Robotic Process Automation

Scrivener Publishing
100 Cummings Center, Suite 541J
Beverly, MA 01915-6106

Publishers at Scrivener
Martin Scrivener (martin@scrivenerpublishing.com)
Phillip Carmical (pcarmical@scrivenerpublishing.com)

Robotic Process Automation

Edited by

Romil Rawat

Rajesh Kumar Chakrawarti

Sanjaya Kumar Sarangi

Rahul Choudhary

Anand Singh Gadwal

and

Vivek Bhardwaj

This edition first published 2023 by John Wiley & Sons, Inc., 111 River Street, Hoboken, NJ 07030, USA
and Scrivener Publishing LLC, 100 Cummings Center, Suite 541J, Beverly, MA 01915, USA

For more information about Scrivener publications please visit www.scrivenerpublishing.com.

Wiley Global Headquarters
111 River Street, Hoboken, NJ 07030, USA

For details of our global editorial offices, customer services, and more information about Wiley products visit us at www.wiley.com.

Library of Congress Cataloging-in-Publication Data

ISBN 978-1-394-16618-3

Front cover images supplied by Pixabay.com
Cover design by Russell Richardson

Set in size of 11pt and Minion Pro by Manila Typesetting Company, Makati, Philippines

Contents

Preface

The book discusses advanced working with digital systems and software. The book's RPA (Robotic Process Automation) software approach makes it straightforward to build, use, and manage software robots that replicate human motion. Software robots are like humans in that they can read what's on a screen, input the right keys, navigate systems, find, and extract data, and do a wide range of specified tasks. Without the need to get up and stretch or take a coffee break, software robots can complete the task faster and more accurately than people. Robotics encompasses a variety of disciplines, including mathematics, computer science, electrical engineering, information technology, mechatronics, electronics, bioengineering, and command and software engineering.

The science of building robots that can stand in for people and replicate their behavior is known as robotics. Robots are applicable to a wide range of situations and for a wide range of purposes, although many are currently used in dangerous environments (such as particle inspection, bomb detection, and disposal), manufacturing processes, or in environments where humans cannot survive (e.g., in space, underwater, in high heat, and for the clean-up and containment of hazardous materials and radiation). Robots try every human behavior, including moving, lifting, talking, thinking, and cognition. Bio-inspired automation is a developing field since so many of today's robots are affected by nature. Some of the tasks include disarming explosives, searching for people amid unstable ruins, and looking into mines and wrecks.

The book's primary objective is:

- To present current trends and implications for robotics, big data, cloud computing, virtual reality, and digital communication technologies using robotic process automation systems.
- To go through the current state of advanced modelling and the RPA project structure.
- To analyse and offer innovative models, strategies, and initiatives for hardware and digital platforms.

1

A Comprehensive Study on Cloud Computing and its Security Protocols and Performance Enhancement Using Artificial Intelligence

Srinivasa Rao Gundu[1], Charanarur Panem[2]* and J. Vijaylaxmi[3]

[1]*Department of Computer Science, Government Degree College-Sitaphalmandi, Hyderabad, Telangana, India*
[2]*School of Cyber Security and Digital Forensic, National Forensic Sciences University, Goa Campus, Goa, India*
[3]*PVKK Degree & PG College, Anantapur, Andhra Pradesh, India*

Abstract

Since cloud computing is becoming an increasingly vital component of both big and small businesses, ensuring its integrity and confidentiality has emerged as a top priority in this space. There are a few different approaches that may be used to secure the cloud. Techniques are implemented through protocols. The protocols that are used in cloud computing may also be used in other types of security systems, such as authentication systems, mailing systems, and cryptonet systems. Cloud computing has challenges in the areas of security on-demand application resource management, and self-monitoring without delay. These challenges arise because of the massive amount of data that is made accessible via cloud computing. When it comes to improving the capabilities of security and privacy in cloud storage, Artificial Intelligence (AI) and machine learning have the potential to play a pivotal role. Therefore, incorporating methods for machine learning into the cloud that already exists with the potential to give enhanced efficiency.

***Keywords*:** Simple Storage Service (S3), Elastic Compute Cloud (EC2), artificial neural network, cryptography, information and communication technology

**Corresponding author*: panem.charanarur_goa@nfsu.ac.in

Romil Rawat, Rajesh Kumar Chakrawarti, Sanjaya Kumar Sarangi, Rahul Choudhary, Anand Singh Gadwal and Vivek Bhardwaj (eds.) Robotic Process Automation, (1–18) © 2023 Scrivener Publishing LLC

1.1 Introduction

When it comes to cloud computing, several factors are working together to speed up its arrival and make it a reality sooner rather than later. Because of the development of more reasonably priced and powerful processors and the software as a service (SaaS) computing architecture, conventional data centres are being transformed into massive computing service pools (MCS). Because of the network's increased capacity and robust and flexible connections, users may now subscribe to high-quality services that employ data and software solely hosted in faraway data centres.

When data is moved to the cloud, users save time and money by not having to worry about maintaining local hardware. Moving data to the cloud is beneficial to users. There are two well-known organizations that provide cloud computing services: Amazon S3 and Amazon Elastic Compute Cloud (EC2) [1].

Additionally, this shift in computing platforms frees local computers from the burden of storing their own data, making it feasible for modern internet-based online services to supply enormous amounts of storage space and computer capabilities that may be customized.

Consequently, the cloud service providers of the users control both the availability and the integrity of the users' data as a direct consequence of this. There are several reasons why cloud computing presents substantial security issues. Data security is one of these issues, which has long been recognized as an important part of service quality. To begin, because users no longer have authority over their data, traditional cryptographic primitives cannot be directly employed in a cloud computing context [2].

It is because of this that cloud storage verification must be done without direct access to all of the data that is stored in the cloud. It becomes increasingly difficult to prove the authenticity of cloud-stored data when you consider the variety of data types that each user has access to, as well as their need for long-term confidence that their data is safe. Second, cloud computing is more than just a third-party data storage facility.

1.2 Aim of the Study

The objective of this project is to examine whether or not Artificial Intelligence (AI) and machine learning can play a major role in enhancing cloud storage facilities in terms of both security and privacy.

1.3 Architecture of Cloud Computing

A cloud computing architecture is a collection of cloud components that are only loosely linked to one another. The two components that may be separated to construct the cloud architecture are the front-end and the back-end.

A network, typically the Internet, serves as the linking mechanism between the system's two ends. Figure 1.1 depicts a graphical representation of cloud computing architecture.

The "front end" of a cloud computing system is the element that interacts with users, or clients. It is made up of the interfaces and programmes needed to access cloud computing platforms, such as a web browser, and its primary component is known as the software stack.

In the background, cloud computing is often referred to as a system's "back end." It has all of the resources necessary to deliver cloud computing services. It includes, among other things, massive amounts of data storage, virtual machines, security measures, services, deployment methods, and servers [3].

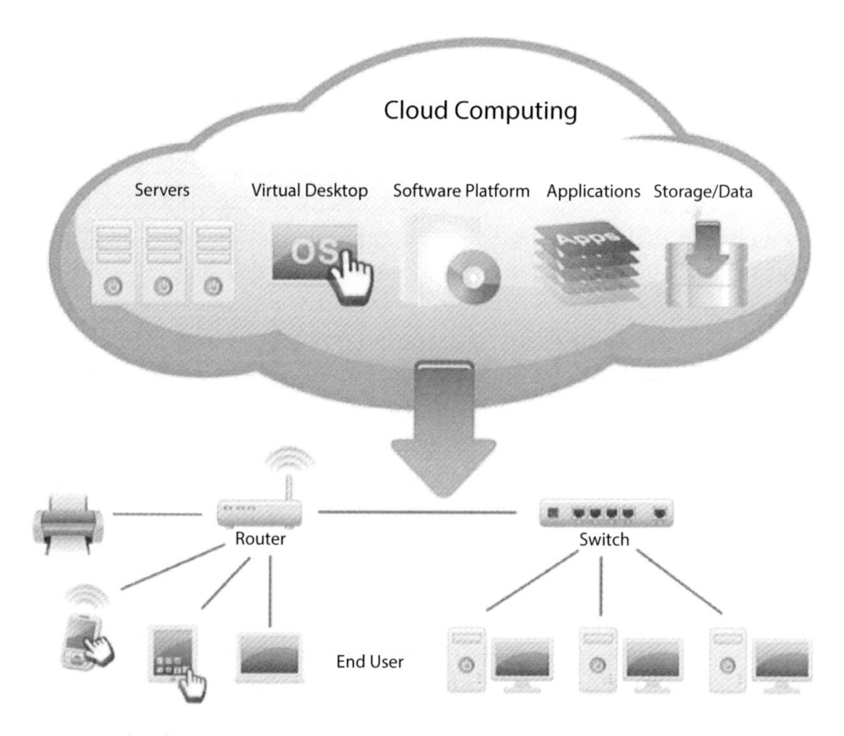

Figure 1.1 Cloud computing.

1.4 The Impact of Cloud Computing on Business

In the world of information technology and cloud computing has become indispensable. Despite its popularity, many firms are reluctant to implement and utilize cloud computing for commercial and operational reasons due to uncertainty about its cost and security consequences. The fundamental appeal of cloud computing for organizations is its cost-effectiveness, whereas the most serious concern is security risks. Cloud computing security is a major concern in the workplace. In addition, various critical implications that enterprises should consider while using cloud computing are explored, as well as security methods for avoiding the highlighted cost and security difficulties [4].

Cloud computing has become a widely accepted and ubiquitous paradigm for service-oriented computing in which computer infrastructure and solutions are provided as a service. Through its characteristics (e.g., self-service on-demand, wide network access, resource pooling, and so on), the cloud has revolutionized the abstraction and use of computer infrastructure, making cloud computing popular. Security, on the other hand, is the most pressing issue and worries about cloud computing are growing as we see more and more innovative cloud computing platforms [5].

Cloud services, software, and infrastructure are clearly becoming more popular in the post-COVID-19 environment, since they can be accessed at any time and from any location. Several research projects and advancements have been suggested to address security threats. Nonetheless, new approaches to make the cloud safer have yet to be discovered. The majority of current cloud security approaches do not address the new sorts of security concerns that cloud computing infrastructure may encounter. As a result, they are unable to identify attacks or vulnerabilities that may originate from either the cloud service provider or the customer [6].

Cloud computing is based on the notion of providing all feasible services, such as software, IT infrastructure, and services to clients over the internet. Cloud computing systems are heterogeneous, large-scale groupings of autonomous computers with a flexible computational architecture. This technology is on the rise since it is becoming the preferred option for firms who do not want to deal with system maintenance or a development team in-house. Many companies, including Amazon Web Services (AWS), Google, IBM, Sun Microsystems, Microsoft, and others, are creating effective cloud products and technologies. Customers and the enterprise exchange data through virtual data centres with cloud technologies [7].

Furthermore, just a few previous studies have looked at the many tiers of cloud architecture. This study did an exhaustive assessment on the challenges that the cloud computing infrastructure encounters at various levels due to the relevance of examining such concerns (application, host, network, and data level). It also discusses the current solutions that have been employed to address these concerns. In addition, this report identifies certain outstanding difficulties that need to be addressed as well as future research prospects [8].

The term "cloud computing" may apply to both a platform, as well as a specific kind of application. The data of both users and corporations is kept safe on servers located in the cloud. The term may also apply to applications that may be downloaded from many websites on the internet. Cloud applications are stored on robust servers that are located in large data centres. These servers also host websites and online services. Several businesses in the information technology sector, including Google, Amazon, Microsoft, and others, are developing and marketing products and services that are hosted on the cloud [9].

1.5 The Benefits of Cloud Computing on Business

Cloud computing has various benefits for enterprises, but a few stick out:

Some of the cost advantages include decreased investment on technology infrastructure, lower capital costs, related savings, and convenience savings.

Some of the technological benefits include reducing maintenance, innovating in technology, devising diversity, and so on. Implementation is straightforward, modification is straightforward, and storage is expanded. Advantages for businesses, customization, adaptability, agility, creativity, selection, and service quality are all important considerations in maintaining vigilance and keeping ahead of the curve [11].

1.6 Generic Security Protocol Features

Generic security protocols as security system enablers offer network security services to various components of a cloud computing environment. These services are provided using generic security protocols. Initial local user authentication protocol, remote user authentication protocol, Single-Sign-On protocol, consuming protocol, cloud trust protocol, secure sessions protocol, and file transfer protocol are used.

Figure 1.2 Cloud computing and security protocols.

The protocols are based on the concepts of generic security objects and a modular security strategy. Each protocol is functionally complete, easy to integrate with other components, and transparent in terms of security credentials and characteristics. They also provide the same set of secure network services to all components of the cryptonet system. The Figure 1.2 shows about the cloud computing and security protocols architecture details.

1.7 Cloud Computing Security Protocol Design

Security protocols are designed in a modular fashion, with each module using a distinct implementation of the idea of generic security objects. This section provides an overview of general cloud computing protocols, as well as various security measures such as the cryptonet system.

1.7.1 Protocol for File Transfer (FTP)

The cloud may be used to store and access a broad range of file types inside an organization. Because a file may include any kind of digital information, such as a written document, image, artwork, video, sound, or piece of software, files can be transferred via the File Transfer Protocol (FTP) service. The File Transfer Protocol (FTP) is a standard protocol for sending data from one computer to another via the Internet. FTP is used to allow users to move files to the cloud. The user information comprises of the username and password, as well as the home directory. The File Transfer Protocol (FTP) is a file distribution protocol that may retrieve data both locally and remotely. FTP was formerly utilised to provide a locking scheme for both cloud and local data, but that practice has since been abandoned.

1.7.2 Local User Authentication Protocol (LUAP)

Users may authenticate themselves on a local level using the LUAP protocol. The local user authentication technique is used in cloud computing and other security systems. The local user authentication method is included within the system's login module. It works with both the username/password authentication mechanism and the certificate-based authentication approach.

The workstation does an automatic examination of the installation environment and settings as soon as the operating system is initiated and then determines the suitable protocol. In order to activate, our system needs a user's Personal Identification Number (PIN) and if set for username/password authentication, it may additionally require the user's fingerprint in addition to the PIN. It communicates with the Security Applet to get the username and password, then provides those credentials to the operating system's native login mechanism. In order to accomplish authentication, the login module will query the user accounts database.

1.7.3 Protocol for Consumption

This protocol was developed with the supplier and the client in mind from the beginning. This protocol addresses concerns about privacy, reliability, and security. Computing in the cloud provides consumers and businesses with delivery platforms that are cost-effective, scalable, adaptable, and have a track record of success. The software-as-a-service model, more often known as SaaS, has the potential to save costs associated with the creation and maintenance of both hardware and software. Platform-as-a-service, also known as PaaS, has the potential to lower costs and simplify the process of acquiring, storing, and administering the software and hardware components of a platform. As a consequence of this, the consumption protocol is a key component to consider in designing the architecture of cloud computing. Cloud computing is presently being used by a very diverse assortment of companies, ranging in size from very small to very big.

1.7.4 Remote User Authentication Protocol

The RUAP protocol allows users to authenticate themselves from a remote location. Cloud computing is a newer kind of computer system that allows users to keep their work on the cloud and then upload and retrieve it as required. Cloud computing is rapidly gaining popularity, but there are still substantial challenges to overcome, such as security and privacy concerns.

Cloud computing is a resource that is available to everybody, which means that a user must connect to the cloud server using open networks in order to utilize it. In the event that a secure system is not built, a number of potential vulnerabilities might be exposed. It is possible for an attacker to have access to the personal information of a user. Because of this, one of the most significant challenges in cloud computing is maintaining the anonymity of users. The remote user authentication protocol is used as a security technique in cloud computing environments in order to ensure that data transmissions are kept secure.

1.7.5 Secure Cloud Transmission Protocol

SCT is a protocol that provides secure data transfer over the internet. The usage of cloud computing has grown in popularity as a result of the many advantages it provides. Small and large businesses alike are increasingly turning to cloud computing services to save costs and streamline operations. Encrypted data transport is required. There are several security concerns to be worried about in the context of cloud computing.

It is vital to interact in a secure way in order to resolve these challenges. Both the user datagram protocol (UDP) and the transmission control protocol (TCP) employ the secure transmission cloud protocol (UDP). It is commonly regarded as a cutting-edge protocol due to its ability to handle a broad range of infrastructure requirements for optimal data transfer over high-speed networks. UDP, or the User Datagram Protocol, is being used to build secure cloud communication protocols (UDT).

1.7.6 Protocol for Cloud Trust

The Cloud Trust Protocol (CTP) is a way that cloud clients may use to request digital data from cloud service providers and subsequently receive that data. This strategy is used to build a user's faith in the owner of the cloud. Using this protocol, you may be able to create a secure method in a cloud computing environment. The usage of this protocol may improve the field of digital trust. Digital trust is the system that assures the security of digitally recorded information. Transparency in information is the foundation for digital trust, and as such, it is the key driver of value collection and return. CTP is used to establish digital trust between a cloud computing client and the provider, as well as to give transparency about the provider's configuration vulnerabilities, authorization, accountability, and operational status conditions. This is performed by using cloud computing.

1.7.7 Protocol for Secure Single-Sign-On

Cloud computing and cloud protocol are two of today's most well-known and fastest evolving technologies. They offer a highly reliable service architecture that can deliver cloud-based integrated services such as on-demand resource computation, resource or data storage or cumulative storage, and exceptionally fast network connectivity. It is probable that it will create a system that is quick and unquestionably successful, needing little in terms of resource management activities and providing an interface for service providers, allowing one to get effective cloud-based services via the usage of internet services.

Each time the client accesses one of today's modern applications, they must memorize and use a new set of credentials. It may be difficult for a company to keep track of the various authentication methods and databases, particularly when each one is utilised by a different kind of organization. As a consequence, we want a dependable protocol that permits speedy sign-on, which will eventually result in a protocol for single sign-on. Clients may utilize a single-on protocol to perform a single sign-on to an identity provider that is trusted by the application that they want to access. Customers no longer need to confirm their identities to a variety of different applications several times, nor do they need to utilize various authentication techniques for each of their apps thanks to the single sign-on protocol.

1.7.8 Secure Session Protocol (SSP)

After the single-sign-on protocol has been successfully established, the secure session protocol is used to offer a session in cloud computing and other security systems. Other security systems may also utilize this protocol. The implementation of the single-sign-on protocol was fruitful, therefore this is the natural next step. The usage of a key exchange certificate may allow for the construction of the protocols necessary for a secure connection. After the key exchange certificate has been installed on both the client and the cloud server, it will be possible for the client and the cloud server to securely swap session-keys and session-ids with one another. A session key is used by the application server's session protocol in order to facilitate the process of controlling the characteristics of encrypted sessions.

A session key is a randomly generated encryption and decryption key that is used to guarantee the security of a communications session between a user and another computer or between a client and a server. This might happen between a client and a server. Session keys are also known as symmetric keys. This is because the same key is used for both encryption and decryption.

This approach, known as session key derivation, is used to generate a session key from a hash value. To do this, the cryptderive key function is used. The key is encrypted using the recipient's public key throughout the session and then included in each message transmitted. Because the amount of security given by session keys is related to how often they are used, those keys are renewed on a regular basis. A distinct session key may be used for each individual interaction.

1.7.9 Protocol for Authorization

In the context of cloud computing, the authorization protocol allows the customer, as well as the service provider to have more control over the level of permit security. The standard known as XACML is used as the foundation for our security system's authorization rules. We used a system called Function-Based Access Control, which denotes that an authorized person, such as a Security Administrator, is the one who organizes a group and decides the access level, role, and permissible activities for each individual member of the group.

A Policy Token with a Target object, which is used to determine the role that each individual member of the group plays, is produced by the Security Administrator. The target consists of the name of a group member, the name of a resource, and the activities that a group member is permitted to do in accordance with the resource authorization policy that has been set.

1.7.10 Protocol for Key Management

Cooperative cryptonet and cloud applications employ key exchange protocols to share group keys. Generic Key Distribution (GKD) complies to the GSAKMP standard. GKD manufactures, distributes, and rekeys keys.

Key exchange protocols are used to exchange group keys between cryptonet and cloud computing applications that operate in a cooperative environment. Generic Key Distribution (GKD) was developed to fulfill this need and adheres to the GSAKMP standard. GKD is accountable for all facets of key production, distribution, and rekeying.

GKD supports both Push and Pull operations, enabling the deployment of shared keys. In addition, it distributes keys in combination with the Secure Application Server. This module uses the PEP component of an application server to create shared-key authorization constraints since it acts as a component. When a group member requests a group-key, he or she establishes a secure connection to the Secure Application Server using Single Sign-On.

The group member used a smart card to access the Secure Application Server's PEP in order to acquire a SAML ticket. After a successful authorization, GKD will transmit the group-key across a secure communication channel to the authorized member of the group. The two kinds of cryptographic keys are secret keys and pairs of public/private keys. The safest method is using secret keys.

A number of other keys are used in cloud computing, including public/private authentication and signature keys, public/private key establishment pairs, symmetric encryption and decryption keys, and symmetric message authentication codes. Public/private authentication, private/private signature, public/private key establishment, and symmetric key wrapping are all instances in which these pairs of keys are used in this way.

The group member uses a smart card to access the Secure Application Server's PEP to acquire a SAML ticket. After successful authorization, GKD sends the group-key via a secure channel to the member. Cryptographic keys may be secret or public/private. Secret keys are the most secure.

Cloud computing uses public/private authentication and signature keys, public/private key establishment pairs, symmetric encryption/decryption keys, and symmetric message authentication codes. This includes public/private authentication, private/private signature, public/private key establishment, and symmetric key wrapping. The Figure 1.3 shows about the cloud computing structure with sharing platforms.

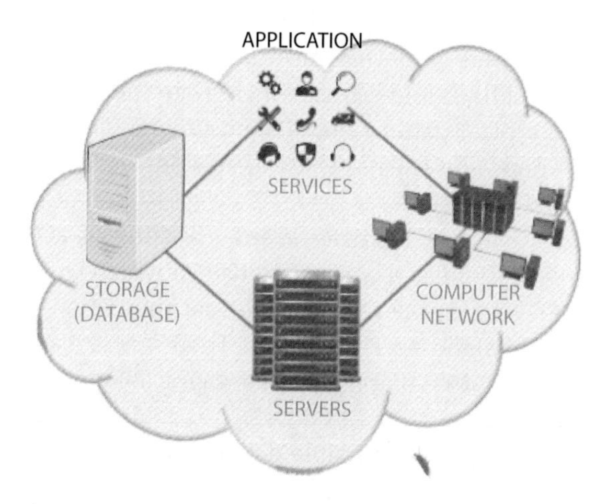

Figure 1.3 Cloud computing structure.

1.8 AI Based Cloud Security

Artificial intelligence (AI)[10] has shown to be an essential component in the disciplines of cyber security and cloud computing security as we go ahead into the era of automation[12]. In light of the fact that AI is capable of rapid learning, it is of the utmost importance to concentrate on finding ways that AI may both improve security and specify how standards can be set around the appropriate use of it. This will ensure that businesses are prepared for the further development of AI.

The term "artificial intelligence" (AI) refers to software developed for computers that has the ability to solve problems and reason in the same way that a human would. The great majority of productive research and development that has been accomplished up to this point may be attributed to the field of machine learning (ML), which is a subfield of artificial intelligence that focuses on training computers to learn by applying algorithms to data.

The phrases "machine learning" (ML) and "artificial intelligence" (AI) are sometimes used interchangeably. AI stands for "artificial intelligence." An issue must be able to be solved using data and there must be a sufficient amount of relevant data that can be acquired for it to be evaluated for a solution utilising artificial intelligence or machine learning.

In addition, there must be access to a large enough amount of computer power in order to complete the essential processing in a time frame that is acceptable.

The volume of data [13, 14] generated by cyber security [15, 16] systems is so large that no human team could possibly hope to process and evaluate it all. In order to identify potentially hazardous scenarios, machine learning algorithms examine all of this information. The more data it examines, the more patterns it finds and learns, which it can then use to identify deviations from the typical pattern flow. The more data it analyses, the more patterns it [17] finds and learns. These alterations have the potential to be seen as cyber threats [18].

For instance, machine learning keeps a record of activities that are regarded as typical, such as the time and date at which workers check in to their respective systems, the information that they often view, and many other traffic patterns and user behaviors. There are several circumstances in which these limits do not apply, such as being able to log on in the middle of the night.

As a direct result of this, possible dangers may be identified and dealt within a far shorter amount of time. Artificial intelligence may be used, using a strategy that is more data-driven, to uncover flaws and vulnerabilities that are now being exploited or that may be exploited in the future and to provide proactive warnings on such weaknesses and vulnerabilities. In order for this to operate, data that is entering and leaving protected endpoints is analyzed and known behaviors and predictive analytics are used to locate and identify potential threats.

1.8.1 Event Detection and Prevention

When AI and machine learning technologies analyze data produced by systems and discover abnormalities, they may take a number of different actions in response, like alerting a person, blocking a specific user, or doing something else entirely. These tactics often result in events being noticed and halted within a matter of hours, therefore putting a stop to the transmission of potentially dangerous code and averting a data breach. Businesses may be able to gain days of notice and time to respond in advance of security issues by using this method, which involves reviewing and integrating data across geographies in real-time.

1.8.2 Delegate to Automated Technology

There are several security systems that can alert you to possible dangers or abnormalities. but automated solutions may eliminate a significant portion of this background noise, allowing you to concentrate on what really matters. When technologies like AI and machine learning are used to perform regular duties and first-level security assessments, security personnel are freed up to concentrate on more major or complicated risks.

This does not imply that these technologies can take the position of human analysts, however, since cyber assaults are often the result of a mix of human and machine activity, these attacks need reactions from both humans and machines. On the other hand, it enables analysts to prioritize their workload and complete their jobs in a more timely manner. To facilitate their business activities, corporations often make use of hundreds or even thousands of interconnected programs. Traditional computer systems store data in a variety of locations, which makes it difficult to ensure that all of those locations are in sync with one another.

With multitenancy Software as a Service (SaaS), which saves human resource, financial, and planning data in one application, all of this is much easier. This central design has many advantages, including the fact that all systems function under the same framework, which eliminates data inconsistencies. It also bridges the divide between the system and those who use it. Access control, on the other hand, must be prioritized.

Because today's workforce is equipped with a range of devices, data is dispersed across several access points, increasing the risk of vulnerability. By prioritizing an access solution that incorporates inspection applications, establishing permissions, and defining rules, the appropriate individuals may have access to the tools they need to operate efficiently.

1.8.3 Machine Learning Algorithms in Cloud Computing

The ever-increasing volume of data that is being collected and processed at ever-increasing rates is a direct result of the ever-increasing number of internet-connected gadgets. This is especially critical in the case of an emergency requiring an instant reaction. When it comes to processing large amounts of data, the length of time it takes has become longer and longer.

The current cloud architecture is not the ideal solution for dealing with these circumstances since the data is routed to a variety of faraway cloud centres. An increase in productivity may be realized by adding machine learning algorithms to an already existent cloud. A huge amount of data is also available on the cloud that may be fed into machine learning algorithms.

In machine learning, clustering is a fundamental technique for organizing and categorizing data. "Clustering" is a term used to describe this basic process, which may subsequently be refined with additional cognitive and predictive algorithms. Machine learning and artificial intelligence techniques have just recently been used by data scientists for cloud computing.

Examples of Keras-based services include Amazon Web Services, IBM Watson, and Microsoft Cognitive AI. Machine learning and artificial intelligence also play a crucial role in addressing the requirements for computing that is both effective and efficient in this day and age of Internet of Things (IoT), Big Data Analytics, and Blockchain.

1.9 Various Neuronal Network Architectures and Their Types

There are three distinct categories of network architectures, which are as follows:

1. Single-Layer Feed-Forward Networks: The input layer of a network with a single layer of feed-forwarding is composed

of source nodes and the neurons that are produced are the output. This is a network that uses feed forwarding.

2. Multilayer Feed Forward Networks: This network simply adds an additional layer that is concealed from view. A higher degree of statistic may be accomplished thanks to the hidden layer that is being used.
3. Recurrent Network: This network has at least one feedback loop in its structure. This loop, which boosts a neuron's ability for learning by feeding its output back into its own input, is shown in Figure 1.1. Additionally, it results in enhanced performance.

1.10 Conclusion

Security methods used for authentication, secure communication, and permissions were examined as part of this study. Along with established technical and legal security requirements, the protocols are built on generic security objects as well. In the context of cloud computing, they maintain security credentials and protocol-specific features in a completely transparent way. In addition, the same attributes might be extended to other well-established methods. Examples of generic security protocols include initial user authentication, remote user authentication, Single Sign-On (SSO), secure sessions and file transfer, cloud transmission and key management protocols.

Acknowledgement

We would like to express our gratitude to Sri Panem Nadipi Chennaih for his assistance and encouragement during the process of writing this book chapter, which we have decided to dedicate to him.

References

1. A. A. Md Shoeb, R. Hasan, M. Haque and M. Hu, "A Comparative Study on I/O Performance between Compute and Storage Optimized Instances of Amazon EC2," 2014 IEEE 7th International Conference on Cloud Computing, 2014, pp. 970-971, doi: 10.1109/CLOUD.2014.146.

2. D. Yang, Y. -C. Chen, S. Ye and R. Tso, "Privacy-Preserving Outsourced Similarity Test for Access Over Encrypted Data in the Cloud," in IEEE Access, vol. 6, pp. 63624-63634, 2018, doi: 10.1109/ACCESS.2018.2877036.
3. Y. Gong, F. Gu, K. Chen and F. Wang, "The Architecture of Micro-services and the Separation of Frond-end and Back-end Applied in a Campus Information System," 2020 IEEE International Conference on Advances in Electrical Engineering and Computer Applications(AEECA), 2020, pp. 321-324, doi: 10.1109/AEECA49918.2020.9213662.
4. D. Berzano *et al.*, "On-demand lung CT analysis with the M5L-CAD via the WIDEN front-end web interface and an OpenNebula-based cloud back-end," 2012 IEEE Nuclear Science Symposium and Medical Imaging Conference Record (NSS/MIC), 2012, pp. 978-984, doi: 10.1109/NSSMIC.2012.6551253.
5. G. Kousiouris, G. Vafiadis and T. Varvarigou, "A Front-end, Hadoop-based Data Management Service for Efficient Federated Clouds," 2011 IEEE Third International Conference on Cloud Computing Technology and Science, 2011, pp. 511-516, doi: 10.1109/CloudCom.2011.76.
6. He, Qinlu, Zhanhuai Li, and Xiao Zhang. "Study on cloud storage system based on distributed storage systems." 2010 International Conference on Computational and Information Sciences. IEEE, 2010.
7. V. Debroy, A. Mansoori, J. Haleblian and M. Wilkens, "Challenges Faced with Application Performance Monitoring (APM) when Migrating to the Cloud," 2020 IEEE International Symposium on Software Reliability Engineering Workshops (ISSREW), 2020, pp. 153-154, doi: 10.1109/ISSREW51248.2020.00046.
8. V. Debroy, A. Mansoori, J. Haleblian and M. Wilkens, "Challenges Faced with Application Performance Monitoring (APM) when Migrating to the Cloud," 2020 IEEE International Symposium on Software Reliability Engineering Workshops (ISSREW), 2020, pp. 153-154, doi: 10.1109/ISSREW51248.2020.00046.
9. M. Bahrami, "Cloud Computing for Emerging Mobile Cloud Apps," 2015 3rd IEEE International Conference on Mobile Cloud Computing, Services, and Engineering, 2015, pp. 4-5, doi: 10.1109/MobileCloud.2015.40.
10. C. Prakash and S. Dasgupta, "Cloud computing security analysis: Challenges and possible solutions," 2016 International Conference on Electrical, Electronics, and Optimization Techniques (ICEEOT), 2016, pp. 54-57, doi: 10.1109/ICEEOT.2016.7755626.
11. C. Prakash and S. Dasgupta, "Cloud computing security analysis: Challenges and possible solutions," 2016 International Conference on Electrical, Electronics, and Optimization Techniques (ICEEOT), 2016, pp. 54-57, doi: 10.1109/ICEEOT.2016.7755626.
12. V. Podolskiy, A. Jindal and M. Gerndt, "IaaS Reactive Autoscaling Performance Challenges," 2018 IEEE 11th International Conference on Cloud Computing (CLOUD), 2018, pp. 954-957, doi: 10.1109/CLOUD.2018.00144.

13. Rawat, R., Logical concept mapping and social media analytics relating to cyber criminal activities for ontology creation. *International Journal of Information Technology*, 15, 2, 893-903, 2023
14. Rawat, R., Mahor, V., Álvarez, J. D., and Ch, F., Cognitive systems for dark web cyber delinquent association malignant data crawling: A review. *Handbook of Research on War Policies, Strategies, and Cyber Wars*, 45-63, 2023.
15. Rawat, R., Chakrawarti, R. K., Vyas, P., Gonzáles, J. L. A., Sikarwar, R. and Bhardwaj, R., Intelligent fog computing surveillance system for crime and vulnerability identification and tracing. *International Journal of Information Security and Privacy (IJISP)*, 17, 1, 1-25, 2023.
16. Rawat, R., Sowjanya, A. M., Patel, S. I., Jaiswal, V., Khan, I. and Balaram, A. (Eds.). *Using Machine Intelligence: Autonomous Vehicles,* Volume 1, John Wiley & Sons, 2022.
17. Rawat, R., Bhardwaj, P., Kaur, U., Telang, S., Chouhan, M. and Sankaran, K. S., *Smart Vehicles for Communication*, Volume 2. John Wiley & Sons, 2023.
18. Mahor, V., Bijrothiya, S., Rawat, R., Kumar, A., Garg, B. and Pachlasiya, K., IoT and artificial intelligence techniques for public safety and security. *Smart Urban Computing Applications*, 111, 2023.

2

The Role of Machine Learning and Artificial Intelligence in Detecting the Malicious Use of Cyber Space

Charanarur Panem[1], Srinivasa Rao Gundu[2*] and J. Vijaylaxmi[3]

[1]School of Cyber Security and Digital Forensic, National Forensic Sciences University, Goa Campus, Goa, India
[2]Department of Computer Science, Government Degree College-Sitaphalmandi, Hyderabad, Telangana, India
[3]PVKK Degree & PG College, Anantapur, Andhra Pradesh, India

Abstract

Cyber security has become a major cause of concern in today's digital world. Data breaches, identity theft, captcha decoding, and other situations that affect millions of individuals and businesses are all too common. When it comes to cyber attacks and crimes, it has always been tough to come up with effective rules and procedures and then put them into action with pinpoint precision. Because of recent developments in artificial intelligence, cyber attacks and criminal behaviour are becoming more common. Research and engineering, including medicine, have all benefited from its utilization.

Artificial intelligence has ushered in a new age in everything from healthcare to robots. When hackers couldn't resist this hot commodity, they turned what were formerly "regular" computer attacks into more "intelligent" forms of crime.

The authors of this chapter examine a variety of AI techniques that they feel have significant promise. Among other things, they go through how to use these tactics in cyber security. They wrap off their discussion with a discussion on the future uses of artificial intelligence and cyber defense.

***Keywords*:** Cyber security, digital world, identity theft, captcha, artificial intelligence, cyber attacks, robots, hackers

**Corresponding author*: srinivasarao.gundu@gmail.com

Romil Rawat, Rajesh Kumar Chakrawarti, Sanjaya Kumar Sarangi, Rahul Choudhary, Anand Singh Gadwal and Vivek Bhardwaj (eds.) Robotic Process Automation, (19–32) © 2023 Scrivener Publishing LLC

2.1 Introduction

To a large extent, hacking has developed from simple theft or vandalism to well-organized and financially sponsored criminals that want to benefit on a global scale because of fast technological advancement. In this case, organized crime's aims might range from personal gain to political development.

As the world becomes more reliant on technology, businesses of all kinds must act faster than ever to protect themselves against cyber assaults and criminal activity. Predicting and detecting an attack before it happens is one of the most critical and hardest parts of Cybersecurity. Cyber assaults may take a number of forms, with varying degrees of sophistication, scope, and objectives. Because of the vast range of threats, organizations and governments must prioritize cyber security as a top priority [1].

Firms must utilize cutting-edge strategies to stay competitive in the face of today's cyber threats. For academics and security professionals alike, Cyber Threat Intelligence (CTI) is becoming an increasingly sought-after tool (CTI). In CTI, you'll find evidence-based information about cyber security risks. The information gleaned from this study improves an organization's capacity to make cyber security choices.

Due to the COVID-19 pandemic outbreak, there is pressure on businesses throughout the globe to establish work-from-home policies without implementing the required and sufficient procedures to counteract these assaults. As COVID-19 spreads, hacker groups are discussing how to use it to launch attacks, including those that target remote work tools and scam job seekers of their money [2].

Hackers regard the social networks of the Deep and Dark Webs as essential to their efforts to learn new tactics and improve their skills. Many hacking-related resources such as compromised data, stolen credit card information, and system flaws may be exchanged on these networks Criminal groups grow and expand as a result of social ties. Internet forums and social networks, in particular, offer criminals like hackers a similar intellectual affinity for exchanging information, organizing attacks, and planning crimes [3].

On these sites, users may follow the postings of other members who they trust or who they feel are experts in certain areas of expertise. Forum users have various specialties and rankings based on their actions and contributions to the community. Because of this, the networks of cybercriminals may grow and new opportunities for cybercrime preparation and execution can be discovered on forums. In order to detect assaults and notify businesses, researchers and cybersecurity professionals depend on these places. It is also possible to design new security technologies by studying the Dark Web's hacker networks [4].

Detecting and preventing cybercrime requires analyzing information on Dark Web sites and this article outlines how to do so. A comparison of current studies, including the study aims, methodology and instruments employed, applicable case studies and results, is included below, as well as any shortcomings that may have been identified. Other topics covered include future trends in Dark Web content analysis, as well as important issues. This list does not include any studies on the Dark Web or how to get data from it. In our research, we are focusing on the content of Dark Web platforms such as websites, forums, and so on.

2.2 Aim of the Study

The aim of the study is to find the role of Machine Learning and Artificial Intelligence in detecting the malicious use of Cyber Space using Dark Web Pattern Recognition and Crime Analysis Using Machine Intelligence.

2.3 Motivation for the Study

Instead of going into great detail on Internet architecture, let us start with a basic explanation. All websites that can be accessed by search engines and are accessible to the general public are included in the "Surface Web" (also called "Open Web" or "Clear Web" in different contexts).

Because they aren't included in search engine indexes, sites on the Deep Web can't be located (or the Invisible Web). When a website is password-protected, it can only be accessed by entering the URL directly into the address bar of the web browser. Otherwise, the website cannot be seen.

It is imperative that academics make a difference between "the black web" and the "deep web". For a variety of technical reasons, search engines cannot index portions of the Deep Web. As a subsection of the Deep Web that employs advanced encryption software to hide users' identities and IP addresses, experts believe that the Dark Web accounts for more than 90% of the web [5].

The most difficult part of the Deep Web to navigate is the Dark Web, often known as the Dark net. Criminals may take advantage of the current circumstances thanks to encryption's guarantee of anonymity. Pedophilia, child pornography, illicit drug and weapon trafficking, human trafficking, terrorist recruiting, preparing terrorist attacks, hiring murderers, and hacking into websites are all frequent in this area of the internet. For a variety of reasons, both professional and amateur hackers participate in these activities. They may extort money from their victims or they may impair the networks of their victims. In certain cases, they may even steal data from businesses themselves [6].

In the Dark Web, privacy, anonymity, and secrecy are all offered. Anyone outside of the Dark Web must utilize special software, such as an Onion Router (TOR), I2P, and Free net, to access the site. For a number of criminal operations, dark networks are the preferred site. There are several examples of CaaS on the Dark Web markets, which provide a wide range of illegal goods and services. Members of Dark Web markets may deal anonymously by using cryptocurrencies like Bitcoin and Monero. Some cybercriminals use Bitcoin suppliers to make it simpler for their colleagues to carry out unlawful activities. When it comes to cyber assaults, organized criminal syndicates often use the Dark Web to disseminate information and hacking tools like malware and ransomware, as well as compromised data [7].

There's no shortage of attack-planning materials on the Dark Web for those who choose it. Payment card and bank account information, as well as PINs and other types of personally identifiable information (PII) that have been exposed may all be purchased by customers (PII). Many more fraud and spam services like email lists for spear phishing may be rented on these markets. DDoS botnets are also accessible.

The degree of technical competence of buyers and sellers on the dark web varies widely. While an elite number of hackers and cybercriminals design and offer advanced hacking tools, the general public either purchases or works with them to execute large-scale cyber attacks and exploit hacked data using the "Crime as a Service" model [8].

There is no longer a need for technological skills to commit cybercrime. It is for this reason that some businesses are now offering security services to their customers as an extra layer of protection. Once a cyber assault is uncovered, we don't know who is responsible for the attack.

In many successful cyber assaults, interpersonal links between the hackers, regardless of their varied degrees of experience, have played a key role. They must work together at these levels to carry out the assaults and achieve their intended objectives. As a result, these networks and markets form networks that mimic friendships or business ties.

Other than cybercrime markets, hacking forums exist. Product features, price, payment choices, and terms of service are all included in product advertisements on these forums. In place of this, buyers and sellers are more likely to utilize private messaging applications or the direct message tool incorporated within the forum to communicate with one other. Markets on the dark web are important in the hacker community. It's plausible to conclude that the Dark Web's hacker population is driven by the possible financial gain, given the prominence of black-market platforms for hackers [9].

Reputation systems are in place on certain forums to discourage intruders and researchers from getting their hands on material. User's level of professionalism and trustworthiness are rewarded with extra privileges such as higher reputation points and access to more portions of the forum as their reputation increases.

To mask the location of hosting servers or TOR Hidden Services, TOR may also be used to host websites. In recent years, access to dark nets has become more difficult. In 2017, TOR added a new layer of secrecy to its service, making it easier for website owners and users to be identified. Sites on the Dark Web will be harder to get to for the broader public without our help. Sites and forums that are only open to those who have been invited are on the rise [10].

However, even though endpoint devices of any sort may be safeguarded for a limited period of time, they are not a long-term solution.

For individuals who work in the area of cybercrime prevention, this may be bad news, but for others, it's good news. In order to track down and shut down websites on the Dark Web, law enforcement agencies may make use of these tools. As a result, criminal proceedings may be brought against individuals responsible. Not all activities on the Dark Web are illegal; many respectable organizations use encryption software for good aims including journalists, political activists, whistleblowers, law enforcement agencies, and researchers [11].

2.4 Detecting the Dark Web

Cyber Threat Intelligence is a term used to describe an information system used to detect, identify, monitor, and react to cyber threats (CTI). Collecting data and analyzing it helps us learn more about potential threats and the strategies and techniques used by those who could pose them. As a bonus, it provides organizations with timely security alerts and other information relevant to the CTI system's kind and function.

With CTI, you'll have the answers to the usual set of five questions. For CTI, it's feasible to gather information from a wide range of sources.

Internal sources include log files, firewall logs, alarms, prior events, malware employed in assaults, and network traffic. Reports from other organizations or governments, as well as specialized blogs, are examples of external sources. A successful CTI, according to Sari, should have five critical characteristics: it should be relevant and timely, information should be accurate and thorough, and it should be actionable.

CTI may be broken down into a number of distinct subcategories, each with its own set of goals and data sources. Only a few examples are given here: open sources, social media, measurement and signatures, human intelligence, and technology (TECHINT).

2.5 Preventing the Dark Web

Three types of Cyber Intelligence are available:

It is the role of strategic cyber intelligence to locate, analyze, and categorize prospective threats.

Cyber intelligence may be divided into three categories:

Risks can be identified in three ways: first, by looking at their sources, aims, and likely consequences; second, operationally, by learning about the attackers' resources, capabilities, and tactics; and third, tactically or technically, by learning about the countermeasures and defense plans that businesses can employ in response to attacks.

In the ever-changing world of technology and computer science, AI continues to provide a wide range of possible applications for organizations and individuals alike. Even while automation and autonomy seem like a good thing, they aren't without their dark side. Unfortunately, this is the situation.

In order to achieve a complex goal, AI systems are software (and possibly hardware) designed by humans that act in the physical or digital dimension by acquiring, interpreting, and rationalizing structured or unstructured information and then deciding on the best actions to take in order to accomplish the given goal. There are two methods in which AI systems may learn: via symbolic rules or through mathematical models. They can also alter their behaviour by analyzing how their previous actions have affected the surroundings.

Cybercrime may be aided by AI in several ways, including speeding up attacks and exploiting new victims. AI can also be used to develop more imaginative illicit business models, all while decreasing the likelihood of being caught. A lesser degree of technical ability and experience may be needed to use IaaS as the service becomes more broadly available, decreasing the barrier to entry [12].

Even though AI is only getting started, it is already being utilized to improve the performance of viruses. At the university level, security researchers are still doing research and their attacks are mostly theoretical and meant to demonstrate a theory of action. Researchers have found evidence that hackers are already planning to use AI in novel ways, even if it is just to assist or augment existing hacking techniques. In order to stop the current attacks and prepare for future ones before they become regular, more research is required.

It is possible that malware authors are using AI in more disguised ways than researchers and analysts are aware of at this moment. Consequently, it is only possible to search for observable signs of AI virus activity. Actually, malware-related artificial intelligence exploits make "traditional" invasions more successful via the usage of artificial intelligence-based approaches.

In 2015, for example, it was shown how to build email messages to bypass spam filters. This system uses a generative language to generate a large database of email messages with excellent semantic quality. The anti-spam system will be able to identify information that is normally invisible by spam filters after these sentences have been altered.

An information security conference called Black Hat USA disclosed how to analyze years of data on BEC attacks using machine learning (ML).

Data dumps and publicly available social media postings are both used in this technique. In addition, the system is capable of accurately predicting whether or not an attack would be effective.

AVPASS18, a tool that aims to derive the detection features and detection rule chain of any antivirus engine, was launched at the same conference. As a result, the software uses this inference to make Android malware seem like a genuine app. AVPASS has a zero percent detection rate on Virus Total, an online malware analysis service, with over 5,000 Android malware samples. It was not possible to identify real-time malware from AVPASS.

Machine learning (ML) is rapidly being used by malware researchers and anti-virus organizations to build more efficient detection techniques for the constantly evolving threats they confront. AI agents taught to explore and seek weak places have shown that machine learning detection systems may be deceived. Even ML-based antivirus engines have been found to be ineffective in the detection of malicious software that has been designed with qualities that allow it to remain undetected.

Reinforcement learning is used by anti-virus detection systems to build a game-based strategy against the system. Among other things, it selects destructive Windows file features that sustain functioning and introduces modifications that enhance the probability of a malware sample being found. Lastly, AI may give new attack ways to enhance old hacking techniques.

We mere humans would be unable to predict such a thing. For the first time ever, we were able to attend DEF CON 2017, the world's largest underground hacker event.

Convention attendees Dan Petro and Ben Morris created Deep Hack, an open-source artificial intelligence tool. Penetration testing online without changing the target system with no prior knowledge is possible.

An artificial neural network is used in conjunction with the Deep Hack online database hacking tool to generate SQL injection strings based on responses from a target server.

DeepExploit22, a tool for automated penetration testing, employs a similar approach. The system utilizes Metasploit to obtain data, build exploits, and conduct tests. Therefore, it employs a reinforcement learning approach called "AC323," which begins by learning from publicly vulnerable services like Metasploit and then tests these situations on the target server to evaluate which attack is most appropriate for the particular scenario.

Criminals might benefit from the use of artificial intelligence (AI) in a variety of ways. Another kind of attack is to interfere with AI systems that are already in operation.

It's possible that artificial intelligence (AI) might be abused, as IBM researchers have suggested in a groundbreaking method.

A malware evasion technique called DeepLocker24 is used to boost the virus's own evasion abilities. With the help of an inherent defect in most AI algorithms, the system is able to do this.

In contrast to traditional approaches, which multiply input data by one or more matrices, machine learning algorithms are defined as nothing more than multiplication operations on factors and matrices. During the algorithm's training phase, these matrices are employed to determine the algorithm's output value. When it comes to supervised machine learning, there aren't many outliers. Making sense of the decisions made by a machine learning model trained on one set of data might be tough. It would be difficult to explain why certain information is labeled as spam if we used a spam filter that relied on deep learning to identify spam. False positives, on the other hand, may be difficult to troubleshoot and resolve, making it much more difficult to identify and remove them from the system.

Algorithms are sometimes referred to as "black boxes" and system design revolves on choosing the right algorithm, fine-tuning its operating parameters, and finding the right training set for the given problem.

For Deep Locker, the lack of information about the product is an asset. Payload obfuscation and encryption are prominent techniques used by malware to evade security scanners' detection. Investigations are often hampered by the use of specialized testing. Malware used in targeted attacks frequently undertakes detailed checks to guarantee that it only runs on the target system as an instance of this. Checks like this may be the only indication that a piece of software is infected with malware and they may serve as a hallmark for the infection itself.

For example, the "inexplicable" nature of neural networks may be used to make these tests seem to be simple mathematical operations between two matrices.

It is more difficult to reverse-engineer the malware's behaviour or even activate it in a test environment since the trigger conditions are never explicitly stored in the infection. Consequently, data analysis is more difficult. In order to identify the target system, Deep Locker leverages a broad variety of data that includes the target machine's software environment, its geolocation, and even audio and video, making it more difficult to predict and detect the trigger conditions for malware to execute.

As a result of the lack of a distinct activation value, Deep Locker's neural network obscures the encryption process even more. Similarly, the question, "Are you on the target machine?" does not yield a yes or no.

Decryption key is provided when input conditions are satisfied and it may be used to decode the remaining malware payload.

It's almost impossible to reverse-engineer a virus and establish its actual identity when data regarding the infection target is concealed using matrix factors. The rest of the malware's code cannot be deciphered until it reaches its victim.

Consider how difficult it is to stop attacks like Deep Locker since the malware cannot be decoded until the defender knows where to look. Deep Locker, like the target system, can only be stopped by identifying the attack after it has been encrypted and is about to execute. A behavioral rule or an antiviral signature in memory might also detect an assault in progress. If a test machine or an engineer believes a dataset is hazardous, things might become much worse. Because Deep Locker doesn't decrypt effectively, it's sometimes hard to identify its damaging actions. Consequently, there would have been no in-memory or behavioral restrictions.

Assaults such as Deep Hack and Deep Exploit are almost hard to detect. As a result, a well-configured application firewall may help defend a network from these attacks and even block subsequent connections from the AI-enabled attack.

ML may also be used to improve password-guessing systems. The fact that this application is still being investigated should not take away from the fact that it has already shown to be more successful than other approaches. Trying to see well on your own is also a challenge. People and organizations that may abuse AI-supported algorithms and approaches are constantly developing them, according to this evidence.

A huge number of potential variations of a password hash is often compared by classic password guessing programs like HashCat29 and John the Ripper30 in order to derive the password. As a starting point, popular passwords are used to generate guesses and subsequently, variations based on the password's actual composition are created, if necessary. There are several ways to say "password," such as "password12345" and "p4ssw0rd."

GANs, for example, may be used to analyze a large database of passwords and generate statistically consistent variations (generative adversarial networks). Password guesses become more targeted and successful as a consequence.

A generative adversarial network consists of a generator and a discriminator (GAN). These models are used by the GAN to distinguish between real samples and simulated ones.

A post on an underground forum in February 2020 reveals an early attempt at this. GitHub has a repository from the same year with software that can filter through 1.4 billion credentials and construct password variation rules based on what it discovers, as referenced in this article.

2.6 Recommendations

There was a question posed by Trend Micro, UNICRI, and Europol: "Has anybody experienced any incidents of criminals utilizing artificial intelligence?" AI will be abused in the future to perpetrate crimes against humanity, as the authors of this research seem to agree.

Among the possible advantages of artificial intelligence are increased productivity, automation, and self-reliance. There are several cross-disciplinary linkages between AI and the Fourth Industrial Revolution, making it a crucial part of this new era. To put it another way, AI may be both a benefit and a curse since it might open the door to new digital, physical, and political threats for civilization. It is necessary to detect the hazards and potential illegal exploitation of AI systems in order to protect society and critical industries and infrastructures.

In this study, existing insights, research, and a systematic open-source analysis were used to examine AI malware, AI-supported password guessing, and AI-aided encryption and social engineering attacks. From automated video creation and processing to AI-enabled reconnaissance and smart and connected technologies such as drones and autonomous cars, AI was utilized to show a broad spectrum of probable future scenarios.

Using so-called deepfakes, artificial intelligence (AI) may be used to modify or create visual and auditory material that is difficult for humans or even technology to distinguish from the real thing.

One theory put out in this article is that criminals would use artificial intelligence (AI) to help and increase the harm they do by increasing the number of victims they may exploit, while also reducing the likelihood that they will be caught. By making it easier for attackers to get started, the bar for entrance will be lowered as "AI-as-a-service" becomes more popular. To put it another way, criminals are more likely to take advantage of AI and use it as a tool for illegal conduct in the future.

Though these assaults are still theoretical, it is possible that malware manufacturers are already using AI in more veiled ways that have yet to be found by researchers and analysts, despite the theoretical nature of these attacks. Malware creators may already be using AI to bypass spam filters, elude detection by antivirus software, or impair the study of malware. Consider this as an illustration. This analysis shows that IBM's recently disclosed Deep Locker technology already has attack capabilities that would be difficult to counter.

As a final point, AI may bring new means of hacking to the table that are difficult for humans to anticipate. Automatic penetration testing, improved password guessing algorithms, and CAPTCHA security system breaching tools, for example, are all forms of penetration testing. Open-source tools like Deep Hack, Deep Exploit, and XEvil were examined for their capabilities as part of the research.

Fraudsters will have additional opportunities to prey on naive victims as AI assistants grow more popular. To sneak into a smart home, criminals might target audio equipment that is left accessible.

Increasingly, cybercriminals will leverage AI methodologies as detailed in this article in order to improve the breadth and scope of their attacks, which will be fueled by the previously stated service-based criminal business model. The second part of the report examined how AI technology may be used by criminals to assist them in carrying out their assaults.

In a recent assessment, thieves are expected to employ AI to automate the early stages of an assault by using AI-aided content generation and increasing corporate information gathering.

Criminals [13, 14] are expected to either directly attack [15, 16] or employ AI for user behaviour modeling in order to overcome AI-based security measures as they become more popular. When more and more aspects of everyday life are automated using AI-based technology [17], there is the possibility for a loss of control over those same areas. Because of this, there is a chance that new types of attacks may be developed. Using artificial intelligence to manipulate the stock market in the area of high-frequency trading is a one-of-a-kind example.

But, as the industry moves toward automation and smart technology, upcoming technologies like 5G and AI will have a huge influence. It's a reasonable bet that criminals will also be using these technologies.

In response to the various conditions below, a number of cures and approaches to reduce the risk of harm were suggested.

A significant partnership between industry and academia is also necessary to establish a body of knowledge and raise awareness about the probable use and abuse of AI by criminals. As a result of this partnership, AI will be able to foresee and anticipate malicious and unlawful behaviours that may be carried out by humans.

Despite the report's focus on misuse, it is clear that AI has a broad range of useful applications, such as assisting law enforcement agencies in identifying crimes of various types and completing important jobs for international organizations. We're excited to see what the corporate and academic communities have to say about AI's limitless potential for good in the world.

Only a thorough understanding of AI's potential, as well as its varied capabilities, scenarios, and attack vectors, can lead to improved preparedness and resilience.

2.7 Conclusion

Cyber security is a major issue in today's digital environment. Data breaches, ID theft, captcha cracking, and other instances that affect millions of individuals and businesses are often reported. Many challenges stand in the way of implementing effective controls and systems for preventing and dealing with cyber risks and crime. The danger of cyber attacks and other crimes has increased dramatically as artificial intelligence has advanced. Every scientific and technological field has found an application for it. AI has ushered in a new age in a variety of industries, including healthcare, robotics, and more. They couldn't resist this roaring fireball and thus, what were once "regular" hacks have developed into "intelligent" ones. There are a variety of AI techniques mentioned in this chapter. Using these tactics in the sphere of cyber security is explained. We end the discussion on AI and cyber security with a look into the future.

Acknowledgement

The authors of this book chapter would like to express our thanks to the late Mr. Panem Nadipi Chennaih for the support and development of this book chapter, it is dedicated to him.

References

1. R. Sabillon, J. Serra-Ruiz, V. Cavaller and J. Cano, "A Comprehensive Cybersecurity Audit Model to Improve Cybersecurity Assurance: The CyberSecurity Audit Model (CSAM)," 2017 International Conference on Information Systems and Computer Science (INCISCOS), 2017, pp. 253-259, doi: 10.1109/INCISCOS.2017.20.
2. H. F. Al-Turkistani and H. Ali, "Enhancing Users' Wireless Network Cyber Security and Privacy Concerns during COVID-19," 2021 1st International Conference on Artificial Intelligence and Data Analytics (CAIDA), 2021, pp. 284-285, doi: 10.1109/CAIDA51941.2021.9425085.
3. M. K. J. Kannan, "A bird's eye view of Cyber Crimes and Free and Open Source Software's to Detoxify Cyber Crime Attacks - an End User Perspective," 2017 2nd International Conference on Anti-Cyber Crimes (ICACC), 2017, pp. 232-237, doi: 10.1109/Anti-Cybercrime.2017.7905297.
4. A. Dalvi, S. Paranjpe, R. Amale, S. Kurumkar, F. Kazi and S. G. Bhirud, "SpyDark: Surface and Dark Web Crawler," 2021 2nd International Conference on Secure Cyber Computing and Communications (ICSCCC), 2021, pp. 45-49, doi: 10.1109/ICSCCC51823.2021.9478098.
5. T. Aoki and A. Goto, "Graph visualization of the dark web hyperlink," 2020 Eighth International Symposium on Computing and Networking (CANDAR), 2020, pp. 89-94, doi: 10.1109/CANDAR51075.2020.00018.
6. H. Chen, "Dark Web: Exploring and Mining the Dark Side of the Web," 2011 European Intelligence and Security Informatics Conference, 2011, pp. 1-2, doi: 10.1109/EISIC.2011.78.
7. M. F. B. Rafiuddin, H. Minhas and P. S. Dhubb, "A dark web story in-depth research and study conducted on the dark web based on forensic computing and security in Malaysia," 2017 IEEE International Conference on Power, Control, Signals and Instrumentation Engineering (ICPCSI), 2017, pp. 3049-3055, doi: 10.1109/ICPCSI.2017.8392286.

8. C. A. S. Murty, H. Rana, R. Verma, R. Pathak and P. H. Rughani, "A Review of Web Application Security Risks: Auditing and Assessment of the Dark Web," 2021 International Conference on Electrical, Computer, Communications and Mechatronics Engineering (ICECCME), 2021, pp. 1-7, doi: 10.1109/ICECCME52200.2021.9591031.
9. A. Biryukov, I. Pustogarov, F. Thill and R. Weinmann, "Content and Popularity Analysis of Tor Hidden Services," 2014 IEEE 34th International Conference on Distributed Computing Systems Workshops (ICDCSW), 2014, pp. 188-193, doi: 10.1109/ICDCSW.2014.20.
10. C. C. Yang, X. Tang and X. Gong, "Identifying Dark Web clusters with temporal coherence analysis," Proceedings of 2011 IEEE International Conference on Intelligence and Security Informatics, 2011, pp. 167-172, doi: 10.1109/ISI.2011.5983993.
11. T. Fu, A. Abbasi and H. Chen, "Interaction Coherence Analysis for Dark Web Forums," 2007 IEEE Intelligence and Security Informatics, 2007, pp. 342-349, doi: 10.1109/ISI.2007.379495
12. Rawat, R., Chakrawarti, R. K., Vyas, P., Gonzáles, J. L. A., Sikarwar, R., & Bhardwaj, R. (2023). Intelligent Fog Computing Surveillance System for Crime and Vulnerability Identification and Tracing. *International Journal of Information Security and Privacy (IJISP)*, 17(1), 1-25.
13. Rawat, R., Mahor, V., Álvarez, J. D., & Ch, F. (2023). Cognitive Systems for Dark Web Cyber Delinquent Association Malignant Data Crawling: A Review. *Handbook of Research on War Policies, Strategies, and Cyber Wars*, 45-63.
14. Rawat, R., Sowjanya, A. M., Patel, S. I., Jaiswal, V., Khan, I. and Balaram, A. (Eds.). Using machine intelligence: Autonomous vehicles, Volume 1. John Wiley & Sons, 2022.
15. Rawat, R., Bhardwaj, P., Kaur, U., Telang, S., Chouhan, M. and Sankaran, K. S., Smart vehicles for communication, Volume 2. John Wiley & Sons, 2023.
16. Mahor, V., Bijrothiya, S., Rawat, R., Kumar, A., Garg, B. and Pachlasiya, K. IoT and artificial intelligence techniques for public safety and security. Smart urban computing applications, 111, 2023.
17. Rawat, R., Logical concept mapping and social media analytics relating to cyber criminal activities for ontology creation. International Journal of Information Technology, 15, 2, 893-903, 2023.

3

Advanced Rival Combatant LIDAR-Guided Directed Energy Weapon Application System Using Hybrid Machine Learning

Srinivasa Rao Gundu[1*], Charanarur Panem[2], J. Vijaylaxmi[3] and Aishvarya Dave[2]

[1]*Department of Computer Science, Government Degree College-Sitaphalmandi, Hyderabad, Telangana, India*
[2]*School of Cyber Security and Digital Forensic, National Forensic Sciences University, Goa Campus, Goa, India*
[3]*PVKK Degree & PG College, Anantapur, Andhra Pradesh, India*

Abstract

Artificial Intelligence (AI) has been employed in a variety of industries and academic institutions to tackle particular challenges. In the same way that electricity and computers are universally applicable, artificial intelligence is. A wide range of areas have benefited from the advancement of machine translation and image recognition technology. Artificial intelligence is being used in the military to enhance, among other things, command, and control, communications, sensing, integration, and interoperability. Everything is being researched, from supply chain logistics to cyber operations to information operations to semi- and self-driving cars. Artificial intelligence in the military might help with sensor and effector coordination, threat detection and identification, enemy position marking, target acquisition, and scattered Joint Fires coordination across networked combat vehicles, including human and unmanned teams. A LASER is a device that causes particular wavelengths of light to be created by activating atoms or molecules. It usually only covers a narrow range of visible, infrared, and ultraviolet wavelengths. There are a lot of environmental factors that affect the propagation of a laser through the atmosphere. In military battles, lasers are used to damage targets. The functioning

**Corresponding author*: srinivasarao.gundu@gmail.com

Romil Rawat, Rajesh Kumar Chakrawarti, Sanjaya Kumar Sarangi, Rahul Choudhary, Anand Singh Gadwal and Vivek Bhardwaj (eds.) Robotic Process Automation, (33–46) © 2023 Scrivener Publishing LLC

of such a direct energy weapon is simply the production of a bigger, more powerful laser.

Keywords: Hybrid Machine Learning (HML), Convolutional Neural Network (CNN), Deep Learning (DL), R-CNN, YOLO, DenseNet

3.1 Introduction

Artificial intelligence can be put to good use. It can help with any intellectual job. There are so many modern artificial intelligence approaches that it's impossible to name them all. The AI affect occurs when a technology becomes commonplace and no longer qualifies as artificial intelligence.

These commercially successful AI applications have subsequently become a part of our everyday lives and have become a central part of our lives. Artificial intelligence (AI) is used in search engines, online advertising targeting, recommendation systems, driving internet traffic, and targeted advertising. Games have been used to evaluate AI power since the 1950s.

The Dartmouth Workshop was founded in 1956 by two senior IBM scientists, Claude Shannon and Nathan Rochester. Ray Solomon and Oliver Selfridge were also there, as were Herbert A. Simon and Arthur Samuel. During the meeting, the term "artificial intelligence" was coined to distinguish itself from cybernetics and its most prominent proponent, Norbert Wiener, who was in attendance.

Artificial Intelligence is a computer that can do a job in lieu of human intellect in a certain situation. Databases are now an essential part of every enterprise-level software's architecture. Artificial intelligence is predicted to be responsible for the bulk of new software value added in future decades.

In the 1980s, academics were interested in teaching computers to learn from data as a natural extension of artificial intelligence (AI). When machine learning was re-established in the 1990s, its goal changed from developing artificial intelligence to solving real-world problems. According to some sources, machine learning will still be a subset of AI in 2020, but it isn't the whole field of artificial intelligence.

During the period from 2010 to 2015, the annual worldwide military robotics budget climbed from $5.1 billion to $7.5 billion. In the military, drones that can take action on their own are commonplace. Many scientists avoid working on military projects.

3.2 Aim of the Study

The aim of the study is to find a way to establish an improved enemy detection system with machine learning in the war field.

3.3 Motivation for the Study

Archimedes' story inspired Direct Energy Weapons. In 214 BC, Archimedes employed mirrors to focus the sun's beams to burn the Roman invaders' ships.

A LASER is a device that causes particular wavelengths of light to be created by activating atoms or molecules. It usually only covers a narrow range of visible, infrared, and ultraviolet wavelengths. There are a lot of environmental factors that affect the propagation of a laser through the atmosphere.

It is crucial to consider diffraction, turbulence, scattering, and absorption. Atmospheric particles may contribute to scattering through the incoming laser's refraction, reflection, or diffraction. Absorption is a phenomenon that reduces a beam's power output by removing energy from it after it has been transformed into another form, most often thermal energy.

The use of lasers for airborne military applications is one of the fastest-growing industries in the defense sector. For high irradiance, laser beams may be focused to extremely tiny regions or they can have very low divergence for concentrating power over a long distance.

Impulsion-driven flight and ballistic orbital flight may be possible in the future as technology progresses. The notion of employing light as a weapon is referred to as laser weapons. A multitude of laser applications are now being used in the industry for simple and precise applications.

In military battles, lasers are used to damage targets. The functioning of such a direct energy weapon is simply the production of a bigger, more powerful laser.

Directed energy weapons have the potential to fundamentally alter the balance of power in the twenty-first century because they are new. As a result of extensive research and development across the world, HPL–DEW has shown its promise in the fields of defense and security. High-powered lasers and high-powered microwaves have emerged as the two principal DEW choices. Another option for a DEW weapon system is HPM or particle beams, although neither has reached the degree of maturity required for direct use in modern conflict.

Next-generation laser technology was used to transport 1.8 Gbps across 45,000 kilometers of GEOLEO lines using a laser terminal (laser terminal). The ultimate objective of the project was to create a fully functional HEL weapon for use in naval operations in the future. A 30kW electric fiber laser prototype system was exhibited by Lockheed Martin in 2015. In 2016, the goal of SL-TM was to design, build, and test a naval ship.

Today's weaponry is more accurate and precise than ever before, eliminating the need for human troops in modern conflicts. In an effort to make sense of the data flow, humans are using additional sensors on the battlefield, which are sending back vast volumes of information to analysts. With AI, the more learning algorithms that live on massive amounts of data, the more accurate these systems may be. In fact, the more data these systems study, the better they may be. A military artificial intelligence arms race occurs when two or more countries compete to build the greatest AI for their military (AI). For many, the increasing geopolitical and military tensions that some have labeled a "Second Cold War" are fueling a worldwide competition among major countries to develop stronger AI for their militaries. Artificial Intelligence disputes between the United States and China sparked the AI Cold War, which has led to an arms race in AI. The term "arms race" is used to describe the quest towards ever-increasing levels of artificial intelligence. AI superiority in the military may conflict with a country's ambition for dominance in other sectors when it seeks both advantages. The AI race, which will be won by one of two parties, is the second threat. Because of the excitement and clear benefits of being the first to develop new AI technology, there are enormous temptations to disregard safety risks, possibly missing crucial features, such as prejudice and fairness. Other teams are motivated to take shortcuts and install an AI system that isn't ready because they fear that another team may break through. This puts other teams and the owner of the AI system at risk.

The purpose of this project is to develop a LIDAR-guided Directed Energy Weapon Application System using artificial intelligence. Next Generation Offensive Precision Weapon Technology might be developed as a result of this idea. When a laser is used to focus on a specific item, LIDAR counts the time it takes for the reflected light to return to the receiver. Long-range weapons that employ highly concentrated energy to harm their target are called directed-energy weapons (DEWs). DEWs may include lasers, microwaves, particles, and sound. However, in order to do this, all major technologies, such as AI, Big Data, Data

Mining, and others, must be extensively engaged. As a consequence, it is recognized as a polygon known as the Dodecahedron. This project provides a full evaluation of the advantages and disadvantages of a fascinating technology.

3.4 Nature of LASERS

In 1917, Albert Einstein provided the first explanation for the laser. In 1960, Theodore H. Maiman created the first LASER device. The achievement was made at the Hughes Laboratory and the invention's success was supported by the ideas of Charles Hard Townes and Arthur Leonard Schawlow. Since then, the LASER business has grown steadily.

The principles of the equipment may be used to grasp the process of working on a LASER device. A LASER's operation is regulated by quantum mechanics. Each atom and molecule, according to the idea, has a set quantity of energy that varies depending on the molecules and atoms employed. The right use of this concept aids in the development of a wide range of lasers.

It must have three essential components in order to work properly. The lasing material or active medium, the external energy source, and the optical resonator are the three essential components of any LASER.

Because of tiny air particle edges, laser beams are bent or dispersed. This phenomenon is unavoidable in laser systems since light's spreading nature allows energy to be scattered as the beam moves even in a perfect vacuum. Laser aiming applications are limited in their ability to achieve high levels of precision and focus by the diffraction effect.

Because it scatters light and reduces the amount of energy it can transfer to its target, turbulent air reduces the amount of energy that can be effectively supplied to a target. Wind speed, height, and other weather factors all affect the degree of turbulence a laser beam experiences.

Scattering occurs when the environment's nature hinders the beam from transmitting accurately, causing it to alter direction. Atmospheric particles may contribute to scattering through the incoming laser's refraction, reflection, or diffraction. The quantity of scattering is determined by the size of the particle impeding the laser's passage. As a result, higher-wavelength lasers' scattering effects are less noticeable. Absorption is a phenomenon that reduces the power output of a beam by removing energy after it has been converted to another form, most often thermal energy. Different atmospheric components have different impacts.

The laser is the surgical weapon of choice due to its precision in determining the precise target location. Laser weapons, despite their high initial costs, are very cost-effective in the long run. It is possible to dynamically regulate the laser deployments so that they may be tuned to have non-fatal, destructive and disruptive effects. Lastly, weapons have relied on rapid energy delivery to kill targets almost exclusively throughout military history. Impulsion-driven flight and ballistic orbital flight may be possible in the future as technology progresses.

3.5 Ongoing Laser Weapon Projects

Since 2010, a timeline of military laser projects has been created.

1. NASA's Blue Beam Project: This project began with the goal of creating holograms from space that might be utilized for clandestine operations on the ground. NASA's space stations use relay satellites or laser projectors to deliver data to aerial vehicles for holographic displays.
2. Robust Electric Laser Initiative, Lockheed Martin Military Weapons, US Department of Defense: The project's objective was to enhance efficiency by more than 30% while maintaining beam quality. Using fiber laser technology, this new weapon boosts the 25kW laser weapon's output to 100kW, a significant increase in power. Successful testing of a 60kW laser was completed in 2017.

An attempt was made to create the first laser communication system that could operate in zero-gravity as part of the Laser Communication Demonstration on the Moon (LLCD) project. Laser communications between a spacecraft in orbit around the moon and a receiver on Earth were the primary goal. Downlink data rates of 620 Mbps and uplink data rates of 20 Mbps were reached with the LLCD. On October 18, 2013, a successful test run was completed.

Boeing built the High Energy Laser Mobile Demonstrator to show that the military can destroy threats like rockets, mortars, and unmanned aerial vehicles (UAVs) at light speed. When it comes to creating an advanced laser weapon, solid-state lasers like the 10kW model were utilized in 2014. As far back as 2011, this has been occurring.

3.6 Directed Energy Weapons (DEWs)

Directed energy weapons (DEWs) are being researched as conventional weapons reach their limits (DEWs). By rethinking how we fight, we want to increase operational effectiveness and combat effectiveness as a whole. An operational framework has left DEW technologies vulnerable to high prices and poor performance.

Software meta-modeling environment GINA allows users to articulate the semantic behavior of systems rather than writing code. A self-descriptive [1, 2] reflective modeling paradigm is used to combine observed defined connection components [3, 4] in the project data [5] environment. The Figure 3.1 shows about the communication links among different objects with reference to the data is conveyed by a narrow beam of laser [6, 7], while the Figure 3.2 highlights about the faster R-CNN model for army position locating, whereas the Figure 3.3 focused for YOLO [8–10] model for helmet and safety vest classification and the Figure 3.4 is the flow of information for DenseNet model for fastness of chin strap on helmet recognition. The Defence Capability features [11–13] are also compared in the Table 3.1.

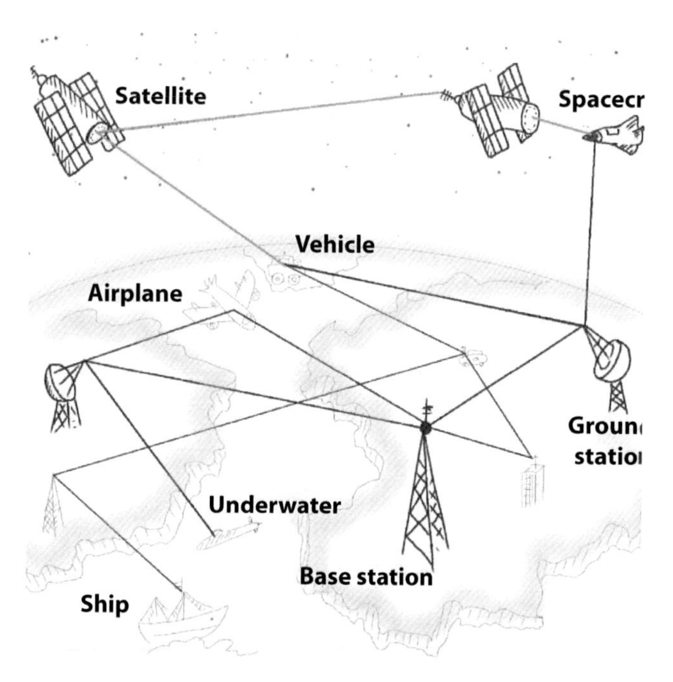

Figure 3.1 Communication links between different objects with reference to the data is conveyed by a narrow beam of laser light and it is highlighted in red color.

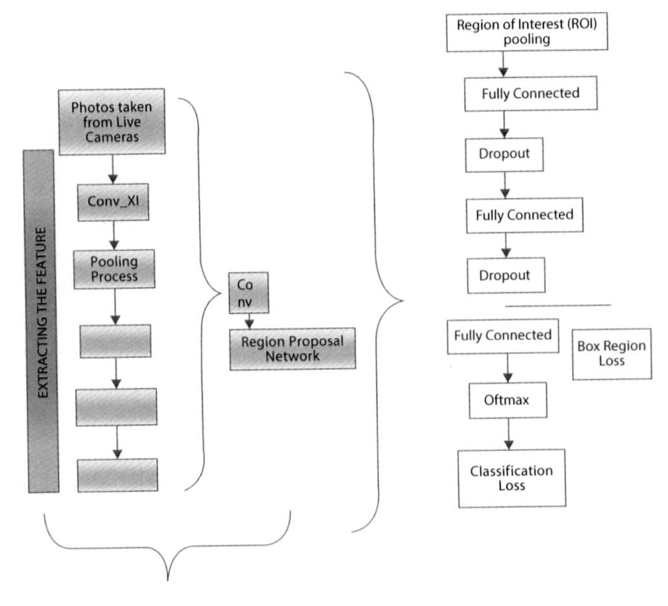

Figure 3.2 Faster R-CNN model for army position locating.

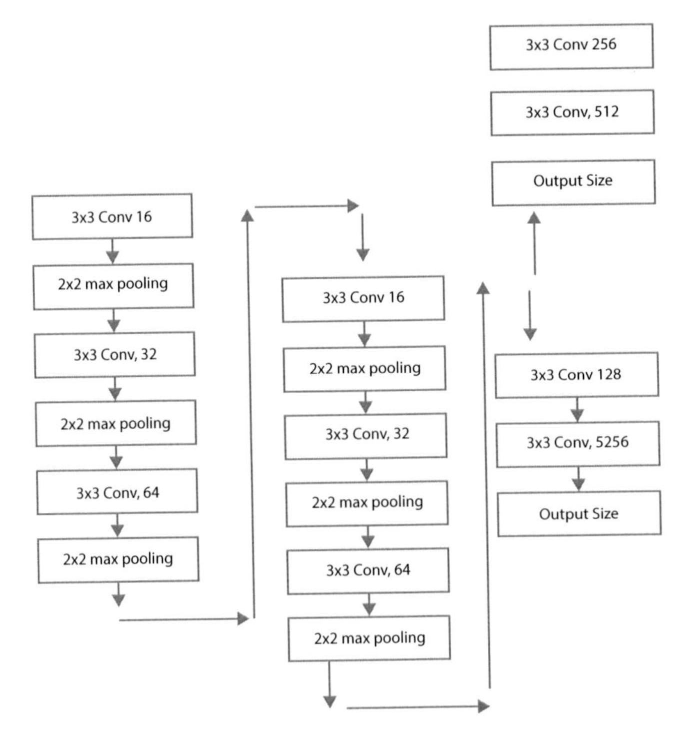

Figure 3.3 YOLO model for helmet and safety vest classification.

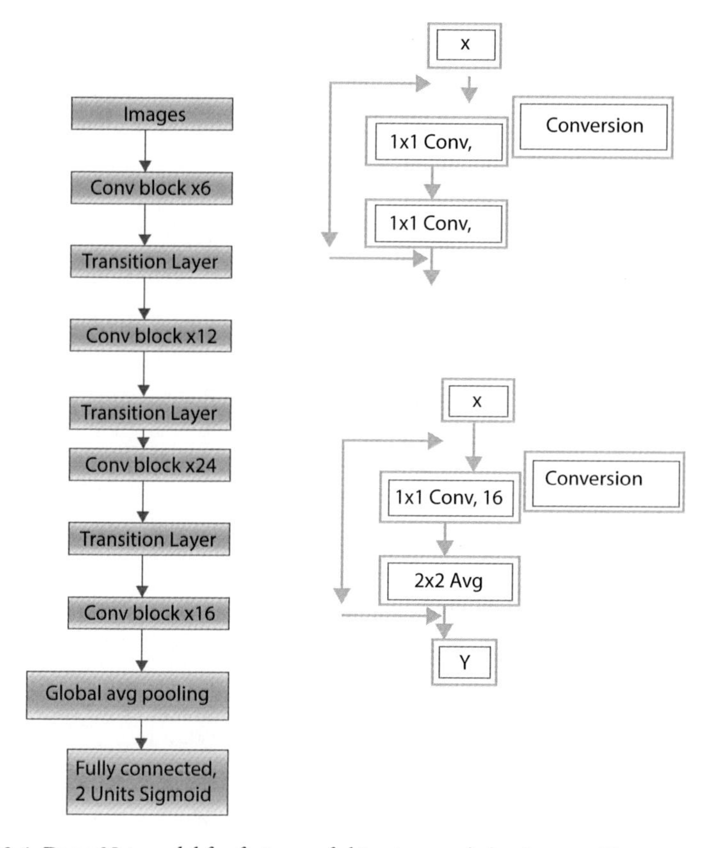

Figure 3.4 DenseNet model for fastness of chin strap on helmet recognition.

3.7 LIDAR Guided LASER Weapon System (LaWS) Requirements

Shown in Table 3.2 are the LIDAR guided LASER Weapon System requirements.

3.8 Methodology

Real-time [14, 15] visual detection [16, 17] of adversaries is much more difficult on the battlefield because of the chaos. A hybrid ML [18] model combining the following three CNN-based DL approaches is proposed in this study to address these issues [19].

Table 3.1

Defense capability features				Target/attacker features	Environment features
Missile	**Gun**	**Laser**	**Sensor**		
Probability of Intercept	Probability of Intercept	Probability of Intercept	Detection Range (in KM)	Target Material Specific Heat Capacity	Weather Dwindling
Maximum Effective Range	Maximum Effective Range	Power		Operating Temperature	
Minimum Engagement Range	Minimum Engagement Range	Aperture Diameter		Target Material Melting Point	
Average Speed	Rate of Fire	Wavelength		Thermal Material Thickness	
Launch Interval	Engagement of Duration	Beam Quality		Target Material Density	
No. of Missile Launchers	No of Guns	Number of LASER Weapons		Thermal Diffusivity of Target Material	
No. of Missiles on Vessel				Average Speed	
				No of Attackers	
				Heat of Fusion	
				Thermal Conductivity	

Table 3.2 LIDAR guided LASER weapon system requirements.

SL. no.	Factors	
1	Design	Mean Allocated Percentage
2	Hardware	20%
3	Support	5%
	(i) Supporting Equipment	5%
	(ii) Tools	5%
	(iii) Testing Equipment	5%
	(iv) System Level Test	5%
	(v) System Level Evaluation	5%
	(vi) Training	5%
	(vii) Data	15%
	(ix) System Engineering	15%
	(x) System Program Management	5%
	(xi) Other	5%
4	Software	5%
5	Integration	20%
	Total	100%

Data Acquisition

The training image datasets were collected from real-world projects via the camera of a mobile phone, videos of IP Cams, CCTV, and the installed PTZ devices. Totally, 3,108 clear helmet images were collected, with 83% (2,639 images) used for training and the other 17% (528 images) for testing. 1,173 clear safety vest images were collected with 65% ages) used for training and the other 35% (411 images) for testing.

The laser weapons [20] are constantly vulnerable [21, 22] to hacks from terrorist [23] groups and hackers. In 2008, a hack caused an oil pipeline in Turkey to burst; it was over pressurized, and the alarms had been turned off. Attackers (cyber criminals) [24] were able to conceal the explosion from the command centre by compromising surveillance cameras, leaving it incapable of reacting quickly. Another threat to the steel industry showed how hackers [25] and terrorists might easily wreak significant harm by entering the knowledge management system powering the facility.

3.9 Conclusion

Currently, the military prefers to have a decision-maker on board. In conflict, however, these communication links may be targeted since the head can be removed and the body is rendered unable to think. In most cases, the majority of drones now in use throughout the globe would cease to function if the data connection between them and their human operator was severed. Since AI algorithms and systems are always evolving due to their quick reflexes and data processing capabilities, they may constantly improve the jobs they have been trained to do. A breakthrough in Artificial Intelligence (AI) was in the works for years and has finally come. The integration of machine intelligence with manned systems has already begun and this trend is projected to grow.

Acknowledgement

The authors of this book chapter would like to express our thanks to the late Mr. Panem Nadipi Chennaih for the support and development of this book chapter, it is dedicated to him.

References

1. V. Varatharasan, A. S. S. Rao, E. Toutounji, J. Hong and H. Shin, "Target Detection, Tracking and Avoidance System for Low-cost UAVs using AI-Based Approaches," 2019 Workshop on Research, Education and Development of Unmanned Aerial Systems (RED UAS), 2019, pp. 142-147, doi: 10.1109/REDUAS47371.2019.8999683.
2. A. Meliones and I. Plas, "Developing video games with elementary adaptive artificial intelligence in unity: An intelligent systems approach," 2017 Intelligent Systems Conference (IntelliSys), 2017, pp. 104-111, doi: 10.1109/IntelliSys.2017.8324230.
3. K. Uchiyama, Y. Iwai, T. Ichinoseki and K. Kayama, "Applying artificial intelligence theory to helicopter SAF simulation based on HLA/RTI (1)," Proceedings Fourth International Conference on Computational Intelligence and Multimedia Applications. ICCIMA 2001, 2001, pp. 210-214, doi: 10.1109/ICCIMA.2001.970470.

4. J. Ai, L. Hua, J. Liu, S. Chen, Y. Xu and C. Hou, "Radar Signal Separation Recognition Method based on Semantic Segmentation," 2021 8th International Conference on Dependable Systems and Their Applications (DSA), 2021, pp. 302-306, doi: 10.1109/DSA52907.2021.00046.
5. R. Edmundson, R. Danby, K. Brotherton, E. Livingstone and L. Allcock, "Investigating combinations of machine learning and classification techniques in a game environment," 2016 12th International Conference on Natural Computation, Fuzzy Systems and Knowledge Discovery (ICNC-FSKD), 2016, pp. 1306-1311, doi: 10.1109/FSKD.2016.7603367.
6. A. A. Hidayat, S. Wasista and Y. P. Pratiwi, "Development of fighting genre game(boxing)using an accelerometer sensor," 2016 International Conference on Knowledge Creation and Intelligent Computing (KCIC), 2016, pp. 201-206, doi: 10.1109/KCIC.2016.7883647.
7. Joseph Redmon, Santosh Kumar Divvala, Ross B. Girshick and Ali Farhadi, "You only look once: Unified real-time object detection", CoRR, vol. abs/1506.02640, 2015.
8. V. L. Shirokov, "The basics of methodology for estimation and choosing parameters of multi-service telecommunication systems," 2nd International Conference on Dependability of Computer Systems (DepCoS-RELCOMEX '07), 2007, pp. 191-197, doi: 10.1109/DEPCOS-RELCOMEX.2007.47.
9. K. Trapezon, T. Gumen and A. Trapezon, "Basics of using connected me system in the transmission of information signals," 2018 14th International Conference on Advanced Trends in Radioelecrtronics, Telecommunications and Computer Engineering (TCSET), 2018, pp. 229-232, doi: 10.1109/TCSET.2018.8336192.
10. T. B. Minde, S. Bruhn, E. Ekudden and H. Hermansson, "Requirements on speech coders imposed by speech service solutions in cellular systems,"1997 IEEE Workshop on Speech Coding for Telecommunications Proceedings. Back to Basics: Attacking Fundamental Problems in Speech Coding, 1997, pp. 89-90, doi: 10.1109/SCFT.1997.623910.
11. M. Bahrami, "Cloud Computing for Emerging Mobile Cloud Apps," 2015 3rd IEEE International Conference on Mobile Cloud Computing, Services, and Engineering, 2015, pp. 4-5, doi: 10.1109/MobileCloud.2015.40.
12. S. Li and J. Gao, "Moving from Mobile Databases to Mobile Cloud Data Services," 2015 3rd IEEE International Conference on Mobile Cloud Computing, Services, and Engineering, 2015, pp. 235-236, doi: 10.1109/MobileCloud.2015.33.
13. M. M. Fuad and D. Deb, "Cloud-Enabled Hybrid Architecture for In- Class Interactive Learning Using Mobile Device," 2017 5th IEEE International Conference on Mobile Cloud Computing, Services, and Engineering (MobileCloud), 2017, pp. 149-152, doi: 10.1109/MobileCloud.2017.15.
14. L. Shankar, "Bigonet Mobile Cloud Framework," 2014 Sixth International Conference on Communication Systems and Networks (COMSNETS), 2014, pp. 1-3, doi: 10.1109/COMSNETS.2014.6734915.

15. S. D. A. Shah, M. A. Gregory and S. Li, "Cloud-Native Network Slicing Using Software Defined Networking Based Multi-Access Edge Computing: A Survey," in IEEE Access, As vol. 9, pp. 10903-10924, 2021, doi: 10.1109/ACCESS.2021.3050155.
16. B. Chihani, E. Bertin and N. Crespi, "Enhancing M2M Communication with Cloud-Based Context Management," 2012 Sixth International Conference on Next Generation Mobile Applications, Services and Technologies, 2012, pp. 36-41, doi: 10.1109/NGMAST.2012.17.
17. O. Segal, N. Nasiri, M. Margala and W. Vanderbauwhede, "High level programming of FPGAs for HPC and data centric applications," 2014 IEEE High Performance Extreme Computing Conference (HPEC), 2014, pp. 1-3, doi: 10.1109/HPEC.2014.7040979.
18. E. El-Araby, I. Gonzalez and T. El-Ghazawi, "Virtualizing and sharing reconfigurable resources in High-Performance Reconfigurable Computing systems," 2008 Second International Workshop on High-Performance Reconfigurable Computing Technology and Applications, 2008, pp. 1-8, doi: 10.1109/HPRCTA.2008.4745683.
19. Zhixin Ba, Haichang Zhou, Huai Zhang and Zhenxiao Yang, "Performance evaluation of some MPI implementations on workstation clusters," Proceedings Fourth International Conference/Exhibition on High Performance Computing in the Asia-Pacific Region, 2000, pp. 392-394 vol.1, doi: 10.1109/HPC.2000.846584.
20. Mahor, V., Bijrothiya, S., Rawat, R., Kumar, A., Garg, B., and Pachlasiya, K., IoT and artificial intelligence techniques for public safety and security. Smart Urban Computing Applications, 111, 2023.
21. Rawat, R., Logical concept mapping and social media analytics relating to cyber criminal activities for ontology creation. International Journal of Information Technology, 15, 2, 893-903, 2023.
22. Rawat, R., Mahor, V., Álvarez, J. D., and Ch, F., Cognitive systems for dark web cyber delinquent association malignant data crawling: A review. Handbook of Research on War Policies, Strategies, and Cyber Wars, 45-63, 2023.
23. Rawat, R., Chakrawarti, R. K., Vyas, P., Gonzáles, J. L. A., Sikarwar, R., and Bhardwaj, R., Intelligent fog computing surveillance system for crime and vulnerability identification and tracing. International Journal of Information Security and Privacy (IJISP), 17, 1, 1-25, 2023.
24. Rawat, R., Sowjanya, A. M., Patel, S. I., Jaiswal, V., Khan, I., and Balaram, A. (Eds.). Using Machine Intelligence: Autonomous Vehicles, Volume 1, John Wiley & Sons, 2022.
25. Rawat, R., Bhardwaj, P., Kaur, U., Telang, S., Chouhan, M., and Sankaran, K. S. Smart Vehicles for Communication, Volume 2, John Wiley & Sons, 2023.

4

An Impact on Strategical Advancement and Its Analysis of Training the Autonomous Unmanned Aerial Vehicles in Warfare [Theme - RPA and Machine Learning]

Srinivasa Rao Gundu[1], Charanarur Panem[2*] and J. Vijaylaxmi[3]

[1]*Department of Computer Science, Government Degree College-Sitaphalmandi, Hyderabad, Telangana, India*
[2]*School of Cyber Security and Digital Forensic, National Forensic Sciences University, Goa Campus, Goa, India*
[3]*PVKK Degree & PG College, Anantapur, Andhra Pradesh, India*

Abstract
For all of our future needs, unmanned aerial vehicles (UAVs) are viewed as a very promising technology. The use of unmanned aerial vehicles (UAVs) has become commonplace in most parts of the globe. More and more scientists are looking into using drones for anything from inspecting and delivering goods to monitoring and spying on people. AI, according to most analysts, will have a significant impact on the military, if not transform it. According to some, Artificial Intelligence (AI) is one of today's most transformational technologies. Artificial Intelligence (AI) is projected to have a substantial impact on the vast majority of common consumer electronics over the next decade or two. Because of its rapid expansion and introduction into the military, Artificial Intelligence (AI) is already having a significant impact on how militaries operate. Despite AI's immense potential and advantages, the military's employment of the technology has proven controversial. Drone warriors can utilize swarm tactics anywhere in the world, thanks to investments by the US military in technology that makes this possible. The development of unmanned aerial vehicles (UAVs) has the potential to revolutionize military operations in the

**Corresponding author*: panem.charanarur_goa@nfsu.ac.in

Romil Rawat, Rajesh Kumar Chakrawarti, Sanjaya Kumar Sarangi, Rahul Choudhary, Anand Singh Gadwal and Vivek Bhardwaj (eds.) Robotic Process Automation, (47–60) © 2023 Scrivener Publishing LLC

future. In this study, the function of pilots, as well as that of remotely piloted and autonomous vehicles, is studied. Strengthening a nation's military may be easier if artificial intelligence is integrated into its equipment. When it comes to autonomous systems that are used in military systems, the discrepancy between government and private AI R&D expenditure may have a cascade effect.

Keywords**:** Artificial Intelligence (AI), autonomous vehicles, delivery, drone troops, inspection, military applications, military equipment, military systems, research and development, surveillance, Unmanned Aerial Vehicles (UAVs)

4.1 Introduction

There has been a great deal of scientific interest in UAVs, or Unmanned Aerial Vehicles, in recent years because of their mobility, the ease of deployment, and their capacity to communicate with humans. Fixed-wing and rotary-wing unmanned aerial vehicles (UAVs) are the two most common forms of UAVs. Each kind of UAV is designed for a distinct function. Fixed-wing UAVs are better suited for tasks that don't need them to remain stationary. Assault and surveillance are two examples of military uses that fall within this category. Flying machines have a greater aerodynamic complexity because of the rotation of their wings. These vehicles may also be stationed at a certain area, although they are unable to carry out long-distance operations. Temporary wireless coverage may be provided more effectively by rotary-wing UAVs than by other kinds of UAVs. Due to the engagement of several industries in the development of UAVs, the usage of a UAV network is now a reality rather than a concept. There are several applications for wireless networking, including weather forecasting and emergency management, agriculture, distribution, and traffic control. Extreme weather conditions and the necessity for a line of sight link to avoid a potentially deadly loss of control are two of UAVs' many constraints. It is widely accepted that drones' battery life and computing power are their most significant drawbacks. For the most part, commercially available unmanned aerial vehicles (UAVs) are only able to hover for a maximum of two hours before they must land and recharge their batteries. As a result, standard procedures and methodologies may not always be relevant for UAV-related concerns [1].

Artificial intelligence's newest innovation is machine learning (ML) (AI). The black-box technique, which solely examines the inputs and outcomes, is currently widely used in scientific research as a result of this method's widespread adoption. Additionally, the availability of high- performance computing (HPC) and powerful graphics processing units (GPUs) has helped in the development of machine learning [2]. Machine learning is being used by a wide spectrum of industries, maybe more than expected.

If someone is interested in artificial intelligence (AI), there are a number of sub-fields that focus on a certain kind of difficulty. If someone has ever wondered how the human mind works, he or she has probably heard about deep learning. Two of the most popular applications for this technology are in computer vision and speech recognition. In the late 1970s, another AI discipline, known as RL, was established. As part of the process, you'll discover the most effective methods for maximising your gains. The many possible states of learning are employed and examined throughout the process of learning. In recent years, robotics and machine learning (RL) have expanded and progressed rapidly. RL is more often employed than Deep Learning in robotics, route planning, and learning hard tasks. As a result, it may be used for a wide variety of decision-making scenarios, including those involving a goal-oriented agent and a specified environment [3].

Other developments in machine learning include FL, which was unveiled by Google in 2016 and promises to enable decentralized network systems. FL is a machine learning environment for training a centralized model on decentralized data without transferring the data to a local shared unit. Dispersed or scattered data may then be utilized to train machine learning algorithms on. FL is becoming a major subject when it comes to unmanned aerial vehicles (UAVs) [4].

For the most part, artificial intelligence (AI) is a fast-growing field that enables computers to do jobs better than humans. Adding artificial intelligence to UAV networks is considered both tough and exciting. If AI is integrated into the network, drone-based applications may also profit. As previously said, AI may play a critical role in UAV resource management with the goal of optimizing energy efficiency. The capacity of an unmanned aerial vehicle (UAV) to choose its own course leaves it subject to AI improvements that might alter its course and deployment. It is also possible to increase the quality of imagery for UAVs by using computer vision. Surveillance, traffic control, and landing site identification might all benefit from the use of this technology. Using AI algorithms that automate hard activities and boost the overall intelligence of the system may significantly improve UAV-based networks [5].

4.2 Aim of the Study

The aim of this study is to:

(i) Estimate and analyze the training of Autonomous Unmanned Aerial Vehicles in Warfare Development
(ii) Assess the impact on Strategical Advancement and its analysis

4.3 Motivation for the Study

Unmanned aerial systems (UAS) have seen a recent surge in commercial and military use, including for surveillance, special operations, ground troop aerial support, and aerial coverage of public gatherings such as rallies, damage estimation, and relief operation management, telecom relay for coverage in remote areas, such as oil and gas exploration, and geographic surveys [6].

Also, small unmanned aerial vehicles (UAVs) are used for the delivery of purchased items to buyers. It has been shown that manned aircraft have an accident rate of one every 100,000 flight hours, whereas, unmanned aerial vehicles (UAVs) have an accident rate that is closer to one per 1000 flying hours, according to research. Unmanned aircraft have a 100-fold higher accident rate than human-piloted aircraft, according to this research. The most common causes of accidents are mechanical surface issues and data connection failures. Keeping the UAV under control while the control station's data link goes down is possible using a variety of methods. Several UAV systems use circular flying patterns with pre-programmed circular radiuses to reestablish the broken link. Unmanned aerial vehicles (UAVs) are built to autonomously land on a backup strip in the event that their first landing strip loses power. According to post-accident research, data connection failure is responsible for more than a quarter of all UAV accidents [7].

Computer vision and machine learning have been extensively studied in recent years in order to construct emergency or autonomous landing systems. Machine learning techniques like SVM and ANN are used in conjunction with digital image processing technologies to choose an optimal landing spot for emergency landing systems while designing them. If you're interested in studying more about SVM and ANN, keep in mind that they each have their own performance criteria. For example, ANN takes longer to train than SVM since it requires a significantly bigger data set and, as a result, more computational power to process it. Neither technique is able to keep up with the continually changing needs for emergency landing location selection systems because of these constraints [8].

4.4 Supervised and Unsupervised Machine Learning for UAVs

Machine learning is a relatively recent term in the artificial intelligence lexicon. A computer's ability to do tasks accurately is facilitated by this component of AI. With today's computers and the amount of data, machine learning has been a big success in recent years. The current focus of research is on using machine learning for unmanned aerial vehicle (UAV) challenges [9].

4.4.1 Supervised Learning Overview

- For the algorithm to learn how to make a judgment for a new unlabeled item in the future, supervised learning provides a ground-truth value for each data input. According to this example, the cost of a UAV may be estimated. The algorithm needs a collection of training data, which contains information about the different UAVs and the labels that go along with them in order to work properly (the price). It is common for the dataset to have separate training and testing sets [10]. Training and testing sets are used to validate the model's correctness, while the training set is used to learn about the link between input and output. When describing supervised activities, regression and classification challenges are often cited as points of discussion. A price prediction is an example of a continuous output value generated by regression issues. As opposed to the above, categorization issues provide distinct values indicating whether an input is benign or malignant.
- Unsupervised and supervised machine learning approaches will be discussed in detail in the sections that follow. Additionally, the techniques utilized to resolve the issues stated in this survey are reviewed.
- The following are some examples of supervised algorithms and neural network architectures:
  In many applications, it is customary to combine classification with regression. Various supervised approaches may be used to conduct classification and regression. Support Vector Machines (SVMs) may also be utilized to tackle regression and classification problems, depending on the application [11].
- Regression methods come in two flavors:
  The output of continuous data may be predicted using an algorithm to achieve pure regression. For example, well-known machine learning approaches include linear regression and logistic regression.

- Automated Classification Systems
  Machine learning relies on pure classifiers. We only offer the Naive Bayes classifier as an example of a pure classifier since it was originally created for classification using the probabilistic Bayes theorem [12].
- A Perceptron with a Stack of Layers (MLP)
  As a theoretical model for mimicking biological human brain networks, automated neural networks (ANNs) have been proposed in the last several years. ANNs are constructed from a collection of loosely linked nodes using perceptrons and layers. Each Perceptron is tasked with a certain input and output duty. MLP is the most basic ANN type. The system's design includes an input layer, a hidden layer, and an output layer that performs classification or regression.
- Neural Networks using Convolutional Layers (CNNs)
  The CNN is a special kind of ANN created for use in computer vision. For example, a photograph may be sent via an algorithm to a CNN, where weights and biases can be altered. Convolutional layers of the CNN architecture are utilized to extract high-level visual attributes from the image. For example, activation functions, layer pools, or padding are not included in this study. Feature extraction and classification are often performed in the earliest convolutional layers of convolutional neural network (CNN) designs.
- Neural Networks with Recurring Connections (RNNs)
  Recurrent neural networks (RNNs) may be used to tackle problems that need sequential information (recurrent neural networks). Using written speech, a video of the speech, or an audio recording, we may demonstrate our point. Automated creation of visual descriptions, voice recognition, and natural language processing (NLP) are all RNN applications. Because the RNN contains a "loop," the model may use the outcomes of earlier neurons. There are two parts to RNNs: one that predicts the output and one that stores short-term memories.

4.4.2 Overview of Unsupervised Learning

Data that has not been labeled is used in unsupervised learning rather than supervised learning, which uses data that has been labeled. As a general rule, the following unsupervised learning activities are acceptable:

clustering, dimension reduction, and data creation. In the next section, we'll take a look at some of the most often used unsupervised methods.

- Neural Network Design and Unsupervised Learning Methods K-means is a prominent approach for clustering data in machine learning. Based on the distance between their centroids and the cluster's input number K, data points are allocated to the cluster with the closest centroids.
  Probabilistic models include Gaussian Mixture Modeling (GMM) and K-means. The clusters and the Gaussian distribution have little association. Because of the K-means algorithm's close connection strategy, each data point has a possibility of being linked to one of the cluster centres [13].
- Auto Encoding (AE): A "representation learning" neural network is used by an AE to encode the input (RL). In order to keep a dataset as small as possible, this method is often used. An AE's design is surprisingly straightforward. The "bottleneck" pushes the original information via an input layer into a compressed knowledge representation.
- Synthetic data that seems to be genuine is generated using an algorithmic structure known as a Generative Adversarial Network (GAN). Images, films, and music are all produced with them [14].

4.4.3 Supervised and Unsupervised UAV-Based Solution Problems

The installation and deployment of unmanned aerial vehicles (UAVs) should be explored in order to free up terrestrial base stations while reducing the power consumption of the UAVs. Unmanned aerial vehicles are positioned based on wireless network congestion predictions rather than continually adjusting their placements. Wireless traffic forecasting is aided by probabilistic models. The GMM family of unsupervised machine learning models includes these models. A Gaussian distribution is assumed to accurately reflect the data distribution.

After the users have been separated into K clusters using the K-means approach, the best GMM model parameters are calculated using a weighted expectation maximization procedure. In the following step, we need to identify the issue that is reducing the power of the drones. In terms of numerical efficiency, the machine learning-aided strategy outperforms the previous one. There are still problems concerning how to manually pick the number of K clusters and the locations of cluster centres when using a K-means approach to categorize people.

We can create a radio map that shows the optimal placement of UAVs as aerial base stations. Combination grouping and regression challenges were offered by the authors based on terrain complexity and the constraints of radio maps. Machine learning is also used to anticipate the channel in order to reassemble the radio map.

Aerial base station installation may also be studied using UAVs. Weighted expectation maximization is used to predict download traffic using machine learning algorithms.

K-means and baseline expectation maximization methods were utilized to evaluate the performance of this machine learning technology. For each hotspot's downlink requirement, contract theory was used to choose the right UAV. The base station's ability to predict the UAV's position improves communication efficiency. In fact, a UAV's capacity may be reduced if it encounters wind turbulence while removing waste from a terrestrial base station. The authors suggest an RNN-assisted architecture in which the UAV's future elevation and horizontal angles relative to its base station are predicted using the previous angles. This approach may be used to estimate the precise position of a fast-moving UAV. For example, the number of hidden nodes and layers in RNNs has yet to be optimized, but researchers are studying how these factors impact prediction accuracy. Numerical tests have shown the high accuracy of a four-layer RNN with sixteen hidden nodes.

4.5 Unsupervised Solution

4.5.1 Estimation of Channel

The estimate and modeling of complicated UAV and ground-to-UAV communications may benefit from the use of machine learning. This study looked at how to predict UAV-to-UAV route loss. Random Forest and KNN algorithms are used to evaluate predictions against real data. Several characteristics, including propagation distance, transmitter altitude, receiver altitude, and slope, may be used to determine route loss. The comparison of the ray-tracing program's data to machine learning results reveals that certain prediction jobs can be successfully handled by machine learning.

A generative neural network was used by the researchers to predict the channel state for two types of terrestrial base stations: (i) street-level base stations located on the ground and (ii) aerial roof-mounted base stations located on the roof of the UAV. Classifying a connection type (LOS/NLOS/outage) and then creating channel parameters based on that classification is done using neural networks. It has also been used to forecast the quality of a UAV-to-ground channel. Predicting route loss, for example, was done using ANNs. Neural networks are used to predict the UAV's signal intensity and channel propagation (ANNs). The impacts of natural phenomena such as diffraction, reflection, and scattering are studied using a shallow neural network (ANN). The input layer includes the UAV's distance, altitude, frequency, and loss of route. This intriguing study highlights the issue of whether ANNs are suitable for real- time applications because of their lengthy processing times. Analog network theory (ANN) is utilized to examine the relationship between UAV-to-ground node signal intensity and distance. The signal strength data from an urban area was used to train an artificial neural network. It was possible to get quite accurate estimates of the channel characteristics by using this data set. The received signal strength at a flying unit from a cellular base station was predicted using ANN as well as other supervised machine learning methodologies.

Using unsupervised learning, however, the three-dimensional channel between UAVs and mobile users on the ground was built. During this analysis, the linkages were divided into two categories: LOS and NLOS, using K-means. Additionally, the SVM technique may be utilized to do regression. Route loss in an urban outdoor setting is predicted using support vector regression and then compared to empirical results in this study [15].

4.5.2 Detection of UAVs

Due to their widespread usage by both the military and people, unmanned aerial vehicles (UAVs) have the potential to be employed for espionage or even as a fatal weapon. Therefore, the capacity to recognize and monitor UAVs is essential in averting these threats. With so many creative solutions, we've separated them into image-based and sound-based categories to make it easier for you to go through them.

We have created a real-time UAV detection method based on the analysis of the drone's acoustic data. Two machine learning algorithms were put to the test in terms of accuracy. There is a preset drone threshold that must be met in order to identify potential UAVs. As the initial machine learning technique, Plotted Image Machine Learning is used (PIL). Picture similarity is assessed by visualizing an audio Fast Fourier Transform (FFT) graph against a target image.

The second technique determines the average distance to the target by combining the KNN algorithm with the Fast Fourier Transform (FFT).

Simulated results suggest that the PIL strategy outperforms the KNN approach when it comes to predicting outcomes. Ambient noise may have a significant influence on sound data, even if the visual clarity and sharpness of the input image are adequate. Another thing to consider is whether or not all of the unmanned aerial vehicles (UAVs) will utilize the same FFT profile to identify a certain target during an incident like this. When using KNN, it is advisable to experiment with more complicated machine learning methods. An unmanned aerial vehicle's location may be detected and tracked using a spiral microphone array. Several spectrograms and filters are applied to the input sound before it is fed into the architecture of the concurrent neural network to finish the feature recognition process.

Scholarly works often discuss the auditory signature of unmanned aerial vehicles (UAVs). In other investigations, UAVs have been identified using a variety of other indications, such as WiFi traffic or the UAV radio frequency signal.

Using images instead of sound may help overcome the challenge of recognizing unmanned aerial vehicles (UAVs). UAVs may be identified visually using a Pan-Tilt-Zoom camera and a variety of deep learning algorithms.

Several computer vision techniques for UAV recognition were examined in a study released very recently.

A hybrid system that incorporates picture, sound, and radio UAV transmission signals will be necessary in the future. Those who are interested may access the most current findings in this area's study.

4.5.3 Imaging for Unmanned Aerial Vehicles

UAVs are used for imaging despite the fact that computer vision is not the focus of this review. Investigating how to find a safe landing spot in an emergency landing scenario is one example of this. The detection issue is reframed as a classification task by analyzing two well-known classifiers (GMM and SVM). A classifier, for example, may classify the initial map as safe or harmful. After that, a filter is employed to eliminate potentially dangerous areas while retaining feasible landing locations. These issues aren't addressed in this work since they may be mistaken for problems with computer vision in general. Aside from the fact that the photographs are shot from a fixed height, nothing else impacts their nature. In order to acquire photographs of unmanned aerial vehicles, these algorithms, also known as "CNN," "feature extractor," and "edge detector," are used.

We know that data is essential since it aids a learning algorithm in identifying and collecting information from the data set. There is a lack of high-quality data for wireless purposes compared to computer vision operations. We feel that the wireless communication industry should place a greater priority on open source, high-quality data.

Some articles compare the performance of machine learning algorithms. This is an important flaw. Compare two machine learning models to see whether there's a logical reason why one is better than the other, or if one NN design is better than NN architecture, for example. Defining parameters and assessing the outcome without further explanation is an example of "black box" machine learning. No one can predict whether a model will succeed or fail in advance.

Attacks [16, 17] by hackers on UAVs that a company owns and employs are quite distinct. Other factors that need to be taken into account include protecting [18] the UAV's onboard storage, guaranteeing that the paths UAVs travel are reasonably safe (i.e., devoid of hazards, thinly populated, etc.), and making sure that the Wi-Fi or radio frequency (RF) signals used by drone platforms are properly encrypted to prevent eavesdropping [19] or alteration. For local storage, the majority of UAV platforms have an onboard storage disc interface. Platform takeovers, where an attacker [20, 21] uses RF, Wi-Fi, or a subscription service like Aerial Armour to detect the flight paths of a UAV in a specific area, undertake de-authentication attacks, seize command of the UAV, and land the misappropriated UAV in a specific location, are frequent attacks and threats by cyberterrorists [20, 21] against venture drones UAV.

4.6 Conclusion

Using machine learning to solve problems that might be dealt with more simply and deterministically is prevalent, creating the illusion that machine learning is not fully justified and typically leads to ML abuse.

Machine learning has consistently beaten numerical simulations in all of the studies we've looked at. Machine learning techniques may not always outperform more traditional ones. The huge success of machine learning in solving a wide range of problems leads us to believe that data selection is a key consideration when determining whether an ML model is appropriate.

For the sake of illustration, let's look at an actual situation. You're working on a computer vision object identification challenge with the objective of recognizing a UAV in a photograph. In this way, it is feasible to understand that the CNN performs well with photos obtained by a UAV, but not so well with images taken by other means. There will be no problem with performance, but it will be unable to reliably forecast future samples if the training set is heavily biased. When data quality and quantity aren't taken into consideration, a model's accuracy suffers, resulting in artificially high ML accuracy. Unmanned aerial vehicle (UAV) problem-solving frameworks, both supervised and unsupervised, have traditionally surmounted a number of hurdles.

4.7 Scope for the Future Work

Machine learning may be able to help with unmanned aerial vehicles in particular. As a result, we expect a wide range of ideas to emerge in the future. UAV-related tasks, such as route loss prediction, may be assessed using better machine learning models employing novel regression methodologies. By employing sound detection or transforming the task to a computer vision challenge, UAVs may be detected. A hybrid system using several inputs, such as sound FFTs and pictures and radio broadcast, where a suitable NN offers a score for each kind of input (e.g., a type of CNN for images and a particular RNN for sound or radio) is used to categorize the output.

Acknowledgement

The authors of this book chapter would like to express our thanks to the late Mr. Panem Nadipi Chennaih for the support and development of this book chapter, it is dedicated to him.

References

1. S. U. Jan and H. U. Khan, "Identity and Aggregate Signature-Based Authentication Protocol for IoD Deployment Military Drone," in IEEE Access, vol. 9, pp. 130247-130263, 2021, doi: 10.1109/ACCESS.2021.3110804.
2. I. C. Freeman, A. J. Haigler, S. E. Schmeelk, L. R. Ellrodt and T. L. Fields, "What are they Researching? Examining Industry-Based Doctoral Dissertation Research through the Lens of Machine Learning," 2018 17th IEEE International Conference on Machine Learning and Applications (ICMLA), 2018, pp. 1338-1340, doi: 10.1109/ICMLA.2018.00217.
3. F. H. Zunjani, S. Sen, H. Shekhar, A. Powale, D. Godnaik and G. C. Nandi, "Intent-based Object Grasping by a Robot using Deep Learning," 2018 IEEE 8th International Advance Computing Conference (IACC), 2018, pp. 246-251, doi: 10.1109/IADCC.2018.8692134.

4. B. H. -Y. Lee, J. R. Morrison and R. Sharma, "Multi-UAV control testbed for persistent UAV presence: ROS GPS waypoint tracking package and centralized task allocation capability," 2017 International Conference on Unmanned Aircraft Systems (ICUAS), 2017, pp. 1742-1750, doi: 10.1109/ICUAS.2017.7991424.
5. S. A. Al-Ahmed, M. Z. Shakir and S. A. R. Zaidi, "Optimal 3D UAV base station placement by considering autonomous coverage hole detection, wireless backhaul and user demand," in Journal of Communications and Networks, vol. 22, no. 6, pp. 467-475, Dec. 2020, doi: 10.23919/JCN.2020.000034.
6. S. Savazzi, U. Spagnolini, L. Goratti, D. Molteni, M. Latva-aho and M. Nicoli, "Ultra-wide band sensor networks in oil and gas explorations," in IEEE Communications Magazine, vol. 51, no. 4, pp. 150-160, April 2013, doi: 10.1109/MCOM.2013.6495774.
7. J. Shmelev, "Simulator training optimization of UAV external pilots," 2014 IEEE 3rd International Conference on Methods and Systems of Navigation and Motion Control (MSNMC), 2014, pp. 75-78, doi: 10.1109/MSNMC.2014.6979734.
8. A. E. Klochan, A. Al-Ammouri, M. M. Dekhtyar and N. M. Poleva, "Fundamentals of the Polarimetric UAV Landing System," 2019 IEEE 5th International Conference Actual Problems of Unmanned Aerial Vehicles Developments (APUAVD), 2019, pp. 153-156, doi: 10.1109/APUAVD47061.2019.8943830.
9. R. L. Mallors, "Autonomous systems: Opportunities and challenges for the UK," IET Seminar on UAVs in the Civilian Airspace, 2013, pp. 1-18, doi: 10.1049/ic.2013.0067.
10. H. Shakhatreh *et al.*, "Unmanned Aerial Vehicles (UAVs): A Survey on Civil Applications and Key Research Challenges," in IEEE Access, vol. 7, pp. 48572-48634, 2019, doi: 10.1109/ACCESS.2019.2909530.
11. H. Zheng, X. Gan and S. Ren, "Equipment Development Risk Assessment Based on Least Square Support Vector Machine," 2019 2nd International Conference on Safety Produce Informatization (IICSPI), 2019, pp. 66-68, doi: 10.1109/IICSPI48186.2019.9096009.
12. X. Deng, J. Guo, Y. Chen and X. Liu, "A Method for Detecting Document Orientation by Using NaÏve Bayes Classifier," 2012 International Conference on Industrial Control and Electronics Engineering, 2012, pp. 429-432, doi: 10.1109/ICICEE.2012.120.
13. X. Xie, W. Huang, H. H. Wang and Z. Liu, "Image de-noising algorithm based on Gaussian mixture model and adaptive threshold modeling," 2017 International Conference on Inventive Computing and Informatics (ICICI), 2017, pp. 226-229, doi: 10.1109/ICICI.2017.8365343.
14. Prabhat, Nishant and D. Kumar Vishwakarma, "Comparative Analysis of Deep Convolutional Generative Adversarial Network and Conditional Generative Adversarial Network using Hand Written Digits," 2020 4th International Conference on Intelligent Computing and Control Systems (ICICCS), 2020, pp. 1072-1075, doi: 10.1109/ICICCS48265.2020.9121178.

15. R. M. Esteves, T. Hacker and C. Rong, "Competitive K-Means, a New Accurate and Distributed K-Means Algorithm for Large Datasets," 2013 IEEE 5th International Conference on Cloud Computing Technology and Science, 2013, pp. 17-24, doi:10.1109/CloudCom.2013.89.
16. Rawat, R., Logical concept mapping and social media analytics relating to cyber criminal activities for ontology creation. International Journal of Information Technology, 15, 2, 893-903, 2023.
17. Rawat, R., Chakrawarti, R. K., Vyas, P., Gonzáles, J. L. A., Sikarwar, R. and Bhardwaj, R., Intelligent fog computing surveillance system for crime and vulnerability identification and tracing. International Journal of Information Security and Privacy (IJISP), 17, 1, 1-25, 2023.
18. Rawat, R., Sowjanya, A. M., Patel, S. I., Jaiswal, V., Khan, I. and Balaram, A. (Eds.). Using machine intelligence: Autonomous vehicles, Volume 1, John Wiley & Sons, 2022.
19. Rawat, R., Bhardwaj, P., Kaur, U., Telang, S., Chouhan, M. and Sankaran, K. S., Smart vehicles for communication, Volume 2, John Wiley & Sons, 2023.
20. Mahor, V., Bijrothiya, S., Rawat, R., Kumar, A., Garg, B. and Pachlasiya, K., IoT and artificial intelligence techniques for public safety and security. Smart Urban Computing Applications, 111, 2023.
21. Rawat, R., Mahor, V., Álvarez, J. D. and Ch, F., Cognitive systems for dark web cyber delinquent association malignant data crawling: A review. Handbook of Research on War Policies, Strategies, and Cyber Wars, 45-63, 2023.

5

FLASH: Web-Form's Logical Analysis & Session Handling Automatic Form Classification and Filling on Surface and Dark Web

Ashwini Dalvi[1]*, Viraj Thakkar[2], Smit Moradiya[2], Aditya Vedpathak[2], Irfan Siddavatam[2], Fark Kazi[1] and S.G. Bhirud[1]

[1]Veermata Jijabai Technological Institute, Mumbai, India
[2]K. J. Somaiya College of Engineering, Mumbai, India

Abstract

Data collection and mining has quickly established itself as a topic of interest, with a focus on openly available data already used extensively. The recent years have turned attention to the dark web due to its unmonitored nature, especially in regard to data behind sign up or registration pages, access to which is not yet automated. With present work already developed on the structure and semantics of a web form, researchers categorize forms as search forms and non-search forms. This paper furthers this by proposing an automated process that determines the more common types of form (Search, Login, Registration, Other) and further demonstrates an automated method to fill and submit forms on the surface and dark web. To achieve this, we perform the interpretation of the form tags and further maintain a headless session for low system overheads. The approach suggested and demonstrated in this paper has been tested for classification and filling (including login after registration) on over 2000 different forms with 84.8% accuracy for classification and 61.1% accuracy for filling.

Keywords: Security automation, robotic process application, robotic process for automated form filler, form classification, query interface, login, registration, search

**Corresponding author*: aadalvi_p19@ce.vjti.ac.in

Romil Rawat, Rajesh Kumar Chakrawarti, Sanjaya Kumar Sarangi, Rahul Choudhary, Anand Singh Gadwal and Vivek Bhardwaj (eds.) Robotic Process Automation, (61–100) © 2023 Scrivener Publishing LLC

5.1 Introduction

Data is nothing less than a resource; right from detecting patterns from past events to the prediction of future events, data is the key. With the exponential growth and adoption of the Internet, researchers have found an increasing interest in knowing and understanding the data on the web. The increasing interest in this subject and the methods to access data on the internet has started to migrate towards a registration/query-based structure. In order to obtain this data from any online service, usually one must create an account with the organization. Only on filling the form is data access provided and this is quite common on the WWW (World Wide Web). Only users who are registered to the particular site or are manually put in a particular query can access these webpages. The problem that lies here is for those who need to access huge amounts of data from various sites at the same time. It is not viable for them to log in and then extract all the data. The prior solution was to take aid of a web crawler with the introduction of forms, but most of the crawlers fail.

This forms an obstacle that needs to be understood to provide a conclusive solution for the automation of data extraction mechanisms on the web. This solution needs to work on all 3 parts of the web - surface web, deep web, and dark web. These 3 combined are the sources of data that are commonly explored in order to access data. Usually, users have to go through an authentication layer consisting of login or registration forms. Filling the required authentication comes naturally when done by humans, but in order to obtain larger amounts of data, automation is important. The ability to bypass this authentication in an automated format proves to be invaluable here, opening the door to whole new webpages and web resources for any automated mechanism to get to. A popular mechanism here would be that of Internet crawling and scraping, which would capture precise information regarding the kind of content websites post behind the authentication armed with this tool.

Noting from prior literature, some researchers study the structure of a web form and there have been studies that focus on the relation between the form fields and the database attributes behind the HTML forms. These researchers have studied the web-forms in great depth and performed classification of the forms but are limited to the determination of search and non-search forms only. Further, other papers were able to classify forms into more categories like request based, feature-based, and AJAX-based. A common trend was observed here: while research would talk about basic classification, there was little research talking about form filling and submission. Select researchers would also only focus on search forms and would not have a methodology associated with themselves. This paper notes the lack of non-search form classification and also a novel approach to fill forms.

This paper develops an automated mechanism to accomplish the method of getting access to the website and its data by clearing the authentication. FLASH (Form Logical Analysis & Session Handling) plays a significant role in allowing a user/program to fill any form. Whether it is a search, login, or registration form, it also maintains the session by allowing the program to browse under the same authentication. Not being limited there, the algorithm can submit forms on different technologies like HTML, JS, and AJAX forms.

All this is possible once the form filling algorithm detects all the forms present on a webpage and fills them. The form filler determines the type of input tags and then appropriately fills data in the fields before automatically submitting it. Additionally, FLASH also detects the specific order in which the forms have to be filled, for example prioritizing the submission of a registration form over a login form for the same web domain. Since the strategy of form filling on the surface web and the dark web is similar, the same algorithm can be used to fill forms on both.

In further chapters, current research towards form filling and the progression of this technology over these papers has been explored. The methodology covers the technique and the results section puts the approach in action and demonstrates it. The results section aims to understand why this proposed method is adopted, using both datasets from the surface and the dark web. This paper ends with concluding remarks that analyze the current state and future work that could be possibly done on this module.

5.2 Literature Review

There is a huge amount of data present on the web and this data is available for everyone to see and use. Some of this data is openly available and some is behind web forms. While it is easy to create an automated approach to access the openly available data, the same cannot be said when it comes to data behind forms. As the importance of data increases, a need for an automated form filling approach also became more prominent. This led to various researchers coming forward with their own proposals and methods. One of them proposed that there are two ways of making deep web data accessible: 1. Virtual web data integration that creates an intermediate form and presents relevant data as entered in the form by a user and 2. Surfacing or indexing the filled forms, i.e., search engine approach which involves first filling the form and getting the data and then indexing it [1].

The aim of this paper lies in being able to understand the first step to filling the forms is to understand the very nature of the HTML forms seen. The literature first presents the more notable works that are currently present and then leads to a more general view of the same.

5.2.1 Notable Approaches

5.2.1.1 Crawling the Hidden Web [14]

This proposed approach focuses on getting the data behind HTML search forms by filling proper keywords to get a better coverage of the data.

Form filling is divided into four parts: Form Analysis, Value Assignment & Submission, Response Analysis, and Response Navigation. Form Analysis is done by parsing the HTML code and getting to know the DOM structure of the page. Value assignment is done by referring to the LVS table, i.e., Label Value Set, which is maintained by HiWE for different domains and based on the method and action attribute of the form, the values are passed. Response Analysis focuses on tuning the value assignment based on the number of results obtained. Response Navigation refers to the URLs that occur on the response page, which are then added in the queue for crawling.

5.2.1.2 Google's Deep Web Crawl [15]

The next increment in the area of filling search forms is Google's deep web crawl paper. This paper focuses on surfacing deep web content. They have devised an approach to get access to the content behind the deep web search forms by using efficient queries. The ideology is that a smaller number of requests made should cover more data. The proposed approach works on the concept of informativeness of a website. Informativeness of a website defines how much the data on that website is distinct from the previously obtained data. This helps in avoiding the wastage of resources on redundant or duplicate data. If a web page has an informativeness value less than a certain threshold, then that webpage is not further accessed. The technique used for search text boxes in this paper is called iterative probing and is based on the tags generated for a certain web form page and its domain. A keyword set is generated and those keywords are used to fill searchable forms to gain more coverage of the data. Other form input types use wildcard, value, placeholder, and name attributes as the value to be filled.

To avoid wasting resources on bad requests, this is a smarter approach of finding keywords based on the content of that domain and the web form itself. This approach was able to get overall more results with a fewer number of requests. But, for a focused crawler, this method cannot be implemented as they are interested in content related to specific topics only. So, researchers interested in a focused approach have to stick with keywords related to their topic only. As for non-search forms, there has not been any concrete approach developed or proposed yet. Until now, this paper has marked the pinnacle of what an automated form filler is capable of achieving if given the proper methodology and attention.

5.2.1.3 Other Approaches

For simplistic understanding, research compared the form interface features with the cosmic universe. They compared fields to individuals (vs. star), as the basic information in a deep web form (vs. in cosmic universe). A semantic entity (vs. group of fields), such as passengers, corresponds to one galaxy. A super-group of fields (and/or concepts) corresponds to a super-galaxy containing stars (and/or other galaxies) [2].

Structures like DOM tree, segment tree, and layout tree are used in the OPAL model for form labelling with the intent of making some inferences and finding relations among form fields and database attributes [3, 4].

A hierarchical ordered schema, such as an ordered tree, can capture the semantics of an interface much better. Furthermore, this paper shows that the structure of interfaces can be exploited to help identify mappings of a field [5].

Researchers then found that the semantic or the meaning of the element is generally associated with the textual labels used along with it, so the layout of the labels was observed [6].

From the source code, there are three important attributes of the form that specify useful information for making an HTTP request: (i) the value of action attribute which corresponds to the URL where the form is to be submitted, (ii) the method attribute which specifies the type of HTTP request to be used, i.e. GET or POST, and (iii) the enctype attribute which specifies the content type that is used for submission of a form [7].

So, to have a better understanding of the search forms, it was essential to understand the search interface present on the dark web. Table 5.1 is from a survey paper based on search interface understanding. It explains the semantic information in forms and how it is present in various approaches (LEX, HSP, HMM, etc.).

Table 5.1 Summary of reductionist analysis [31].

<table>
<tr><th></th><th>Methodology</th><th>Semantic information</th></tr>
<tr><td>CombMatch, LITE LabelEx DEQUE</td><td>The methodology uses CombMatch for Chunk Partitioning and LITE for Pruning. It models in the format Logical Attribute = <text-label, form element(s)></td><td>4 label assignments</td></tr>
<tr><td>HSP</td><td>It parses using Token position and determines the model with the application of Pattern= <attr-name, operator, value></td><td>3 query conditions</td></tr>
<tr><td>LEX</td><td>It uses an interface expression (tt|tete|t|t|ee|t) for parsing and Logical Attribute = <attr-label, element label, domain/constraint element, domain type, default value> for modelling</td><td>2 logical attributes</td></tr>
<tr><td>HMM</td><td>This methodology parses using Pre-order DOM traversal and models with the equation Segment= <attr-name (s), operator(s), operantor(s), misctext(s)></td><td>2 segments</td></tr>
<tr><td>SchemaTree ExQ</td><td>Parses using text and field tokens forming a tree structure as it parses to all levels. It follows the model Tree Node = text-label or form element</td><td>1 tree</td></tr>
</table>

As understanding about the forms started increasing, there were observations made that some forms are similar in some aspects. Firstly, clustering was done with respect to the search interface, i.e., visible form features [8]. Then, there was domain specific clustering being done to make observations. Also, there was segregation of forms done with respect to AJAX, as it handles the forms indifferently [9]. This further lead to the classification of forms into a broad category, i.e., Search and Non-Search forms [10, 11]. Further, some research classified non-search forms into login, subscription, and registration forms [12].

For a search form, it gives the appropriate results only on entering the valid keywords, otherwise it always returns a *404 - Page not found* error.

Finding the valid keywords was the major task then. There was a typical approach of having a static set of keywords and using them, but the search forms generally work on GET HTTP requests. It appends the values assigned to form fields, so there can be an infinite number of distinct URLs, but most of those are going to result in a *404 - Page not found* error.

Even SQL injections were made to find the records of the database behind the search forms. The approach was designed in such a way that whenever a form is detected, an SQL injection attack is done which also helped in getting the useful keywords which can be used to fill forms [13].

Most proposed approaches kept a dataset for keywords with respect to domain. So, if a specific domain was encountered based on the text on the page, those specific keywords are used to fill the forms.

Upon referring to the literature, the following table was made which shows the distribution of progression of features of form filler in accordance to the stages or steps involved in the working of an automated form filler. The papers referred to for the literature are categorized with respect to the work done in the direction of form filling. Each feature is dedicated to a certain process which is required in an automated form filling approach.

- Downloading the Data: To download the form (HTML) and then analyze it or to directly try to analyze it without downloading it, i.e., while the surfing web on a browser
- Creation of Own Query Interface: To create own query interface to get user input for forms
- Form Classifications: Classification of HTML forms into search and non-search forms or some other classifications as different types of forms require different ways to fill them.
- Automated Form Filling of Search Forms: Whether search forms are filled using the mentioned approach
- Automated Form Filling of Non-Search Forms: Whether non-search forms are filled using the mentioned approach.
- GUI Requirement: If there any GUI requirements for using the tool/approach mentioned in the paper

The process starts with downloading of the HTML form to fill out that HTML form. In between, based on the objectives of research, certain features vary. Some papers tried to create their own query interface, whereas some tried to classify the forms into categories like search and non-search forms. Table 5.2 gives a gist of features covered by all the papers containing certain form filling approaches.

Table 5.2 Feature distribution over literature of papers.

Name of paper	Year of publication	Downloading the data	Creation of own query interface	Form classifications	Automated filling of search forms	Automated filling of non-search forms	GUI requirement
Crawling the Hidden Web [14]	2000	Yes	No	No	Yes	No	
DEQUE: Querying the Deep Web [7]	2004	Yes	Yes	No	No	No	Yes
An interactive clustering-based approach to integrating source query interfaces on the deep web [5]	2004	Yes	No	No	No	No	No
A Novel Design of Hidden Web Crawler Using Reinforcement Learning Based Agents [12]	2007	Yes	No	Yes (SNS)	No	No	No

(*Continued*)

Table 5.2 Feature distribution over literature of papers. (*Continued*)

Name of paper	Year of publication	Downloading the data	Creation of own query interface	Form classifications	Automated filling of search forms	Automated filling of non-search forms	GUI requirement
Organizing Hidden-Web Databases by Clustering Visible Web Documents [8]	2007	Yes	No	Yes (C)	No	No	No
A Framework of deep web Crawler [16]	2008	Yes	No	No	Yes	No	No
Advanced deep web crawler based on Dom [9]	2008	Yes	No	Yes (AJAX)	No	No	No
Learning to extract form labels [17]	2008	Yes	No	No	No	No	No
Google's deep web crawl [15]	2008	Yes	No	Yes (GP)	Yes	No	No
Harnessing the deep web: Present and Future [1]	2009	Yes	Yes	No	Yes	No	No

(*Continued*)

Table 5.2 Feature distribution over literature of papers. (*Continued*)

Name of paper	Year of publication	Downloading the data	Creation of own query interface	Form classifications	Automated filling of search forms	Automated filling of non-search forms	GUI requirement
Carbon: Domain-Independent Automatic Web Form Filling [18]	2010	Yes	No	No	No	No	No
Hidden web crawling for SQL injection detection [13]	2010	Yes	No	No	No	No	No
Research on discovering deep web entries Based on topic crawling and ontology [19]	2011	Yes	No	No	Yes	No	No
A New Hidden Web Crawling Approach [20]	2012	Yes	No	Yes (SNS, R)	Yes	No	No

(*Continued*)

Table 5.2 Feature distribution over literature of papers. (*Continued*)

Name of paper	Year of publication	Downloading the data	Creation of own query interface	Form classifications	Automated filling of search forms	Automated filling of non-search forms	GUI requirement
An Ontology-Based Topical Crawling Algorithm for Accessing deep web Content [21]	2012	Yes	No	No	No	No	No
Automatic discovery of Web Query Interfaces using machine learning techniques [22]	2012	Yes	No	Yes (SNS)	No	No	No
VIQI: A New Approach for Visual Interpretation of deep web Query Interfaces [23]	2012	Yes		Yes (D)	No	No	No

(*Continued*)

Table 5.2 Feature distribution over literature of papers. *(Continued)*

Name of paper	Year of publication	Downloading the data	Creation of own query interface	Form classifications	Automated filling of search forms	Automated filling of non-search forms	GUI requirement
HiCrawl: A Hidden Web Crawler for Medical Domain [24]	2013	Yes	Yes	Yes (D)	No	No	No
The ontological key: automatically understanding and integrating forms to access the deep web [3]	2013	Yes	No	Yes (F)	No	No	No
Automation of the deep web with User Defined Behaviours [25]	2013	Yes	No	No	No	No	No

(Continued)

Table 5.2 Feature distribution over literature of papers. *(Continued)*

Name of paper	Year of publication	Downloading the data	Creation of own query interface	Form classifications	Automated filling of search forms	Automated filling of non-search forms	GUI requirement
Extraction of relational schema from deep web sources: a form driven approach [26]	2014	Yes	No	No	No	No	No
Method and system for form-filling crawl and associating rich keywords [27]	2014	Yes	No	No	Yes	No	No
Focused deep web Entrance Crawling by Form Feature Classification [11]	2015	Yes	No	Yes (SNS)	Yes	No	No
Towards XML Schema Extraction from deep web [28]	2016	Yes	No	Yes (D)	Yes	No	No

(Continued)

Table 5.2 Feature distribution over literature of papers. (*Continued*)

Name of paper	Year of publication	Downloading the data	Creation of own query interface	Form classifications	Automated filling of search forms	Automated filling of non-search forms	GUI requirement
Querying deep web Data Bases without Accessing to Data [10]	2017	Yes	No	Yes (SNS)	No	No	NA
A new clustering approach to identify the values to query the deep web access forms [29]	2018	Yes	No	No	No	No	Yes

SNS: Search and Non-Search
D: Domain Based
F: Field Classification
GP: Get and Post based
C: Clustering of similar forms based on visible form features
AIAX: AIAX form and Non-AIAX form

Table 5.2 above is the culmination of all the literature that was referred to in the making of this paper and it simplifies the research in order to check the extent to which the papers demonstrate the process of form filling. As depicted in the table, many pieces of research have already focused on the basic structure of a web form, while some ontological research has studied the relation between the form fields and the database attributes behind the HTML forms. These researchers categorize forms as search forms and non-search forms. Approaches like iterative probing for finding relevant queries on search forms came forward. However, there is still no concrete and detailed research on the further classification, filling, and submission of non-search forms. The focus of this paper is to propose an automated process that determines the type of form (Search, Login, Registration, Other) and then develop and demonstrate an automated method to fill and submit all these forms.

5.3 How FLASH Offers Better Results

The approach proposed in this paper has tried to bridge the gap between the limitations of the above methods and a highly efficient form filler. The developed form filler can clear the authentication for non-searchable forms also. It's able to fill all the forms on a webpage and pass separate requests simultaneously. In a website where you can surf only if you have an account, the form filler is also able to get in using a certain credential and it surfs the entire domain while maintaining a single session with the same credentials.

5.4 Methodology

In this section, the adopted approach for the filling forms has been discussed. There are five major steps and each one is crucial for the proper working of the module. The module finds the forms on a page, categorizes them, then fills the forms in accordance to their type and generates a request that is sent to the server for fulfilment. Focus has been made to ensure the module is completely headless, i.e., there is no need to render the webpage. This special attention is warranted as it greatly improves the performance and also enables everything to be done using raw HTML.

Figure 5.1 gives a general idea of the working of FLASH. Here are two short examples which will help in getting a grasp of the methodology. These examples will cover 2 types of scenarios: single form and multiple forms on the same page

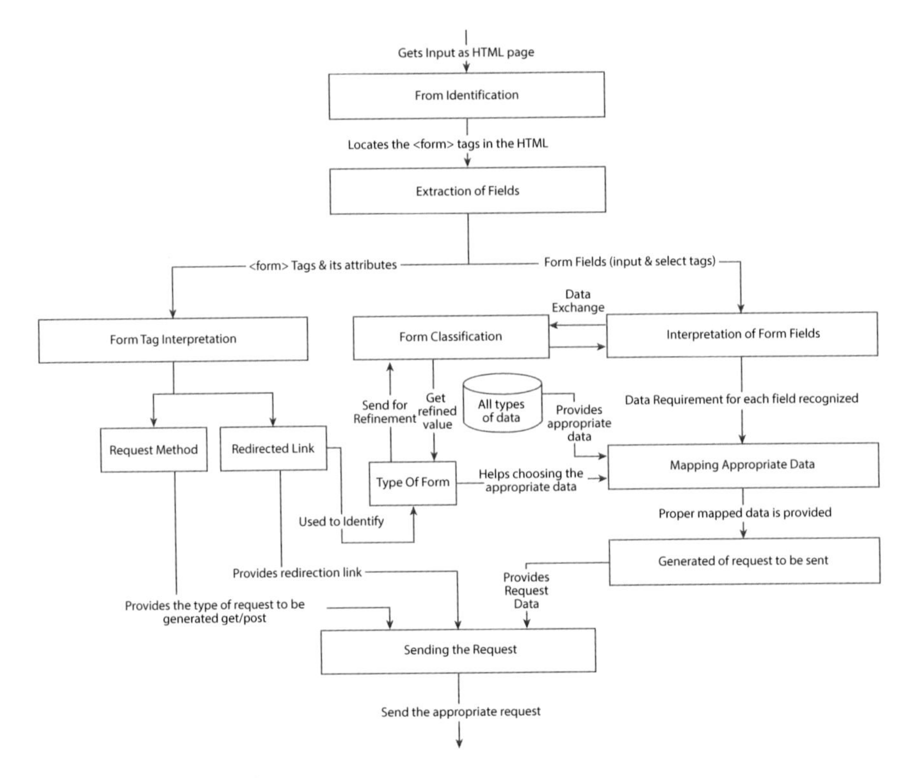

Figure 5.1 Flowchart for FLASH.

Single Form on a Page:

1. The raw HTML page is received.
2. The form identification step retrieves the form HTML tag from the page (including all the content between the start tag and the end tag: data between <form>…</form>).
3. The extracted form is divided into the data (attributes) in the <form> tag and the form fields in that form (input & select tags).
4. The form tag is interpreted to get a request method, as well the redirection link associated with the form.
   4.1 Interpretation of form fields give the type of data required for each type of form field.

4.2 The redirection link and the processing of the interpretation of the form fields are used to identify the type of the current form (Registration, Login, Search, Other).

5. The type of form along with the help of the data from the database are used to map appropriate data with each form field.
   5.1 This mapping is then used to generate the request data which is to be sent.
6. Finally, the redirection link, the request method, and the data are sent in a request and all are used to send the request and retrieve the data beyond the current form.

Multiple Forms on the Same Page:

Consider an example where the search form, registration, and login all are on the same webpage. FLASH is designed to be capable of handling such a situation as well.

1. The raw HTML page is received.
2. All the form tags are retrieved and for each one of the forms, all the analysis is done except the request sending.
   2.1 Extraction of fields, Form Tag Interpretation, Interpretation of Form Fields, Mapping of Appropriate Data, Generation of Request
3. At this point, 3 different requests are present, each one of them for each one of the form types.
4. Now, the request sending is based upon the type of form which is interpreted.
   4.1 The priority order of the forms is as follows:
      4.1.1 Search/Registration (Whichever is encountered first)
      4.1.2 Login
      4.1.3 Other
5. Here, for the login form, the request is regenerated with the data which was used in the sign-up form submission.
6. Then, the requests are sent one after another.

The following are the steps of how FLASH was fabricated and programmed in detail:

5.4.1 Form Identification and Field Extraction

5.4.1.1 *Form Identification*

The first and foremost step in filling is the identification of the form. This is a fairly simple step where the <form> HTML tag is found in the HTML. A page can have multiple forms. The sign-up form is given the max priority in case of multiple login type forms. All the data in the form tag is stored in a form object which will be used to interpret details about the form and be used to generate the request. Figure 5.2 shows about the identifying forms and extracting fields and Figure 5.3 displays about the interpretation of form fields and Figure 5.4 is the framework for generating and sending report.

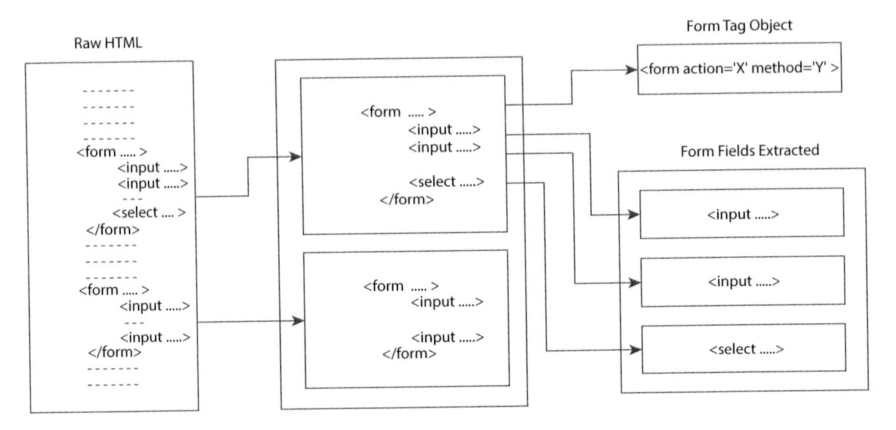

Figure 5.2 Identifying forms and extracting fields.

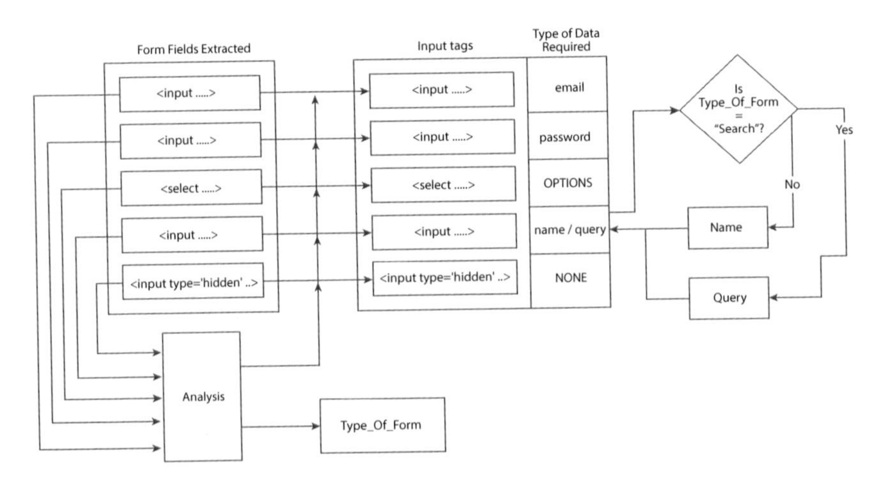

Figure 5.3 Interpretation of form fields.

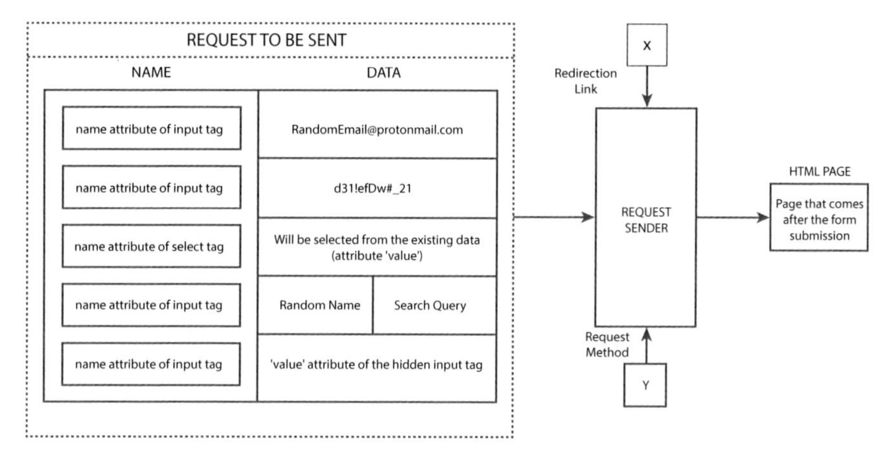

Figure 5.4 Generating and sending report.

5.4.1.2 *Extraction of Fields*

For each form extracted, the data to fill in the forms is to be identified. Hence, the data is extracted in 2 parts:

Firstly, this can be done using the tag attributes of the form HTML tag.

The second way is with the form fields. These fields include normal input tags as text fields for name, search query, email, or input tags as radio buttons, checkboxes, or a drop-down with a select tag. All the types of inputs are managed by FLASH.

5.4.2 Form HTML Tag Interpretation

The form tag object will have its own attributes like action and method. The form tag object is received from the second step. The action gives the page where the request is to be redirected, whereas the method represents the request method that is accompanied with the form.

5.4.2.1 *Identification of Form Type*

If the action redirects to a search page, the type of form is set as search. If the attribute does not exist, the redirection link is then, the domain name. If the redirection link is '#', the redirection is on the same page itself. If the value for the action value starts with a '/', then the current link is appended with the value which will then be used as the redirection link.

5.4.2.2 *Identification of Request Method*

The value of the method attribute of the form gives the type of request which is to be sent. This value usually is 'GET' or 'POST'. The form filling methodology handles both GET as well as POST forms. For search forms, it is usually a GET form and for registration and login forms are usually in POST form.

5.4.3 Interpretation of Form Fields

The extracted form fields are now interpreted to understand what kind of data is to be put into these tags. To get the interpretation, the tags are sent into the analysis function, which analyses the input tags and gives the type of data that is required. If the data required is NONE, then no data has to be given as input from the module side. The data will be readily available on the webpage itself as the 'value' attribute of the input tag.

Each of the form's inputs are classified into different parts to understand which kind of data is to be sent. The inputs are classified upon various levels to ensure proper data mapping.

The classification is based upon the attributes of the tag. Several lists are generated of each type of data to be mapped. Data is mapped in the format {input tag: type of data to be mapped}.

The hierarchy order of the classifying attributes is given below:

0. Type (Hidden): Usually have a predefined value (often dynamic and generated by JS)
1. Type (Everything other than Text)
2. Placeholder
3. Role
4. Name
5. Type (Text)
6. Value (the predefined value if any)

If the tag is not classified according to the above, then it is considered normal text.

Once all tags are mapped according to their type of data, the appropriate data is added to create a request. The request mappings are then re-arranged to match the order of inputs in the original form as the POST request follows a top-down approach, i.e., the first input then the second and similarly for each input in the form.

During the classification, the values of the attributes of placeholder, name all are also considered and analysed at the same time to understand the type of form as a login/registration or a search form. These attributed values are compared to a list of search related words. If the any of the values are present in the list, the form is deemed to be a search form. This is a by-product of the analyser.

If the form is a search form, all the instances of data requirement of type 'name' are converted into a data type requirement of 'query'.

For select tags (drop downs), input types checkbox, radio, or date, the type requirement is set to option.

5.4.4 Form Classification

The classification is in 7 types: Search, Login, Register, Reset Password, Broken Form, No Form, and Other. The other category would include forms like feedback forms, payment forms, etc. For the purpose of this paper, the definitions of each form type were predefined.

While the forms Search, Login, Register, Reset Password are all self-explanatory, the definitions of Not Form and Broken Form were defined.

A form is classified as 'Not Form' if that form tag does not have any sort of input attributes. For form filling, there is no use of this type of form unless the data in the form is generated using JavaScript.

A form is classified as 'Broken Form' if that form tag does not have any user input field. At most, that form would have a submit button for form submission. Examples for this type of form would be the form tags which are used to initialize some browser variables (usually hidden-type input tags). This is a form for the 'add to cart' (submit button) feature for a marketplace type of website. The form which just displays some dynamic value to the user where the user can just click submit.

In this step, the forms are categorized to get the type of the current form. It can be Search, Login, Register, or any other type of form. The type of form received from the interpretation of the form HTML tag gives the baseline for what the type of form will be. This value is now further refined by this step-in conjunction to the analysis provided by the interpretation of the fields. The same has been shown in Table 5.3 where it is seen how the program decides on what type of form is to be assigned in any criteria. The X in the same table is to identify any type of form which was interpreted by analyzing the FORM tag.

To analyze for the type of form, the check will be done by checking the sub-strings for the attributes. For an example, in a search form, words like 'search' or 'find' would be in the form tags or any of the other attributes.

A few of the attributes of the form fields are really important in the classification of the type of form. The most important one is the 'role' attribute. Usually, the role attribute is used to define the role of a particular HTML

Table 5.3 Criteria for form classification.

	Number of fields for each input type							
Type of form received from FORM tag	**Hidden**	**Submit/ button**	**Search**	**Name/ text**	**Email**	**Password**	**Other**	**Type of form assigned**
X	0	0	0	0	0	0	0	Not Form
X	>= 0	>= 0	0	0	0	0	0	Broken Form
X	>= 0	>= 0	>= 1	>= 0	>= 0	>= 0	>= 0	Search
X	>= 0	>= 0	0	1	0	1	>= 0	Login
X	>= 0	>= 0	0	0	1	1	>= 0	Login
Register / Login	>= 0	>= 0	0	1	1	1	>= 0	Register
X	>= 0	>= 0	0	1 OR 2	>= 0	0	>= 0	Reset Password
X	>= 0	>= 0	0	>= 0	1 OR 2	0	>= 0	Reset Password
Register / Login	>= 0	>= 0	0	0	0	3 OR 4	>= 0	Reset Password
Register / Login	>= 0	>= 0	0	>= 1	>= 1	>= 2	>= 0	Register
X	>= 0	>= 0	0	0	>= 2	>= 1	>= 0	Register

tag. The next most important attribute to be analyzed is the placeholder, since it is the data that a user will see to enter the data. If at least 2 attributes from different tags indicate to the same type, the form is considered to be the considered type of form.

5.4.5 Generation and Sending of Request with Appropriate Data

In this step, the data is now mapped per the data requirement. For email, name, password, and search queries, they would all be stored into a database of predefined credentials.

A separate database is also maintained which keeps track of the request sent for a registration form, as well as the domain name to which the request was sent. Then, in the login form for the same domain name, the registration credentials are used to map the data.

In case of the type requirement 'OPTION', data is used which is present in the 'value' attribute of the input/select tag. In case of a select tag, a random number will be picked. If in a select tag the values are numeric values, the value attribute of the data > 18 will be selected. If the data to be selected is a 4-digit number, the number selected will be at least 20 numbers away from the default number. For radio, random is selected. For checkbox, if the option for check box is only 1, then it is selected as it can be agreed to terms and conditions and one is not allowed to log in without it. For multiple options for a checkbox, at least one is selected, then some random amount of checkbox will be selected in the consideration.

5.4.5.1 Generation of the Request

The data is received from the mapping of data. A request is in the format {name1: value1, name2: value2, ...}. Hence, all the name attributes of the input tags are extracted with their mapping with the data. This makes the request which will be sent in the next step.

5.4.5.2 Sending the Request

The details about where and how to send the request are generated from 3.3. Now, for the final step, the request is to be sent. A request will be sent as per the 'request method' to the 'redirection link'. This part is handled by the requests library of python, which generates the request and sends it. For a 'GET' request, the data is appended with the redirection link. In the terms of the requests library, the 'GET' request will be sent to the redirection link with the parameters generating the request data. A similar approach is followed for a 'POST' request. The requests library of python takes care of sending the request in the same way one's browser would.

After the final step, the received HTML page/session after the form submission is sent back to the crawler to continue further crawling.

5.5 Results

In the previous section, the approach to the problem was discussed. In this section, the results of the implementation of FLASH in a real world as well as a control environment are discussed. This will help to analyze how the form filler performs in the real world.

5.5.1 Classification of Forms

All the previous research on filling web forms was mostly based on analysis of the web forms. Few papers were found which discussed the actual automated filling of web forms, but these papers don't clearly mention the method used and their algorithms are not made public. Due to this, the comparative analysis of FLASH was not possible.

5.5.2 Dataset

The dataset selection and results have been divided into two different categories. In order to maintain that FLASH can work under most conditions faced, the researchers have manually created a smaller dataset which is a random set of all kinds of found on the internet. The dark web datasets are used in order to observe the performance of the form filler on the dark web sites, i.e., the onion links.

The categorization is as follows:

5.5.2.1 Manual

In manual types of datasets, each one of the links which is added to the dataset is surely to have at least one form on the webpage. It is manually ensured by the researchers that the dataset contains each one of the main types of forms included. These datasets are used to find the accuracy in terms of identifying the type of form. This allows for a comprehensive testing of the form classification and filling modules in all aspects.

5.5.2.2 *Automated*

Automated datasets have a huge number of links (usually in hundreds or thousands, sometimes even more). This allows for rigorous testing of the form filler module. Since manual verification of all the links of the full dataset cannot be done, the inferences noted by the form classifier and filler have been noted as well as visualized.

5.5.3 Accuracy Measures

For the calculation of accuracy, two crucial aspects, form filling and analysis, were considered. Namely, 'Accuracy of Form Classification' and 'Accuracy of Form Filling'. These accuracy values have only been calculated for the manual datasets of the surface web and dark web, as the calculation of accuracy involves manually verifying the type of form and successful filling for each one of the forms in the dataset.

- Accuracy of Form Classification = Number of forms correctly Identified / Total Number of Forms
  Accuracy of Form Classification signifies the precision of FLASH in categorization/classification of forms. For a form to be correctly identified, it has been classified (by FLASH) into the same category as the form type decided by the researchers.
- Accuracy of Form Filling = Number of forms correctly Filled / Total Number of Forms
  Accuracy of Form Filling is the success measure of FLASH in filling a form. For a form to be correctly filled, the response after filling the form should be acceptable to a human (in this case researcher). Consider these examples:
  - On searching the keyword 'cricket' on a search form, the human would either expect results related to the keyword 'cricket' or a message from the website saying no results found. Any other response would be considered to be incorrect.
  - On a registration form, the expectation is to get an account registered on the website. Any other response would be deemed incorrect.
  - On a login form, the expectation is to either login into the account or receive a message from the website saying incorrect username, password, no account found, etc. Any other response would be considered incorrect.

- For rest of the form types, no general rule can be considered as those forms may be really distinct by themselves and the success or failure of the FLASH approach would be verified by the researcher itself.

5.5.4 Manual Dataset

For a manual dataset, there are 2 types of datasets which were manually collected and verified for their form type. This includes a surface web dataset and dark web dataset, as seen in Table 5.4 and Table 5.5, respectively.

Websites in both of the manual datasets have been verified by the researchers for the classification of the form. Additionally, it was verified that at the time of testing all websites were up and running. It has also

Table 5.4 Manual surface web dataset.

Form type	Number of web forms
Broken Form	3
Login	15
Not Form	24
Other	20
Register	9
Reset Password	6
Search	100

Table 5.5 Manual dark web dataset.

Form type	Number of web forms
Broken Form	54
Login	20
Not Form	28
Other	12
Register	9
Search	43

been ensured that all of the types of forms are covered. Some websites with 'Broken Form' and 'Not Form' type of forms have been intentionally added to the dataset to ensure proper working for each and every type of form.

5.5.4.1 *Form Classification*

For the surface web, some websites consider Sign Up as a Login feature itself. Due to this discrepancy, the form classification is considered combined as 'Login/Register'. The search was dominant in the dataset having more than 50% forms in the dataset being search forms. For the dark web, the concepts of Login and Register both were clearly distinct on the websites. The 'Other' category of forms could be forms like payment forms, feedback forms, etc.

Surface Web

Table 5.6 Accuracy of form classification on manual surface web.

Form type	Count	Correctly identified	Incorrectly identified	Accuracy (in %)
Broken Form	3	3	0	100.00
Login	15	6	9	40.00
Not Form	24	12	12	50.00
Other	20	9	11	45.00
Register	9	9	0	100.00
Reset Password	6	3	3	50.00
Search	100	87	13	87.00
Grand Total	177	129	48	72.88

Dataset

Table 5.6 is the calculation for the accuracy of the identification methodology in FLASH. The approach was 100% accurate in classification of the surface web's search type forms and couldn't correctly identify 9 of the Login/Register forms. FLASH achieved approximately 72.88% accuracy overall in classification of forms from the manual surface web dataset.

In order to understand where the methodology lacks, Table 5.7 presents a correlation matrix which can show which forms were incorrectly identified.

Table 5.7 Correlation matrix for manual surface web dataset.

		FLASH outcomes							
		Broken form	**Login**	**Not form**	**Other**	**Register**	**Reset password**	**Search**	**Grand total**
Actual Value	Broken Form	3	0	0	0	0	0	0	3
	Login	2	6	3	1	3	0	0	15
	Not Form	6	1	12	4	0	0	1	24
	Other	0	1	6	9	1	2	1	20
	Register	0	0	0	0	9	0	0	9
	Reset Password	0	0	1	2	0	3	0	6
	Search	0	2	8	2	1	0	87	100
	Grand Total	11	10	30	18	14	5	89	177

Some clear deductions that are visible: while Register forms are always identified correctly, they are also often confused with Login forms. This suggests that there are some improvements required here and can be analyzed using the data collected for further future improvements.

FLASH achieved 90% accuracy in classifying the 'Login'. The 'Other' type of forms get the advantage of uncertainty, as precise classification for them is not done. An overall accuracy of 97.59% in classification for forms present in the dark web manual dataset.

Dark Web

The dark web form failures were the ones where the form's placeholders for input tags, as well as the website itself, were in a different language. Otherwise, the dark web's forms were easily classifiable as the distinguishing factors among the forms. Complications like using Sign Up forms as Login are not present. Table 5.8 presents the details for accuracy of form classification on manual dark web dataset, while Table 5.9 represents the correlation matrix for manual dark web dataset, whereas Table 5.10 shows the results of form filling for surface web dataset and similar Table 5.11 shows the results of form filling for dark web dataset. Table 5.12 is the presentation of automated surface web dataset and Table 5.13 is presentation for automated dark web dataset.

Table 5.8 Accuracy of form classification on manual dark web dataset.

Form type	Count	Correctly identified	Incorrectly identified	Accuracy (in %)
Broken Form	54	54	0	100.00
Login	20	18	2	90.00
Not Form	28	28	0	100.00
Other	12	12	0	100.00
Register	9	7	2	77.78
Search	43	43	0	100.00
Grand Total	166	162	4	97.59

Table 5.9 Correlation matrix for manual dark web dataset.

		FLASH outcomes						
		Broken form	Login	Not form	Other	Register	Search	Grand total
Actual Values	Broken Form	54	0	0	0	0	0	54
	Login	0	17	1	0	2	0	20
	Not Form	0	0	28	0	0	0	28
	Other	1	0	0	11	0	0	12
	Register	0	2	0	0	7	0	9
	Search	0	0	0	0	0	43	43
	Grand Total	55	19	29	11	9	43	166

Table 5.10 results of form filling for surface web dataset.

	Fail	Success	Grand total
Login	13	2	15
Other	17	3	20
Register	7	2	9
Reset Password	6	0	6
Search	31	69	100
Grand Total	74	76	150

Table 5.11 Results of form filling for dark web dataset.

	Fail	Success	Grand total
Login	5	15	20
Other	3	9	12
Register	6	3	9
Search	3	40	43
Grand Total	17	67	84

Table 5.12 Automated surface web dataset.

	Number of web forms
Total URLs	1055
Total Forms	2290

Table 5.13 Automated dark web dataset.

	Number of websites
Total URLs	10367
URLs Unreachable	10304
URLs without Forms	47
URLs with Forms	16

5.5.4.2 *Form Filling*

Surface Web

FLASH was successfully able to fill each and every search form from the manual surface web forms successfully. Overall, FLASH successfully filled almost 51% of the forms from the manual surface web dataset. Figure 5.5 shows about the surface web results and Figure 5.6 shows the dark web results.

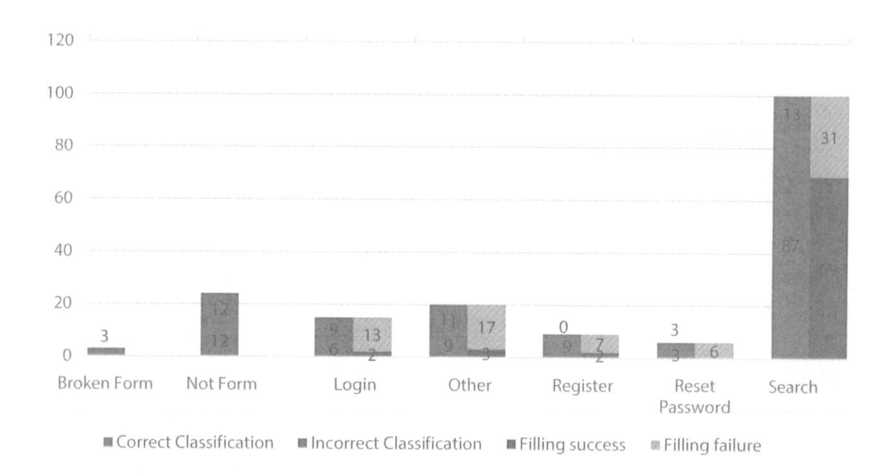

Figure 5.5 Surface web results.

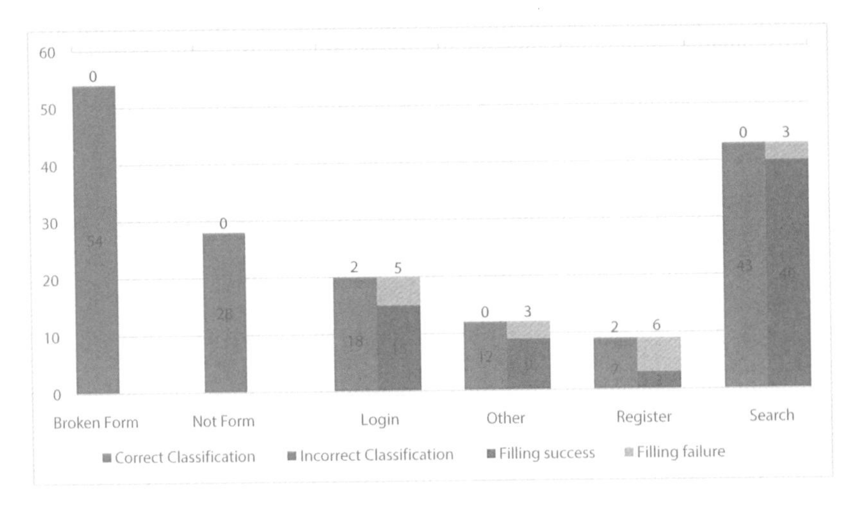

Figure 5.6 Dark web results.

Dark Web
FLASH achieved about 80% successful form fills of the dark web manual dataset. It was able to successfully return with a session from the majority of Login pages (75%) and also manages to Register on 3 websites. On further inspection, it was discovered that the failure to register, both on the surface and dark web, was due to captcha, an element that was not yet worked upon in the FLASH structure.

5.5.5 Automated Datasets

There is an absence of datasets that contain form-field categorizations and this makes it difficult to provide a proper justification, making the selection of automated databases tricky. A total of 2 databases were used to verify this algorithm: one for the surface web and one for the dark web. For a surface web automated dataset, we created the V-10 Database. This dataset is extremely detailed and contains multiple attributes that allow it to provide much more detailed insights into how the FLASH algorithm performs. For the dark web automated dataset, DUTA's 10K has been used. It has over 10,000 dark web links. These massive datasets will be beneficial for rigorous and unbiased testing of form classification and form filling.

Due to the size of the dataset, manual verification of the website's type of form is an extensive task. Thus, inferences have been gathered by form classification and form filling using FLASH. Table 5.14 shows about the classification of forms on automated surface web dataset, and Table 5.15 represents the classification of forms on automated dark web dataset. Table 5.16 shows the

Table 5.14 Classification of forms on automated surface web dataset.

Type of form	Broken form	Login	Not form	Other	Register	Reset password	Search	Grand total
Number of Forms	331	82	89	472	126	34	1156	2290

Table 5.15 Classification of forms on automated dark web dataset.

Type of form	Broken form	Login	Not form	Other	Register	Reset password	Search	Grand total
Number of Form	0	4	2	12	4	1	3	26

input tag field distribution for automated surface web dataset and Table 5.17 represents the input tag field distribution for automated dark web dataset.

5.5.5.1 *Form Classification*

Surface Web (V10 Dataset)
As observed, approximately 50% of the forms in the V-10 Database are search forms, followed by the 'Other' type of forms. It is followed by 'Broken Form' and 'Register' forms. Finally, Login, 'Not Form', and Reset Password types of forms are found.

'Broken Forms' are the forms which do require user input in some way but do not contribute to any type of authentication nor to any type of searching feature. For example, a form which has only an 'Add to Cart' button and nothing else would be considered a 'Broken Form'.

'Not Form' is a form tag which has no effect on the webpage. That form tag is just to add some sort of dynamic content into the webpage.

Dark Web (DUTA Dataset)
In the DUTA dataset, a majority of the forms found were of the 'Other' type, followed by Login and Register Forms, accompanied by Search Forms and 'Not Form' forms. The least frequent was the Reset Password form. Broken Forms did not make any appearance.

5.5.5.2 *Form Input-Tag Inference*

Surface Web (V10 Dataset)
'Other' type forms have the maximum amount of input tags compared to any other type of form in the V-10 Dataset. None of the forms in the V-10

Table 5.16 Input tag field distribution for automated surface web dataset.

Type of form	Hidden	Search	Name	Email	Password	Select	Gender	Radio	Checkbox	Button	Total
Login	217	0	143	58	52	16	0	1	36	0	523
Other	1571	0	1368	295	57	178	0	29	123	3	3624
Register	251	0	110	42	11	8	0	1	10	1	434
Reset Password	91	0	15	74	10	0	0	2	6	1	199
Search	567	827	0	30	48	171	0	8	59	0	1710
Total	2697	827	1636	499	178	373	0	41	234	5	6490

Table 5.17 Input tag field distribution for automated dark web dataset.

Type of form	Hidden	Search	Name	Email	Password	Select	Gender	Radio	Checkbox	Button	Total
Login	1	0	1	4	5	0	0	0	2	0	13
Not Form	0	0	0	0	0	0	0	0	0	0	0
Other	16	0	12	3	0	6	0	2	1	0	40
Register	6	0	5	3	5	2	0	0	5	0	26
Reset Password	0	0	0	1	0	0	0	0	0	0	1
Search	1	2	0	1	0	2	0	0	0	0	6
Total	24	2	18	12	10	10	0	2	8	0	86

Dataset have a gender type of input. Search is the only dataset which has a searchable kind of input.

Dark Web (DUTA Dataset)
'Other' type forms have the maximum amount of input tags compared to any other type of form in the DUTA Dataset [30]. None of the forms in the DUTA Dataset have a gender or button type of input. Search is the only dataset which has a searchable kind of input.

5.6 Limitations and Future Work

This section is dedicated to the limitations of the current form filling approach and the future work which will overcome some of those limitations and will lead to more effective and efficient form filling techniques.

5.6.1 Limitations

The success ratio for filling register forms is lower than expected, which is a limitation. The current form filler is able to clear all the basic HTML forms and is also able to pass keywords to search forms. But there are a few limitations to the current version.

- Forms driven by JavaScript
- Forms containing CAPTCHA

The form filler will face an issue only if one of the above conditions is true, otherwise the form filler is able to do its job perfectly.

5.6.2 Future Work

Web forms nowadays have captchas as one of the form fields. Captchas were made to distinguish between a human accessing a server and a machine accessing a server. Generally used captchas are text-based images. So, a feature can be added to the form filler to break the captcha code. To do that, the methodology should be able to render the image, recognize text, perform segmentation of the text, i.e., recognize each character individually, and pass the data in the post request while maintaining the same session.

Using Machine Learning/AI techniques to aid form classification could lead to even higher form classification accuracy.

Web forms are made up of HTML and JavaScript and these forms generally handle events using JavaScript, so to handle that the methodology must be able to parse JavaScript perfectly in order to deal with forms using JS (JavaScript, jQuery, AJAX).

Also, there are some forms which handle form validation on a very extreme level, for example they pass each character as you type in the form of POST requests in one single session. So, to deal with this kind of form a mechanism that sends individual characters in post requests while maintaining one session is needed.

Weapons smuggling [32, 33], drug trading [34], and the dissemination of unethical content by cybercriminals [35, 36], such as obscenity and violent imagery [37], are some of the most common illicit activities that take place on the dark web. Sites promote the ideologies of white supremacists, neo-Nazis, and other radical and cyberterrorist [35–37] organisations.

5.7 Conclusion

Filling of forms would grant access to immense amounts of resources for a crawler which wouldn't be available otherwise. With this paper, the researchers aim to take a crucial step into Automated Form Classification, Form Filling, and Submission on the Web. The form filler described is successfully able to detect all the web forms on a web page. All the form fields are detected and filled with appropriate data. The steps have been broken down into their simplest components allowing them to put forth the straightforward methodology seen in the paper. While there is always going to be a scope of improvement, the researchers aim to lay the groundwork and also provide a new dataset by the end of this paper. This will allow for filling the gap present in successful comparison and testing of different form filling methodologies.

FUNDING: This study has not been funded by any grants.

CONFLICT OF INTEREST: The authors declare that they have no conflict of interest.

References

1. Jayant Madhavan, Loredana Afanasiev, Lyublena Antova, Alon Halevy. 2009. Harnessing the deep web: Present and Future. arXiv preprint arXiv:0909.1785
2. Radhouane Boughammoura, Lobna Hlaoua and Mohamed Nazih Omri. 2015. G-Form: A Collaborative Design Approach to Regard deep web Form as Galaxy of Concepts. In International Conference on Cooperative Design, Visualization and Engineering (pp. 170-174). Springer, Cham. https://doi.org/10.1007/978-3-319-24132-6_20

3. Tim Furche, Georg Gottlob, Giovanni Grasso, Xiaonan Guo, Giorgio Orsi, Christian Schallhart. 2013. The ontological key: automatically understanding and integrating forms to access the deep web. The VLDB Journal, 22(5), 615-640. https://doi.org/10.1007/s00778-013-0323-0
4. Rui Zhang, Derong Shen, Yue Kou, Tiezheng Nie. 2010. Author Name Disambiguation for Citations on the deep web. International Conference on Web-Age Information Management (pp. 198-209). Springer, Berlin, Heidelberg. https://doi.org/10.1007/978-3-642-16720-1_21
5. Wu, W., Yu, C., Doan, A., & Meng, W.. 2004. An interactive clustering-based approach to integrating source query interfaces on the deep web. Proceedings of the 2004 ACM SIGMOD international conference on Management of data (pp. 95-106). https://doi.org/10.1145/1007568.1007582
6. Lu Jiang, Zhaohui Wu, Qinghua Zheng, Jun Liu. 2009. Learning deep web Crawling with Diverse Features. Proceedings of the 2004 ACM SIGMOD international conference on Management of data (pp. 95-106). https://doi.org/10.1145/1007568.1007582
7. Denis Shestakov, Sourav S. Bhowmick , Ee-Peng Lim. 2004. DEQUE: querying the deep web. Data & Knowledge Engineering, 52(3), 273-311. https://doi.org/10.1016/j.datak.2004.06.009
8. Luciano Barbosa, Juliana Freire, Altigran Silva. 2007. Organizing Hidden-Web Databases by Clustering Visible Web Documents. 2007 IEEE 23rd International Conference on Data Engineering (pp. 326-335). IEEE. https://doi.org/10.1109/ICDE.2007.367878
9. Weicheng Ma, Xiuxia Chen, Wenqian Shang. 2008. Advanced deep web crawler based on Dom. 2012 Fifth International Joint Conference on Computational Sciences and Optimization (pp. 605-609). IEEE. https://doi.org/10.1109/CSO.2012.138
10. Dr. Radhouane Boughammoura, Pr. Mohamed Nazih Omri. 2017. Querying deep web Data Bases without Accessing to Data. 2017 13th International Conference on Natural Computation, Fuzzy Systems and Knowledge Discovery (ICNC-FSKD) (pp. 597-603). IEEE. https://doi.org/10.1109/FSKD.2017.8393338
11. Lin Wang, Ammar Hawbani and Xingfu Wang. 2015. Focused deep web Entrance Crawling by Form Feature Classification. International Conference on Big Data Computing and Communications (pp. 79-87). Springer, Cham. https://doi.org/10.1007/978-3-319-22047-5_7
12. J. Akilandeswari, N.P. Gopalan. 2007. A Novel Design of Hidden Web Crawler Using Reinforcement Learning Based Agents. International Workshop on Advanced Parallel Processing Technologies (pp. 433-440). Springer, Berlin, Heidelberg. https://doi.org/10.1007/978-3-540-76837-1_47
13. Xin Wang, Luhua Wang, Gengyu Wei, Dongmei Zhang, Yixian Yang. 2010. HIDDEN WEB CRAWLING FOR SQL INJECTION DETECTION. 2010 3rd IEEE International Conference on Broadband Network and Multimedia Technology (IC-BNMT) (pp. 14-18). IEEE. https://doi.org/10.1109/ICBNMT.2010.5704860

14. Sriram Raghavan, Hector Garcia-Molina. 2000. Crawling the Hidden Web. Stanford. http://ilpubs.stanford.edu:8090/456/
15. Madhavan, Jayant, David Ko, Łucja Kot, Vignesh Ganapathy, Alex Rasmussen, and Alon Halevy. 2008. Google's deep web crawl. Proceedings of the VLDB Endowment, 1(2), 1241-1252. https://doi.org/10.14778/1454159.1454163
16. Xiang Peisu, Tian Ke, Huang Qinzhen. 2008. A Framework of deep web Crawler. 2008 27th Chinese Control Conference (pp. 582-586). IEEE. https://doi.org/10.1109/CHICC.2008.4604881
17. Nguyen, H., Nguyen, T., & Freire, J. 2008. Learning to extract form labels. Proceedings of the VLDB Endowment, 1(1), 684-694. https://doi.org/10.14778/1453856.1453931
18. Samur Araujo, Qi Gao, Erwin Leonardi, Geert-Jan Houben. 2010. Carbon: Domain-Independent Automatic Web Form Filling. International Conference on Web Engineering (pp. 292-306). Springer, Berlin, Heidelberg. https://doi.org/10.1007/978-3-642-13911-6_20
19. Gang Liu, Kai Liu, Yuan-yuan Dang. 2011. Research on discovering deep web entries Based ontopic crawling and ontology. 2011 International Conference on Electrical and Control Engineering (pp. 2488-2490). IEEE. https://doi.org/10.1109/ICECENG.2011.6057954
20. L. Saoudi, A. boukerram, S. mhamedi. 2012. A New Hidden Web Crawling Approach. International Journal of Advanced Computer Science and Applications (IJACSA), 6(10). http://dx.doi.org/10.14569/IJACSA.2015.061039
21. K. V. Arya, Baby Ramya Vadlamudi. 2012. An Ontology-Based Topical Crawling Algorithm for Accessing deep web Content. 2012 Third International Conference on Computer and Communication Technology (pp. 1-6). IEEE. https://doi.org/10.1109/ICCCT.2012.10
22. Heidy M. Marin-Castro, Victor J. Sosa-Sosa, Jose F. Martinez-Trinidad, Ivan Lopez-Arevalo. 2012. Automatic discovery of Web Query Interfaces using machine learning techniques. Journal of Intelligent Information Systems, 40(1), 85-108. https://doi.org/10.1007/s10844-012-0217-4
23. Radhouane Boughammoura, Lobna Hlaoua, Mohamed Nazih Omri. 2012. VIQI: A New Approach for Visual Interpretation of deep web Query Interfaces. 2012 8th International Conference on Computing Technology and Information Management (NCM and ICNIT) (Vol. 1, pp. 344-349). IEEE. https://doi.org/10.1109/ICITeS.2012.6216656
24. Sonali Gupta, Komal Kumar Bhatia. 2013. HiCrawl: A Hidden Web Crawler for Medical Domain. 2013 International Symposium on Computational and Business Intelligence (pp. 152-157). IEEE. https://doi.org/10.1109/ISCBI.2013.39
25. Vicente Luque Centeno, Carlos Delgado Kloos, Peter T. Breuer, Luis Sánchez Fernández, Ma. Eugenia Gonzalo Cabellos, Juan Antonio Herráiz Pérez. 2013. Automation of the deep web with User Defined Behaviours. International Atlantic Web Intelligence Conference (pp. 339-348). Springer, Berlin, Heidelberg. https://doi.org/10.1007/3-540-44831-4_35

26. Yasser Saissi, Ahmed Zellou, Ali Idri. 2014. Extraction of relational schema from deep web sources: a form driven approach. 2014 Second World Conference on Complex Systems (WCCS) (pp. 178-182). IEEE. https://doi.org/10.1109/ICoCS.2014.7060888
27. Dalvi, N., Ramakrishnan, R., Kakade, V., Choudhury, A. K., Selvaraj, S. K., Bohannon, P.,... & Kirpal, A. S. 2014. Method and system for form-filling crawl and associating rich keywords. U.S. Patent No. 8,793,239. Washington, DC: U.S. Patent and Trademark Office. https://patents.google.com/patent/US8793239B2/en
28. Yasser Saissi, Ahmed Zellou, Ali Idri. 2016. Towards XML Schema Extraction from deep web. 2016 4th IEEE International Colloquium on Information Science and Technology (CiSt) (pp. 94-99). IEEE. https://doi.org/10.1109/CIST.2016.7805022
29. Saissi, Yasser, Ahmed Zellou, and Ali Idri. 2018. A new clustering approach to identify the values to query the deep web access forms. 2018 4th International Conference on Computer and Technology Applications (ICCTA) (pp. 111-116). IEEE. https://doi.org/10.1109/CATA.2018.8398666
30. Mhd Wesam Al-Nabki and Eduardo Fidalgo and Enrique Alegre and Laura Fernández-Robles. 2019. ToRank: Identifying the most influential suspicious domains in the Tor network. Expert Systems with Applications. https://doi.org/10.1016/j.eswa.2019.01.029
31. Ritu Khare, Yuan An, and Il-Yeol Song. 2010. Understanding deep web search interfaces: a survey. SIGMOD Rec. 39, 1 (March 2010), 33–40. DOI:https://doi.org/10.1145/1860702.1860708
32. Rawat, R., Chakrawarti, R. K., Vyas, P., Gonzáles, J. L. A., Sikarwar, R., & Bhardwaj, R. 2023. Intelligent Fog Computing Surveillance System for Crime and Vulnerability Identification and Tracing. *International Journal of Information Security and Privacy (IJISP)*, 17(1), 1-25.
33. Rawat, R., Sowjanya, A. M., Patel, S. I., Jaiswal, V., Khan, I., & Balaram, A. (Eds.). 2022. *Using Machine Intelligence: Autonomous Vehicles Volume 1*. John Wiley & Sons.
34. Rawat, R., Bhardwaj, P., Kaur, U., Telang, S., Chouhan, M., & Sankaran, K. S. 2023. *Smart Vehicles for Communication, Volume 2*. John Wiley & Sons.
35. Rawat, R. 2023. Logical concept mapping and social media analytics relating to cyber criminal activities for ontology creation. *International Journal of Information Technology*, 15(2), 893-903.
36. Mahor, V., Bijrothiya, S., Rawat, R., Kumar, A., Garg, B., & Pachlasiya, K. 2023. IoT and Artificial Intelligence Techniques for Public Safety and Security. *Smart Urban Computing Applications*, 111.
37. Rawat, R., Mahor, V., Álvarez, J. D., & Ch, F. 2023. Cognitive Systems for Dark Web Cyber Delinquent Association Malignant Data Crawling: A Review. *Handbook of Research on War Policies, Strategies, and Cyber Wars*, 45-63.

6

Performance Analysis of Terahertz Microstrip Antenna Designs: A Review

Ram Krishan

Department of Computer Science, Mata Sundri University Girls College, Mansa, Punjab, India

Abstract

Terahertz (THz) technology has gained considerable research attention in recent years as a result of its growing number of applications in spectroscopy, medicine, communication, material characterisation, sensing, and imaging. Terahertz antennas are key devices for sending and receiving THz electromagnetic waves in upcoming THz systems because of their tiny size, wide frequency bandwidth, and high data rate. Between the microwave and infrared frequency bands, the THz is a group of frequencies ranging from 0.1 to 10 THz. The great penetration with minimal attenuation and high-resolution imaging capacity of THz technology is one of the reasons for its popularity. This chapter will compare and contrast the performance of microstrip THz antennas composed of various substrate materials.

***Keywords*:** Terahertz (THz), performance, microstrip antenna, analysis, process, challenges

6.1 Introduction

Terahertz (THz) waves, also known as sub-millimetre waves, are found in the electromagnetic spectrum between the optical and microwave frequency ranges [1]. The useable frequency band is bigger than millimetre waves, the beam direction is stronger, and the secrecy and anti-interference performance are superior. THz waves are more efficient and have greater

Email: ramkrishan@pbi.ac.in

Romil Rawat, Rajesh Kumar Chakrawarti, Sanjaya Kumar Sarangi, Rahul Choudhary, Anand Singh Gadwal and Vivek Bhardwaj (eds.) Robotic Process Automation, (101–112) © 2023 Scrivener Publishing LLC

penetration than light waves [2]. THz antennas have the optimum performance because of their wide operational bandwidth, which is based on the distinctive features of the THz waves. In THz wireless communication systems, THz antennas are necessary for sending and receiving THz waves [3]. The performance of THz antennas has a direct impact on the overall system's quality, particularly the operational bandwidth and gain of the antennas [4, 5]. Furthermore, THz antennas face numerous new obstacles when compared to microwave antennas [6]. The gadget size is substantially decreased because the THz antenna operates in the high-frequency range [7]. Materials and production processes limit the packing of THz antennas [8]. Another issue in terahertz antennas is determining how to make them radiate effectively [9]. As a result, THz antennas must meet higher standards in terms of antenna design, manufacturing materials, and process technology [10]. Figure 6.1 shows about the Terahertz radiation spectrum [3] and Figure 6.2 displays about the structure of microstrip antenna.

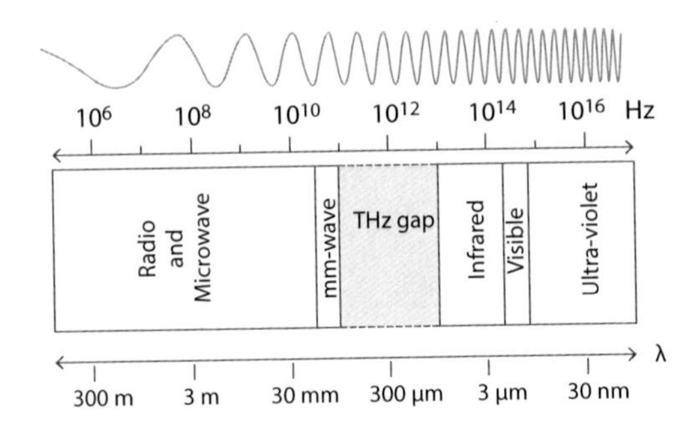

Figure 6.1 Terahertz radiation spectrum [3].

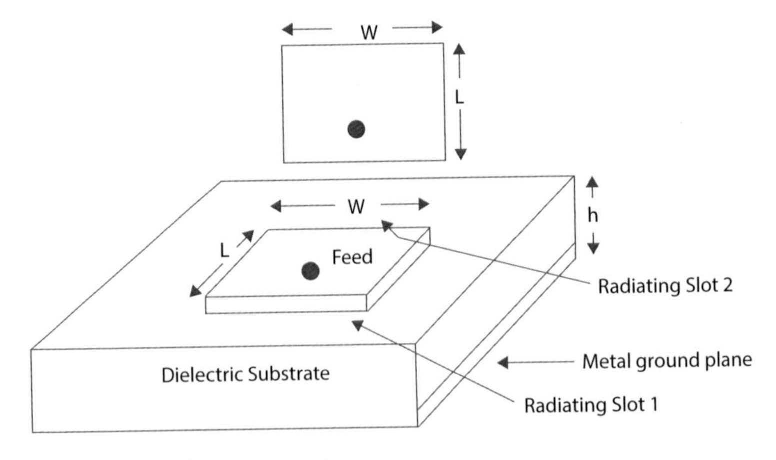

Figure 6.2 Structure of microstrip antenna.

This section provides basic information about THz technology. Section 6.2 presents the microstrip antenna theory with different structures and feeding methods for design concept understanding. Section 6.3 summarizes the challenges of THz antenna development. Section 6.4 discusses the different performance evaluation parameters used for an antenna. Section 6.5 presents a detailed performance analysis of different kinds of microstrip terahertz antennas with references. Finally, the conclusion is presented in Section 6.6.

6.2 Microstrip Antenna Design

Microstrip antennas are used for wireless communication because these antennas are lightweight, low profile and cost, and can be easily integrated into the RF devices [11]. Microstrip patch antennas provide high efficiency and compactness, making them useful for WLAN applications [12]. A microstrip patch antenna is most commonly a 1/2 wavelength construction, however, it may also be built as a 1/4 wave component. Despite the fact that sheet metal patches over a ground plane with an air-dielectric are also widely used, patch antennas are imprinted on the top surface of a 2-layer PCB. Patch antennas are viewed as directional antennas with an essential projection of radiation more than 70 x 70 degrees (when direction away from ground plane). A rectangular patch with measurements for width (W) and length (L) is placed on a ground plane with a substrate thickness (h) and dielectric constant (r) [13]. A number of substrates are available for microstrip antenna design. The typical range of dielectric constants lies between $2.2 \leq \varepsilon_r \leq 12$. The dielectric is used as a substrate of the microstrip antenna and as the relative dielectric constant of the antenna substrate increases the length of the antenna decreases [14].

For better antenna performance, the substrates should be thick and the dielectric constant range needs to be near the lower end which gives good efficiency and larger antenna bandwidth [11]. The dielectric constant of the substrate (ε_r), the height of the substrate (h), and the resonant frequency (fr) are all required for the microstrip patch antenna design. Using values of ε_r, f_r, and h antenna Width (W) and Length (L) can be calculated for its design. Microstrip antenna patches might be square, rectangular, triangular, thin strip (dipole), circular, elliptical, or any other shape [15]. The different forms of patch [16] components used in microstrip antennas are shown in Figure 6.3. Because square, rectangular, strip (dipole), and circular patches are simple to make and analyze [17], they are commonly employed.

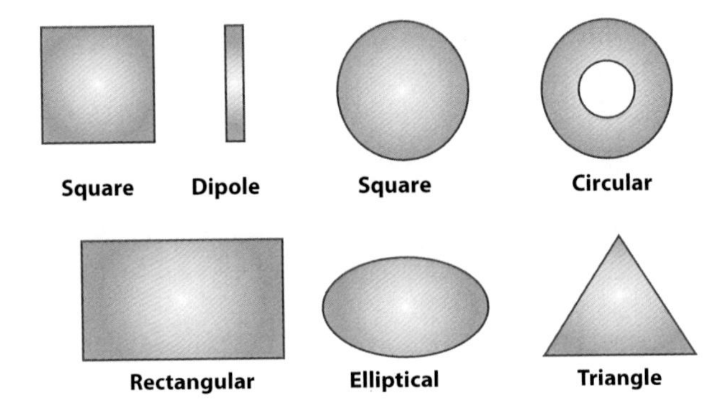

Figure 6.3 Shapes of microstrip patch elements.

6.2.1 Feeding Methods

A feedline is used to activate transmission via direct or indirect contact [10]. There are several dissimilar feeding techniques available, but the main ones used are line feed, coaxial feeding, aperture coupling, and proximity coupling. Figure 6.4 depicts the microstrip antenna's feeding mechanisms.

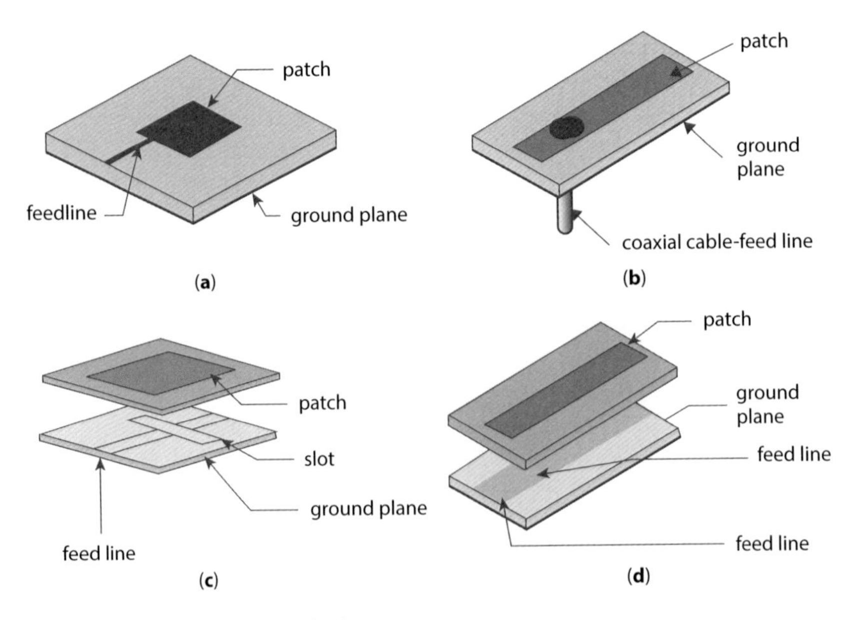

Figure 6.4 Microstrip antenna feeds.

Microstrip Line Feed
Microstrip line feed is one of the most straightforward antenna construction methods, as it is simply a directed strip that connects to the patch and may be regarded as a patch enlargement. Controlling the inset position makes it straightforward to model and coordinate. The hindrance of this feed strategy is that as the thickness of the substrate increments, surface wave and misleading feed radiation increments, restricting the transmission capacity.

Coaxial Feeding
Coaxial feeding is a sustaining technique in which the coaxial's inward conductor is attached to the receiving apparatus's radiation patch, whereas the external conductor is joined to the ground plane. It is simple to plan and construct.

Aperture Coupling
Two unique substrates are separated by a ground plane in aperture coupling. There is a microstrip encouraging a line on the base side of the lower substrate whose vitality is related to the patch via space on the ground plane separating the two substrates. This setup allows for separate feed optimization as well as a radiating element.

Proximity Coupling
The widest bandwidth and lowest spurious radiation are seen in proximity coupling. Fabrication of this feed, on the other hand, is complicated.

6.3 Challenges of Terahertz Antenna Development

This section covers the various challenges faced in the development of the terahertz microstrip antenna. Figure 6.5 represents the Terahertz antenna challenges paradigm. Because the substrate of microstrip antennas is very thin and sensitive to (0.1-1 THz) frequency and the size of antennas is very small, they face some of the challenges listed below.

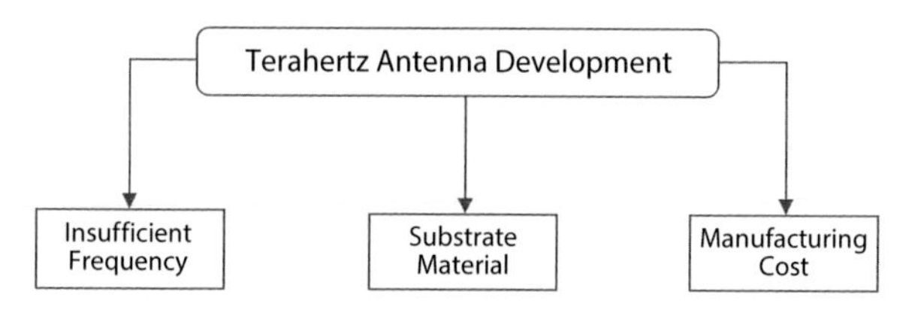

Figure 6.5 Terahertz antenna challenges.

Insufficient Frequency Band
The resonant frequency of the THz antennas built by the majority of researchers normally lies in 0.1 to 1 THz, according to the aforementioned research. Because this interval's frequency band is near the millimetre wave band, this does not completely match the design requirements for the THz antenna's resonance frequency. So, we just need to employ the frequency range of 1 to 10 THz.

Substrate Material Matching
FR-4 or other dielectric polymers are commonly used substrate materials in the design of millimetre-wave antennas. But due to low gain, such substrate material cannot fully fulfil the THz microstrip antenna's resonance frequency requirements. An appropriate and alternative substrate material is required based on the antenna resonance frequency.

Manufacturing Cost
Despite the fact that current modern process technology may match some THz antenna production requirements, the majority of them are inefficient and expensive, owing to the high accuracy of process machines, which results in high equipment costs. The high-cost issue is a pressing issue that must be addressed in the future.

6.4 Antenna Performance Attributes

For accurate evaluation of antenna performance, several antenna parameters are examined. In this work, the following parameters are used for the comparative analysis of various microstrip THz antennas.

Impedance Bandwidth
Impedance bandwidth is a parameter that defines the frequency range over which antenna impedance is matched to the feedline [18]. A radio transmitter may be severely damaged if the antenna is operated outside its operating range. Similarly, to ensure the best reception, antenna design should be tuned within the optimum bandwidth limits only. The percentage refers to a unit of measurement known as fractional bandwidth (FBW). This is the antenna's absolute bandwidth (or impedance bandwidth) divided by the antenna's centre frequency.

$$\mathrm{FBW} = \frac{\mathrm{BW}}{\mathrm{fc}} \tag{6.1}$$

Because it is scale-independent, the fractional bandwidth is a better measure of bandwidth when comparing various antennas.

Antenna Gain

Gain is also an important parameter that signifies the energy radiated by the antenna in the given direction as compared to the reference antenna with the same power input [19]. In order to improve system performance and for efficient, long-distance, and reliable wireless communication link design, antenna gain is to be maximized. It is measured in decibels (dB). Ideally, the reference antenna is an isotropic antenna (gain in dBi), but to obtain a more practical, realizable gain prediction a dipole antenna (gain in dB) is used as the reference antenna. Basically, gain (G) and directivity (D) are the same, however, gain includes radiation efficiency (e_{cd}) of the antenna and takes into account losses within the antenna conductor loss (e_c), dielectric loss (e_d), reflection loss, etc.

$$\text{Gain } G(\theta, \varnothing) = e_{cd} D(\theta, \varnothing) \tag{6.2}$$

where, $e_{cd} = e_c e_d$; $0 < e_{cd} < 1$.

Antenna gain can be increased by increasing substrate height and patch width or by using a larger aperture antenna that could capture more energy from the incident EM wave [14].

$$A_e = e_{cd} \frac{\lambda^2}{4\pi} D_0 \tag{6.3}$$

There is a trade-off situation: when increasing the gain, the beamwidth will also be reduced, hence, this more directional antenna will be critical to orient in the desired direction for proper reception of the power. Therefore, careful design and choice of optimal parameters are critical for a good communication radio link setup.

Return Loss

Return Loss [20] is an important metric to measure the performance of an antenna. The fraction of radio waves arriving at the antenna input that are rejected versus those that are accepted is known as the Return Loss of an antenna. It is expressed in decibels (dB).

6.5 Comparative Analysis of Microstrip THZ Antennas

Table 6.1 compares numerous THz microstrip antennas based on the different performance parameters discussed in Section 6.4. One of the main

Table 6.1 Performance comparison of microstrip terahertz antennas.

Reference	Type of antenna	Frequency (THz)	Gain (dBi)	Return loss (dB)	Substrate material
[21]	Fractal Butterfly Antenna	4.9	16.95	-33.59	Graphene
[22]	Double T-type Microstrip Antenna	0.3 and 0.76	7.13 and 3.71	-29 and -40	Arlon Cuclad 250gt
[23]	Dual Frequency t-type Microstrip Antenna	0.632 and 0.8702	8.2	N/A	Fr-4
[24]	Microstrip Array Antenna	0.1	15.7	-26.05	Liquid Crystal Polymer
[25]	Slotted Rectangular Microstrip Antenna	4.952	4.254	-55.31	Fr-4
[26]	Single-element Dipole Microstrip Antenna	0.29	13.9	N/A	Indium Phosphide and Benzocyclobutene
[27]	Dual-band Slotted Antenna	0.02	5.9	N/A	Dielectric
[28]	Microstrip Antenna with Aircylinder PBG	0.62	9.17	-28.42	Photonic Bandgap (PBG)
[29]	Long Spiral Terahertz Antenna	1 to 6	12	N/A	Quartz
[30]	Terahertz Antenna based on Photonic Crystal	0.630	7.934	-44.71	Polyimide

research objectives for potential THz microstrip antennas is their optimized performance.

The e-healthcare [31, 32] industry is expected to undergo a transformation and switch to terahertz microstrip antennas. Yet, because data traffic is transmitted, they are exposed to security lapses and online attacks [33, 34] from cyberterrorist [35] organisations working at online social network and at dark web channels [36]. The latest developments in quantum computation raise the danger of cyber-security threats and weaknesses in microstrip antennas even more. Valuable patient information must thus be protected from both passive and active threats and vulnerabilities.

6.6 Conclusion

The potential spectrum resources are shifting to the THz band due to the development trend of wireless communications and the building of a THz wireless communication framework can give a greater data transfer rate. THz antennas are critical components in communication systems for sending and receiving THz waves. In the interim, the performance of THz antennas has a significant impact on the quality of communication systems. This chapter examined the THz microstrip antennas in-depth, covering THz fundamental principles, THz antenna development challenges, and analysis of THz microstrip antennas. The majority of THz antennas are in the hypothetical stage, with only a few real items being produced. The THz antenna's future research challenge is enormous. Some of the future research directions are enhancing antenna geometry for downsizing, optimization of antenna radiation performance for high gain, and employing appropriate technology for THz antenna integration.

References

1. S. Koenig *et al.*, "Wireless sub-THz communication system with high data rate," *Nat. Photonics*, vol. 7, no. 12, pp. 977–981, 2013, doi: 10.1038/nphoton.2013.275.
2. R. Zhang, K. Yang, and A. Alomainy, "Modelling of the Terahertz Communication Channel for In-vivo Nano-networks in the Presence of Noise," no. April 2018, 2016, doi: 10.1109/MMS.2016.7803812.
3. R. Krishan, "Terahertz Band for Wireless Communication - A Review," in *Terahertz Wireless Communication Components and System Technologies*, Ist., A. B. y Mohammed El Ghzaoui, Sudipta Das, Trupti Ranjan Lenka, Ed. Springer Singapore, 2022.

4. C. Christodoulou and P. Wahid, "Fundamental Parameters of Antennas," *Fundam. Antennas*, pp. 13–20, 2009, doi: 10.1117/3.416262.ch2.
5. "PART I Fundamental Parameters and Definitions."
6. Y. He, Y. Chen, L. Zhang, S. W. Wong, and Z. N. Chen, "An overview of terahertz antennas," *China Commun.*, vol. 17, no. 7, pp. 124–165, 2020, doi: 10.23919/J.CC.2020.07.011.
7. N. Khiabani, "Modelling, design and characterisation of terahertz photoconductive antennas," no. September, pp. 1–220, 2013, [Online]. Available: http://research-archive.liv.ac.uk/14213/.
8. X. Chen, X. Liu, and C. Science, *Handbook of Antenna Technologies*, no. Schmuttenmaer 2004. 2020.
9. T. Nagatsuma, "Antenna Technologies for Terahertz Communications," in *International Symposium on Antennas and Propagation (ISAP)*, 2018, pp. 1–2.
10. S. K. Vijay, J. Ali, P. Yupapin, B. H. Ahmad, and K. Ray, "A Triband EBG Loaded Microstrip Fractal Antenna for THz Application," pp. 1–11.
11. C. A. Balanis, *Antenna Theory: Analysis and Design*, 4th ed. John Wiley and Sons, 2016.
12. R. Krishan and V. Laxmi, "'X' Shape Slot-Based Microstrip Fractal Antenna for IEEE 802.11 WLAN," in *Advances in Intelligent Systems and Computing*, vol. 553, Springer Nature, 2017, pp. 135–143.
13. R. Krishan and V. Laxmi, "Design of Microstrip Antenna for Wireless Local Area Network," vol. 4, no. 4, pp. 361–365, 2015.
14. R. Krishan, "A Novel Cross-Slotted Dual-Band Fractal Microstrip Antenna Design for Internet of Things (IoT) Applications," in *Advanced Sensing in Image Processing and IoT*, 1st ed., CRC Press, 2022, pp. 245–254.
15. R. Kiruthika and T. Shanmuganantham, "Comparison of different shapes in microstrip patch antenna for X-band applications," *Proc. IEEE Int. Conf. Emerg. Technol. Trends Comput. Commun. Electr. Eng. ICETT 2016*, pp. 3–8, 2017, doi: 10.1109/ICETT.2016.7873722.
16. A. A. K. M. S, "A Review of Design and Analysis For Various Shaped Antenna in Terahertz and Sub-terahertz Applications," vol. 12, no. 8, pp. 2053–2071, 2021.
17. S. Finich, N. A. Touhami, and A. Farkhsi, "Design and Analysis of Different Shapes for Unit-Cell Reflectarray Antenna," *Procedia Eng.*, vol. 181, pp. 526–530, 2017, doi: 10.1016/j.proeng.2017.02.429.
18. R. Poddar, S. Chakraborty, and S. Chattopadhyay, "Improved Cross Polarization and Broad Impedance Bandwidth from Simple Single Element Shorted Rectangular Microstrip Patch: Theory and Experiment," *Frequenz*, vol. 70, no. 1–2, pp. 1–9, 2016, doi: 10.1515/freq-2015-0105.
19. A. D. Yaghjian and S. R. Best, "Impedance, bandwidth, and Q of antennas," *IEEE Trans. Antennas Propag.*, vol. 53, no. 4, pp. 1298–1324, 2005, doi: 10.1109/TAP.2005.844443.

20. Y. Kumar and S. Singh, "A Compact Multiband Hybrid Fractal Antenna for Multistandard Mobile Wireless Applications," *Wirel. Pers. Commun.*, vol. 84, no. 1, pp. 57–67, 2015, doi: 10.1007/s11277-015-2593-x.
21. Y. Shi, X. Zhang, Q. Qiu, Y. Gao, and Z. Huang, "Design of Terahertz Detection Antenna with Fractal Butterfly Structure," *IEEE Access*, vol. 9, pp. 113823–113831, 2021, doi: 10.1109/ACCESS.2021.3103205.
22. A. Hlali, Z. Houaneb, and H. Zairi, "Dual-band reconfigurable graphene-based patch antenna in Terahertz band: Design, analysis and modeling using WCIP method," *Prog. Electromagn. Res. C*, vol. 87, no. October, pp. 213–226, 2018, doi: 10.2528/pierc18080107.
23. M. Khulbe, M. R. Tripathy, H. Parthasarthy, and J. Dhondhiyal, "Dual Band THz Antenna Using T Structures and Effect of Substrate Volume on Antenna Parameters," *Proc. - 2016 8th Int. Conf. Comput. Intell. Commun. Networks, CICN 2016*, pp. 191–195, 2017, doi: 10.1109/CICN.2016.43.
24. M. S. Rabbani and H. Ghafouri-Shiraz, "Liquid Crystalline Polymer Substrate-Based THz Microstrip Antenna Arrays for Medical Applications," *IEEE Antennas Wirel. Propag. Lett.*, vol. 16, no. c, pp. 1533–1536, 2017, doi: 10.1109/LAWP.2017.2647825.
25. Prince, P. Kalra, and E. Sidhu, "Rectangular TeraHertz microstrip patch antenna design for riboflavin detection applications," *Proc. 2017 Int. Conf. Big Data Anal. Comput. Intell. ICBDACI 2017*, pp. 303–306, 2017, doi: 10.1109/ICBDACI.2017.8070853.
26. H. Vettikalladi, W. T. Sethi, A. F. Bin Abas, W. Ko, M. A. Alkanhal, and M. Himdi, "Sub-THz Antenna for High-Speed Wireless Communication Systems," *Int. J. Antennas Propag.*, vol. 2019, 2019, doi: 10.1155/2019/9573647.
27. M. Habib Ullah, M. T. Islam, M. R. Ahsan, J. S. Mandeep, and N. Misran, "A dual band slotted patch antenna on dielectric material substrate," *Int. J. Antennas Propag.*, vol. 2014, 2014, doi: 10.1155/2014/258682.
28. M. N. E. Temmar, A. Hocini, D. Khedrouche, and M. Zamani, "Analysis and design of a terahertz microstrip antenna based on a synthesized photonic bandgap substrate using BPSO," *J. Comput. Electron.*, vol. 18, no. 1, pp. 231–240, 2019, doi: 10.1007/s10825-019-01301-x.
29. A. Yahyaoui *et al.*, "Design and comparative analysis of ultra-wideband and high directive antennas for THz applications," *Appl. Comput. Electromagn. Soc. J.*, vol. 36, no. 3, pp. 308–319, 2021, doi: 10.47037/2020.ACES.J.360311.
30. R. K. Kushwaha, P. Karuppanan, and L. D. Malviya, "Design and analysis of novel microstrip patch antenna on photonic crystal in THz," *Phys. B Condens. Matter*, vol. 545, no. May, pp. 107–112, 2018, doi: 10.1016/j.physb.2018.05.045.
31. V. Mahor, S. Bijrothiya, R. Mishra, R. Rawat, and A. Soni, The Smart City Based on AI and Infrastructure: A New Mobility Concepts and Realities. *Autonomous Vehicles Volume 1: Using Machine Intelligence*, 277-295, 2022.
32. V. Mahor, K. Pachlasiya, B. Garg, M. Chouhan, S. Telang, and R. Rawat, Mobile Operating System (Android) Vulnerability Analysis Using Machine

Learning. In *Proceedings of International Conference on Network Security and Blockchain Technology: ICNSBT 2021* (pp. 159-169). Singapore: Springer Nature Singapore, 2022, June.

33. V. Mahor, B. Garg, S. Telang, K. Pachlasiya, M. Chouhan, and R. Rawat, Cyber Threat Phylogeny Assessment and Vulnerabilities Representation at Thermal Power Station. In *Proceedings of International Conference on Network Security and Blockchain Technology: ICNSBT 2021* (pp. 28-39). Singapore: Springer Nature Singapore, 2022, June.
34. R. Rawat, S. Gupta, S. Sivaranjani, O.K. Cu, M. Kuliha, and K.S. Sankaran, Malevolent Information Crawling Mechanism for Forming Structured Illegal Organisations in Hidden Networks. *International Journal of Cyber Warfare and Terrorism (IJCWT)*, 12(1), 1-14, 2022.
35. R. Rawat, Y.N. Rimal, P. William, S. Dahima, S. Gupta, and K.S. Sankaran, Malware Threat Affecting Financial Organization Analysis Using Machine Learning Approach. *International Journal of Information Technology and Web Engineering (IJITWE)*, 17(1), 1-20, 2022.
36. R. Rawat, V. Mahor, M. Chouhan, K. Pachlasiya, S. Telang, and B. Garg, Systematic literature Review (SLR) on social media and the Digital Transformation of Drug Trafficking on Darkweb. In *International Conference on Network Security and Blockchain Technology* (pp. 181-205). Springer, Singapore, 2022.

7

Smart Antenna for Home Automation Systems

Manish Varun Yadav[1]* and Swati Varun Yadav[2]

[1]*Electronics and Communication Engineering, CHRIST (Deemed to be University), Bengaluru, Karnataka, India*
[2]*Electrical & Electronics Engineering Department, BITS Pilani, K K Birla Goa Campus, Goa, India*

Abstract

A smart antenna for home automation systems is suggested in this design. An antenna is a device that minimizes human movements while dealing with electronics systems and software. This smart antenna helps reduce physically challenged peoples' movement and supports home automation systems. The presented antennas could be used for any small device, like a mobile, tablet, laptop, Wi-Fi, or WiMAX, which are essential in the current scenario. The presented smart antenna is capable of radiating a large frequency band from 3 to 13.8 GHz, which covers the .5-7 GHz (5G(I) Sub-6 GHz band), 'Wimax' 3.5 and 5.5 GHz bands, 'WLAN' 5.2 and 5.8 GHz bands, 'C' 4-8 GHz Band, 'X' 8-12 GHz Band, and other home automation applications with high efficiency. The impedance bandwidth of the smart antenna is 128%, with a size of 15x15x1.5 mm^3. The suggested design includes a modified patch in the shape of a square patch attached with one circular element fed by a microstrip line. Circular pieces have been designed for better resonances at lower modes. The antenna is simulated with an FR4 substrate using a CST Simulator. The design is investigated by simulations and corresponding S-parameter results are presented. The robotics process automation is well described in Table 7.3. The proposed structure also demonstrates stable radiation patterns across the operating bandwidth. The proposed radiator has a high gain of 5.21 dBi and an efficiency of 80%.

**Corresponding author*: manishvarun21@gmail.com

Romil Rawat, Rajesh Kumar Chakrawarti, Sanjaya Kumar Sarangi, Rahul Choudhary, Anand Singh Gadwal and Vivek Bhardwaj (eds.) Robotic Process Automation, (113–124) © 2023 Scrivener Publishing LLC

***Keywords*:** Smart antenna, home automation systems, 5G (Sub-6GHz) band, impedance bandwidth, X-band, slot antenna

7.1 Introduction

Radio communication or wireless signals are only possible with antennas. Advanced equipment in wireless communication requires compact and versatile antennas that can radiate multiple frequency bands and this can be done with the help of planar antennas. Various designs have been presented for home automation such as curvy slots, triangular-shaped, printed circular rings, circular, 'U,' and 'L' shaped, and printed semi-circular slots [1–6]. High frequency is achieved by printing a tiny fractal element in an antenna [7]. By making a U-shaped antenna, bandwidth enhancement is achieved [8]. By making circular trims in the designed radiator, high gain is achieved up to 4.5dB [9]. Two resonant bands are obtained by trim slots in the structure, main patch, and attaching circles in the backplane [10]. Every communication system needs an antenna. This design proposes a slotted radiator fed by a strip line with a patch and partial ground plane. The lower frequencies for the home automation system are excited due to the square main patch.

On the other hand, the higher-order frequencies are excited due to the partial backplane. The proposed design is simulated using CST simulation tools. Upcoming wireless mobile generation needs a 5G mobile and 5G communication. Competing with this upcoming demand requires 5G antennas capable of radiating or receiving 5G (sub 6GHz band) signals. The proposed design antenna concept is useful for 3.5-7GHz (5G-(I) Sub-6GHz band), 'Wimax' 3.5 and 5.5 GHz bands, 'WLAN' 5.2 and 5.8 GHz bands, 'C' 4-8 GHz Band, 'X' 8-12 GHz Band, and other home automation applications.

In Section 7.2, the 5G antenna dimensions are specified. The simulations of various antenna metrics, as well as a related discussion, are covered in Section 7.3. The design comes to a close after a thorough examination of several simulated results for the design.

Figure 7.1 shows the home automation system block diagram using the smart antenna. The proposed antennas can communicate with all the home automation devices visible in the block diagram. This means we can control and send the signals to all wireless devices capable of receiving the wireless communication frequencies. The robotics process automation is defined in Table 7.3.

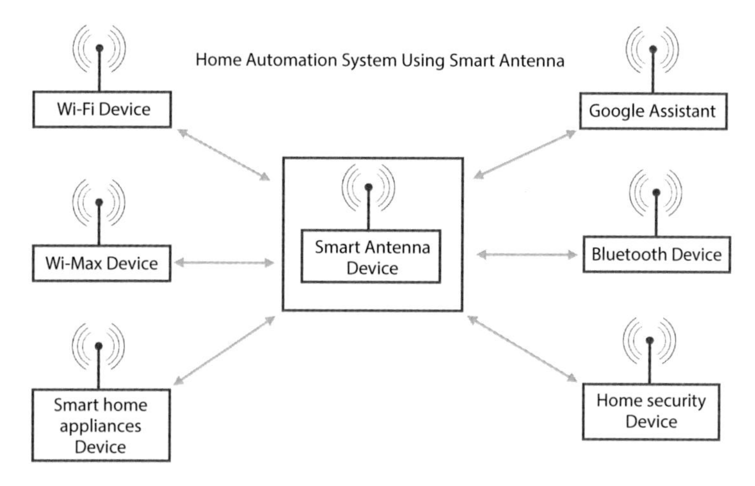

Figure 7.1 Block diagram of home automation system using smart antenna.

7.2 Home Automation Antenna Geometry and Robotics Process Automation

The presented antenna's front view is shown in Figure 7.2, back view in Figure 7.3, and side view in Figure 7.4. The 5G antenna is a compact wide-band radiator simulated using CST software. The FR-4 is used as a substrate to etch-out geometry. Both side layers on the FR-4 are copper with a conductivity of $5.8*10^7$ S/m. Figures 7.2 and 7.3 depict the geometry of numerous

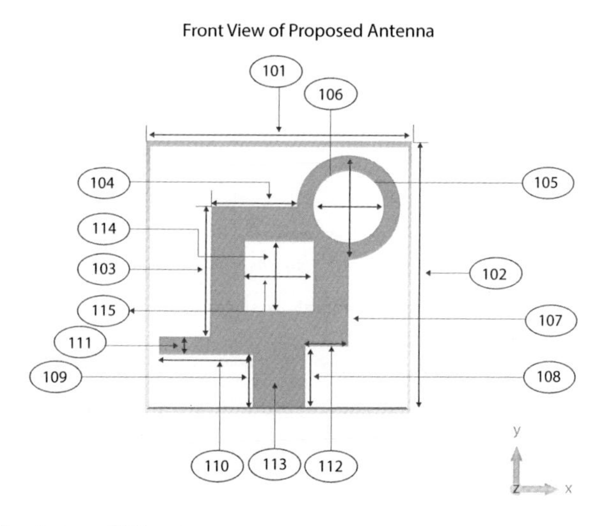

Figure 7.2 Front view of 5G antenna.

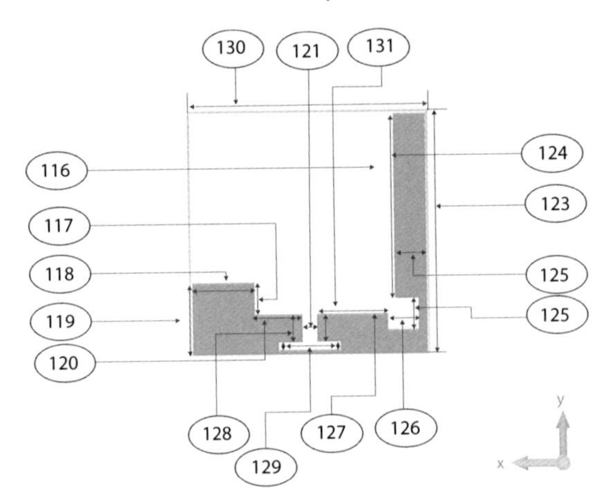

Figure 7.3 Back view of 5G antenna.

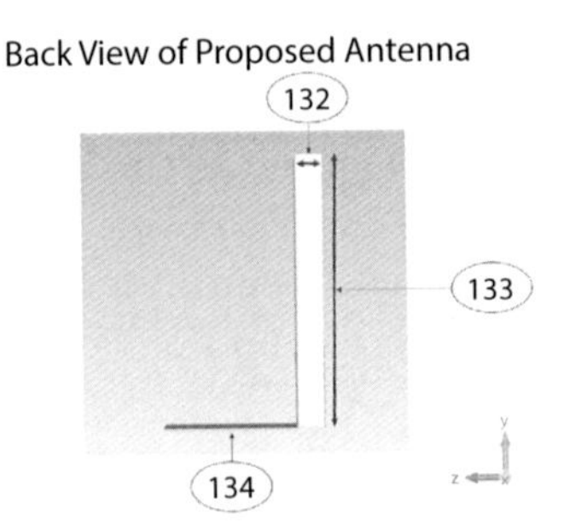

Figure 7.4 Side view of 5G antenna.

elements with a partial backplane and slotted patch on opposite sides, with overall dimensions of the 5G antennas as 15x15x1.5 mm^3. The thickness of the dielectric is 1.5 mm. The patch is fed through a 50Ω strip line with a width of 3 mm. 0.0018 mm is the thickness of the copper [11, 12]. The patch [13, 14] consists of a square and a circular slot. There are a total of 2 slots in the frontside of the patch. The effect of each element's [15, 16] introduction has its relevance in generating the final conclusion we obtained. The impedance B/W of the proposed structure is 128%, with a central frequency [17] of 8.4 GHz (3 GHz to 13.8 GHz). The offered design description is mentioned in Table 7.1. The robotics process automation is expressed in Table 7.3.

Table 7.1 Proposed smart antenna for home automation design description.

Front side			**Back side**		
Figure no.	**Component no.**	**Parameter's values (mm)**	**Figure no.**	**Component no.**	**Parameter's values (mm)**
1	101	15	2	116	11.5
	102	15		117	2
	103	7.5		118	4
	104	5		119	4.5
	105	6		120	3
	106	4		121	1.5
	107	5		123	15
	108	3.5		124	11.5
	109	3		125	2
	110	5.5		126	2
	111	1		127	4.5
	112	2.5		128	1.75
	113	3		129	0.5
	114	4		130	15
	115	4		131	4.5
Side View					
3	132	3.5	3	133	15
	134	Input Port			

For the home automation system, robotics process automation is an essential part. We analyze the radiation process through robotics process automation to compete in this process. Different angles rotate the smart antenna for home automation, as mentioned in Table 7.3. According to the angles, it radiates the different radiation patterns on the E and H planes with the help of robotics process automation.

7.3 Results for Home Automation Smart Antenna

Figure 7.5 shows (S11) the curve of the smart antenna for home automation systems. The return loss curve shows the antenna's good signal transmission capability throughout the entire bandwidth. A smart antenna's loss is defined by following formula:

$$Return\ Loss\ (S_{11}) = -20 \log 10\ (\tau)$$

Figure 7.6 shows the smart antenna's VSWR curve which is defined by the formula:

$$VSWR = \frac{1+|\tau|}{1-|\tau|}$$

where (τ) is the reflection coefficient.

VSWR for any practical antenna should be less than 2. This curve shows a good VSWR value validated antenna that has a good signal transmission capability throughout the entire bandwidth.

Figure 7.7 shows the gain (in dB) of the 5G antenna. We can observe a peak gain of 5.21 dB and we have discussed the S_{11} result. Now, moving forward, we will discuss the other performance metrics. The proposed 5G antenna achieved a peak radiation efficiency of 80%.

Figure 7.8 shows the front view surface current distribution at 5 GHz and 11 GHz frequencies for E-field and H-field current distribution. An equivalent surface replaces an antenna and its radiated current field shows

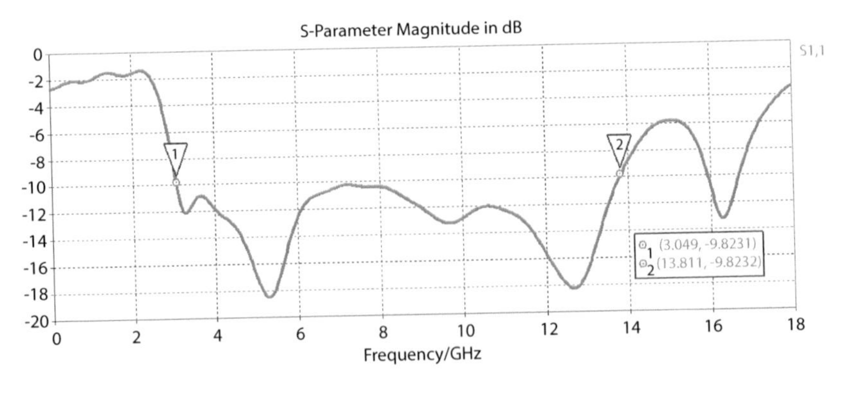

Figure 7.5 Return loss (S_{11}) curve.

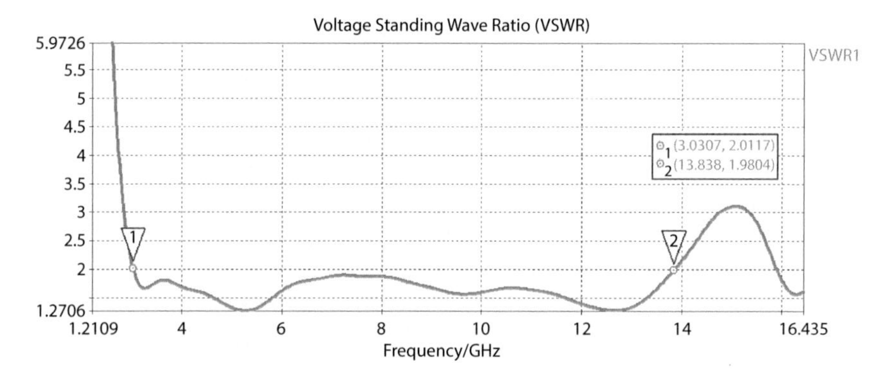

Figure 7.6 VSWR curve.

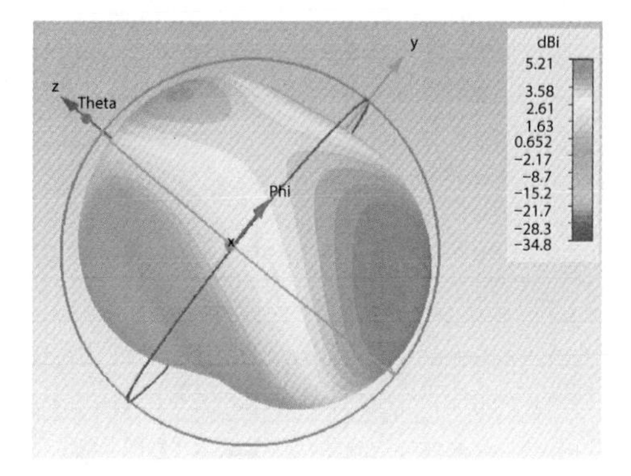

Figure 7.7 Gain of antenna.

its signal strengths. An antenna can generate both an E-field and H-field, which shows the good signal strength of the design. Figure 7.8a shows the current density at 5 GHz frequency with a value of 18925 v/m and Figure 7.8b shows the current density at 11 GHz frequency with a current density of 131 A/m.

The published smart antennas for home automation system comparison is visible in Table 7.2. In Table 7.2, the maximum antenna parameters are compared and define the intelligent antenna properties. After an extensive literature survey, we can say that our proposed smart antenna is capable of advancing home automation systems. The robotics process automation is defined in Table 7.3. With the help of simulations, an antenna rotates and covers all the angles at different frequencies. Table 7.3 shows the robotics process automation, how the antenna is rotated, and how it can radiate the different radiation patterns.

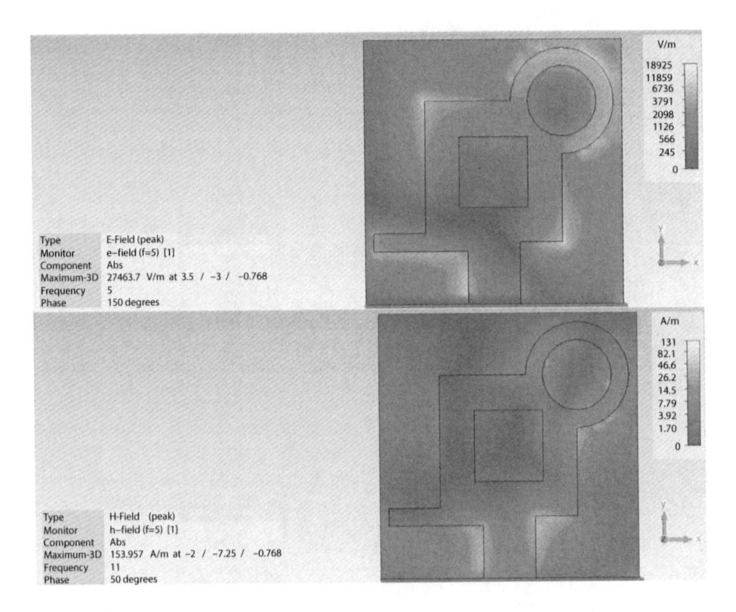

Figure 7.8 Surface current distribution of antenna.

Table 7.2 Comparison of published smart antennas for home automation systems.

Ref.	Overall volume (in λ)	Peak gain (dBi)	Peak (η) (%)	Band (GHz)	Fractional B/W (%)
[2]	.28λ*.25λ*.016λ	1.7	NA	3.1-22	150%
[5]	.20λ*.25λ*.015λ	5.1	89%	3.1-11	110%
[7]	.26λ*.26λ*.019λ	3.5	75%	3.9-14	142%
[9]	.33λ*.22λ*0.1λ	4.5	62%	2-9	127%
[13]	.23λ*.23λ*.015λ	3.2	81%	3.5-19	145%
[16]	.55λ*.41λ*.022λ	2	60%	3.1-11	109%
[18]	.33λ*.24λ*.014λ	5.2	87%	2.9-16	139%
[24]	.18λ*.14λ*.15λ	2.79	72%	2.8-12	122%
[27]	.32λ*.2λ*.014λ	2.3	78.3%	2.7-7.3	108%
[28]	.14λ*.18λ*.015λ	2.2	69%	3.1-11	110%
[29]	.2λ*.3λ*.014λ	2.1	70%	2.3-11	129%
Presented	**.15λ*.15λ*.015λ**	**5.21**	**80%**	**3-13.8**	**128%**

Table 7.3 Robotics process automation 'smart antenna for home automation system' at different degrees (angles).

Theta	4 GHz	4 GHz	6 GHz	6 GHz	9.2 GHz	9.2 GHz
(degree)	E-plane	H-plane	E-plane	H-plane	E-plane	H-plane
30	-13.51	-5.544	-1.607	0.95853	-2.836	-0.0062
60	-11.29	-3.032	-10.63	-3.8778	-16.13	0.4021
90	-3.391	-1.019	-8.601	-3.6596	-9.236	0.672
120	-0.5334	0.02129	-0.861	-2.695	-2.802	0.588
150	0.1572	0.2181	2.015	-2.098	-0.5514	0.1486
180	-0.36	-0.2549	2.092	4.215	-0.1029	-0.1056
210	-2.525	-1.254	0.1033	-0.654	-0.8007	-0.0466
240	-8.822	-2.472	-3.688	-3.222	-3.691	0.2001
270	-13.91	-3.084	-11.55	-4.242	-8.685	0.4306
300	-6.188	-3.345	-8.749	-6.994	-12.83	0.4335
330	-4.49	-4.531	-0.765	-2.191	-3.878	0.2763
360	-5.951	-6.254	1.105	-0.4096	-0.0238	-0.0238

Smart antennas [21, 22] can reduce the risks of harmful attacks [23] and vulnerabilities [24] through remote botnet because of their limited radiated area, immunity to attacks from cybercriminals [25], and node location validation. This directional antenna-equipped node uses the arrival orientation of the acquired signal to compute the position of a sender node and may identify fraudulent nodes by comparing their placement to an authorised list. The nodes with directed antennas may be able to identify, reduce, or even completely remove security threats when discussing eavesdropping, interference, wormhole threats, or Sybil attacks by employing the lines of protection from hostile attacks.

7.4 Conclusion

A smart antenna for home automation systems has been reported. The proposed antenna for home automation appliances, including home appliances, is 15*15*1.5 mm^3 in size and has a bandwidth of 128% (3–13.8GHz).

The compactness of the structure was determined to be responsible for the peak gain of 5.21 dBi and a smart antenna efficiency of 80%. The surface current is also consistent across the whole wideband frequency range. As a result, the antenna is suitable for 3.5-7GHz (5G-(I) Sub-6GHz band), 'Wimax' 3.5 and 5.5 GHz bands, 'WLAN' 5.2 and 5.8 GHz bands, 'C' 4-8 GHz Band, 'X' 8-12 GHz Band, and other home automation applications.

References

1. Mishra B, Verma RK, Yashwanth N, Singh RK (2021). A review on microstrip patch antenna parameters of different geometry and bandwidth enhancement techniques. International Journal of Microwave and Wireless Technologies 1–22.
2. Li G, Zhai H, Li T, Li L and Liang C (2012). A compact antenna with broad bandwidth and quad-sense circular polarization. IEEE Antennas and Wireless Propagation Letters 11, 791-794.
3. Yash, B., Yadav, M. V., Baudha, S. (2020). A compact mace shaped ground plane modified circular patch antenna for ultra-wideband applications. Telecommunications and Radio Engineering, 79(5), 363-381.
4. Khidre, A., Lee, K. F., Elsherbeni, A. Z., & Yang, F. (2013). Wide band dual-beam U-slot microstrip antenna. IEEE transactions on antennas and propagation, 61(3), 1415-1418.
5. S. V. Yadav and A. Chittora, "A High Power Circularly Polarized Antenna with Overmoded Waveguide," 2021 IEEE Indian Conference on Antennas and Propagation (InCAP), 2021, pp. 431-434, doi: 10.1109/InCAP52216.2021.9726452.
6. Srivastava, I., Baudha, S., Yadav, M. V., & (2020). A Novel Approach for Compact Antenna with Parasitic Elements Aimed at Ultra-Wideband Applications. In 2020 14th European Conference on Antennas and Propagation (EuCAP) IEEE, 14(4), 1-5.
7. Shagar, A., & Wahidabanu, S. (2011). Novel wideband slot antenna having notch-band function for 2.4 GHz WLAN and UWB applications. International Journal of Microwave and Wireless Technologies, 3(4), 451-458.
8. Kapoor, K. (2019). U-Shaped Microstrip Patch Antenna with Partial Ground Plane for Mobile Satellite Services (MSS). In 2019 URSI Asia-Pacific Radio Science Conference (AP-RASC) (pp. 1-5). IEEE.
9. Kurniawan, A., & Mukhlishin, S. (2013). Wideband antenna design and fabrication for modern wireless communications systems. Procedia Technology, 11, 348-353.
10. Ghosh, A., Ghosh, S. K., Ghosh, D., & Chattopadhyay, S. (2016). Improved polarization purity for circular microstrip antenna with defected patch

surface. International Journal of Microwave and Wireless Technologies, 8(1), 89-94.

11. Baudha, S., & Yadav, M. V. (2019). A compact ultra-wide band planar antenna with corrugated ladder ground plane for multiple applications. Microwave and Optical Technology Letters, 61(5), 1341-1348.
12. Abdelraheem, A. M., & Abdalla, M. A. (2016). Compact curved half circular disc-monopole UWB antenna. International Journal of Microwave and Wireless Technologies, 8(2), 283-290.
13. Gupta, S., (2019). Parasitic rectangular patch antenna with variable shape ground plane for satellite and defence communication. In 2019 URSI Asia-Pacific Radio Science Conference (AP-RASC) (pp. 1-4). IEEE.
14. Varun Yadav, S., & Chittora, A. (2022). Circularly polarized high-power antenna with higher-order mode excitation. International Journal of Microwave and Wireless Technologies, 14(4), 477-481. doi:10.1017/S1759078721000611.
15. Basak, A., Manocha, M., Baudha, S., & Yadav, M. V. (2020). A compact planar antenna with extended patch and truncated ground plane for ultra wide band application. Microwave and Optical Technology Letters, 62(1), 200-209.
16. Yadav, M. V., & Baudha, S. (2021). A miniaturized printed antenna with extended circular patch and partial ground plane for UWB applications. Wireless Personal Communications, 116(1), 311-323.
17. S. V. Yadav and A. Chittora, "A Compact high Power UWB TEM Horn Antenna," 2020 IEEE International Conference on Electronics, Computing and Communication Technologies (CONECCT), 2020, pp. 1-3, doi: 10.1109/CONECCT50063.2020.9198586.
18. Hota, S., Mangaraj, B. B. (2019, December). Miniaturized planar ultra-wide-band patch antenna with semi-circular slot partial ground plane. In 2019 IEEE Indian Conference on Antennas and Propogation (InCAP) (pp. 1-4). IEEE.
19. S. V. Yadav and A. Chittora, "Cone-Shaped high gain antenna for Ultra-Wideband applications," 2021 IEEE Indian Conference on Antennas and Propagation (InCAP), 2021, pp. 27-30, doi: 10.1109/InCAP52216.2021.9726181.
20. Garg, H., *et al.* (2019). Dumbbell shaped microstrip broadband antenna. Journal of Microwaves, Optoelectronics and Electromagnetic Applications, 18, 33-42.
21. Rawat, R., Bhardwaj, P., Kaur, U., Telang, S., Chouhan, M., & Sankaran, K. S. (2023). *Smart Vehicles for Communication, Volume 2.* John Wiley & Sons.
22. Mahor, V., Bijrothiya, S., Rawat, R., Kumar, A., Garg, B., & Pachlasiya, K. (2023). IoT and Artificial Intelligence Techniques for Public Safety and Security. *Smart Urban Computing Applications*, 111.
23. Mahor, V., Bijrothiya, S., Mishra, R., & Rawat, R. (2022). ML Techniques for Attack and Anomaly Detection in Internet of Things Networks. *Autonomous Vehicles Volume 1: Using Machine Intelligence*, 235-252.

24. Mahor, V., Bijrothiya, S., Mishra, R., Rawat, R., & Soni, A. (2022). The Smart City Based on AI and Infrastructure: A New Mobility Concepts and Realities. *Autonomous Vehicles Volume 1: Using Machine Intelligence*, 277-295.
25. Mahor, V., Pachlasiya, K., Garg, B., Chouhan, M., Telang, S., & Rawat, R. (2022, June). Mobile Operating System (Android) Vulnerability Analysis Using Machine Learning. In *Proceedings of International Conference on Network Security and Blockchain Technology: ICNSBT 2021* (pp. 159-169). Singapore: Springer Nature Singapore.
26. Mazinani, S. M., & Hassani, H. R. (2009). A novel broadband plate-loaded planar monopole antenna. IEEE Antennas and Wireless Propagation Letters, 8, 1123-1126.
27. Thukral, S. (2021). A Compact Hammer Shaped Printed Antenna With Parasitic Elements For Defense And Mobile Satellite Applications. In 2021 IEEE Indian Conference on Antennas and Propagation (InCAP) (pp. 521-524). IEEE.
28. Bansal, Y. (2020). A compact slot antenna for ultra-wideband applications. Telecommunications and Radio Engineering, 79(3).
29. Kim, G. H., & Yun, T. Y. (2013). Compact ultrawideband monopole antenna with an inverted-L-shaped coupled strip. IEEE Antennas and Wireless Propagation Letters, 12, 1291-1294.

8

Special Military Application Antenna for Robotics Process Automation

Manish Varun Yadav[1]* and Swati Varun Yadav[2]

[1]*Electronics and Communication Engineering, CHRIST (Deemed to be University), Bengaluru, Karnataka, India*

[2]*Electrical & Electronics Engineering Department, BITS Pilani, K K Birla Goa Campus, Goa, India*

Abstract

A special military application antenna for robotics process automation is presented in the following chapter. An antenna is a device that uses wireless communication. Wireless communication's main advantage is protecting our soldiers from undefined enemies. To keep this thing in mind, we have designed a special military application antenna. The presented antenna is useful for defense and satellite communication, including wi-fi and Wimax, which is useful for the robotics automation process. Most of the military robotics automation is based on wireless communication. Our proposed antenna is very useful and capable of receiving or transmitting high signals in terms of GHz. The presented geometry can radiate the large frequency band from 2.9 to 11.6 GHz, which covers the 5G-(I) Sub-6GHz band and X-Band Communication, with high efficiency. The impedance bandwidth of the radiator is 120%, with an electrical size of .14λx.14λx0.014λ in lambda. The antenna is simulated with an FR4 substrate using a CST Simulator. Simulations also investigate the 08-stages evolution process and corresponding S-parameter results are presented. The proposed structure also demonstrates stable radiation patterns across the operating bandwidth. The proposed radiator has a high gain of 6.78 dBi and an efficiency of 89%. Therefore, it is useful for 5G-(I) Sub-6GHz band and X-band military applications, including satellite mobile, Radar, and Satellite microwave communication.

**Corresponding author*: manishvarun21@gmail.com

Romil Rawat, Rajesh Kumar Chakrawarti, Sanjaya Kumar Sarangi, Rahul Choudhary, Anand Singh Gadwal and Vivek Bhardwaj (eds.) Robotic Process Automation, (125–138) © 2023 Scrivener Publishing LLC

***Keywords*:** Military application antenna, robotics process automation, impedance bandwidth, radiation efficiency, 5G, X-band

8.1 Introduction

Most of the military robotics automation is based on wireless communication. Wireless communication's main advantage is protecting our soldiers from undefined enemies. An antenna is a device that uses wireless communication and various designs have been presented for defense and military applications, including curved slot, triangular-shaped, circular, 'U', and 'L' shaped, and printed semi-circular slots [1–6]. High frequency is achieved by printing a tiny fractal element in an antenna [7]. By making a U-shaped in antenna, bandwidth enhancement is achieved [8]. By making circular trims in the designed radiator, high gain is achieved up to 4.5dB [9]. Two resonant bands are obtained by trim slots in the structure, main patch, and attaching circles in the backplane [10, 11]. Wide-band is reported by making a corrugated structure and a half-curved element slot in the radiator [11, 12]. Parasitic elements are added to the structure and multiple slots are also introduced to enhance the bandwidth [13, 14]. As reported, the truncated ground has a modified patch for improved performance [15]. Stable radiations have been achieved by extending the circular element with a backplane and cutting a rectangular element [16, 17]. Wideband spectrum applications have been reported due to trimming semi-circular shape slots in the ground plane and making a modified patch element with the defected backplane [18, 19]. Planar radiators with a unique shape like a 'Dumbbell Shaped' planar antenna and a flexible radiator are reported [20, 21]. The signal path is increased by making a long strip and cut slots in a backplane for a stable pattern is reported [22]. A balanced radiation pattern has been achieved by a modified patch with PGP backplane and properly cutting circular slots on the backplane in the planar antenna [23, 24].

As reported, higher frequencies were achieved by adding rectangular plates at the edges [26]. High gain is achieved by designing a slotted circular fractal antenna [27]. Large impedance is reported by cutting multiple slots from front-side antenna [28]. As reported, the lower band starts resonating by making an L-shape strip [29]. In Section 8.2, the antenna dimensions are specified, as well as an eight-stage development method and simulation results. The robotics process automation is also described in Table 8.3. The simulations of various antenna metrics, as well as related discussion, are covered in Section 8.4. The paper comes to a close after a thorough examination of several simulated results for the design.

8.2 Special Military Application Antenna for Robotics Process Automation

The configuration of the special military application antenna is shown in Figure 8.1. The FR4 substrate is used in the proposed unique antenna design, which has μr = 1, loss tangent = 0.025, and εr = 4.3. Figure 8.1 depicts the geometry of numerous elements and the overall dimension is 15x15x1.5 mm^3.

The substrate thickness is 1.5 mm. The patch is fed through a 50Ω strip line with a width of 3 mm. The thickness of the ground plane and of the patch is 0.018 mm. Table 8.1 shows about the geometry parameters of special military application antenna.

The patch consists of multiple circular and rectangular slots, making an antenna very efficient and capable of radiating specific frequencies. The back side of the military antenna has a partial ground plane designed to increase the signals' electrical path. The unique design makes use of the robotics automation process. The surface current is powerful in the proposed antenna, which means an antenna is capable of working in any condition.

Figure 8.2 shows the prospective view of the special military application antenna. The antenna port (red color) is showing at the bottom side of the special antenna, which fed the signal by a 50Ω signal line on the front and back sides.

Figure 8.3 shows the 8 stages of evolution of the special military application antenna. The first stage of the design comprises of a circular patch with a full copper backplane. The second is made by adding circular elements

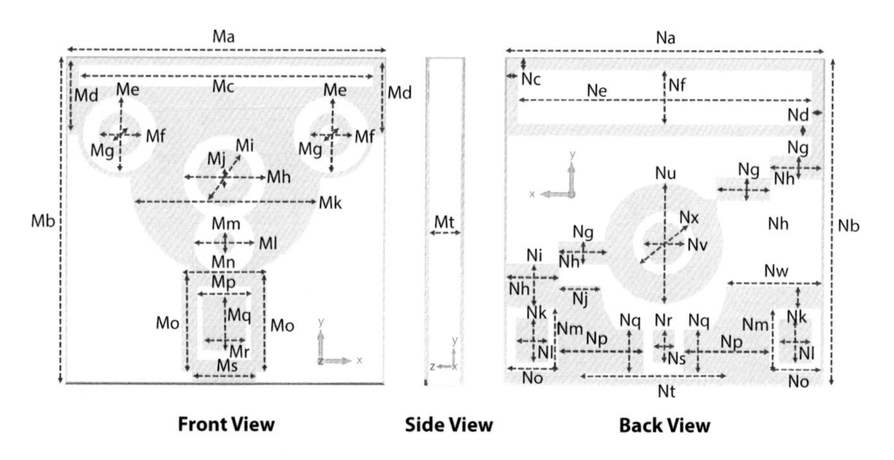

Figure 8.1 Configuration of special military application antenna.

Table 8.1 Geometry parameters of special military application antenna.

Parameters	**Ma**	**Me**	**Mb**	**Mf**	**Mc**	**Mg**	**Md**	**Mh**	**Mj**
Values (mm)	15	4	15	2	14	1	3.5	4	1
Parameters	**Mi**	**Mo**	**Mk**	**Mp**	**Ml**	**Mq**	**Mm**	**Mr**	**Mn**
Values (mm)	2.6	4.5	9	2.6	3	2.5	1	2	4
Parameters	**Ms**	**Mt**	**Na**	**Ne**	**Nb**	**Nf**	**Nc**	**Ng**	**Nd**
Values (mm)	3	1.5	15	14	15	2.5	.6	1	14
Parameters	**Nh**	**Nm**	**Ni**	**Nn**	**Nj**	**No**	**Nk**	**Np**	**Nl**
Values (mm)	2.5	3	2.03	2	2	2.5	2	4.1	1.5
Parameters	**Nq**	**Nv**	**Nr**	**Nw**	**Ns**	**Nx**	**Nt**	**Nu**	-
Values (mm)	2	2.6	1.5	4.5	1	3	7	5.6	-

Figure 8.2 Prospective view of special military application antenna.

on top of the main patch. Stage 3 is made by adding a rectangular element at the top of the main patch. Stage 4 is developed by cutting a slot in the ground plane along with a parasitic circular element. Stage 5 is done by making a ladder structure and Stage 6 is made by cutting a circular slot and the rectangular slot on the front side of the special antenna. Stage 7 is developed by cutting a circular and rectangular slot on the back side of the antenna. Stage 8 is made by a partial ground plane and modified main patch. All stages are simulated on CST tools. Each element's introduction's effect has relevance in generating the conclusion we obtained.

The proposed special military application antenna is a low-profile wideband radiator designed and simulated. The impedance B/W of the

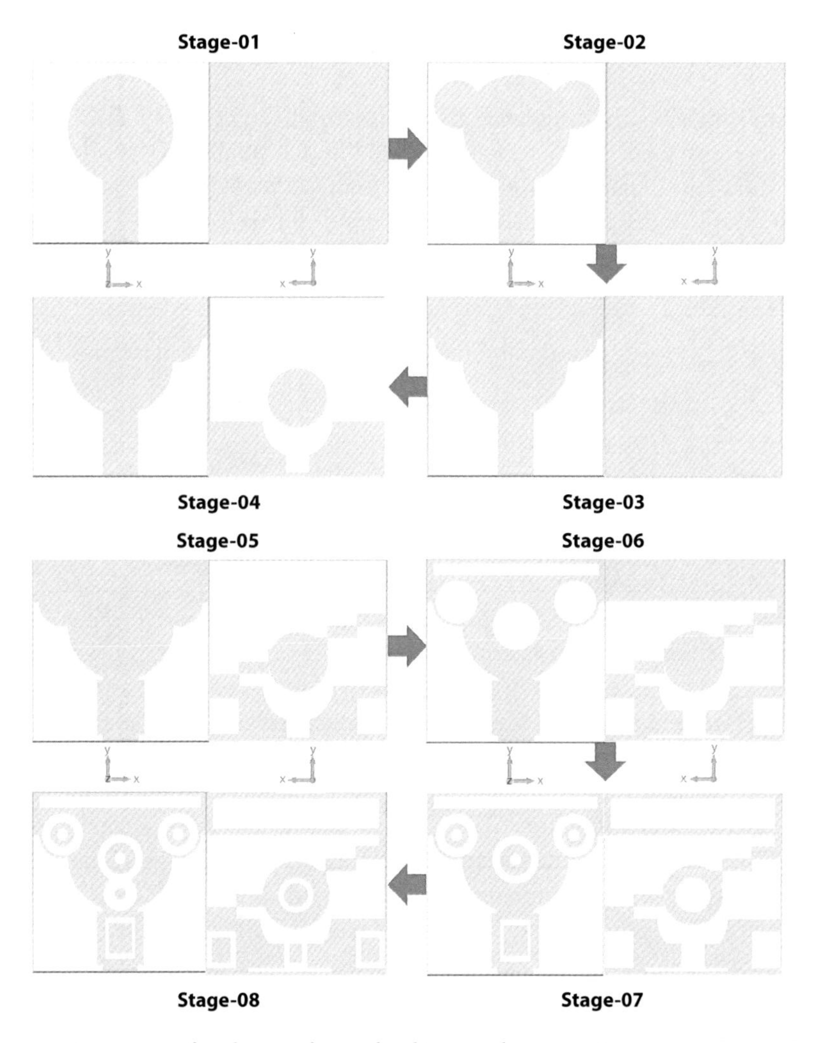

Figure 8.3 8 Stages of evolution of special military application antenna.

proposed structure is 120%, with a central frequency of 7.28 GHz (2.9 GHz to 11.6 GHz). The antenna's width is represented by 'Ma & Na', length by 'Mb & Nb', and 'Mt' is the height of the FR-4 substrate. The robotics process automation is defined in Table 8.3. For the home automation system, robotics process automation is an essential part. We analyze the radiation process through robotics process automation to compete in this process. Different angles rotate the smart antenna for home automation, as mentioned in Table 8.3. According to the angles, it radiates the different radiation patterns on the E and H plane with the help of robotics process automation.

8.3 Results for Special Military Application Antenna

Figure 8.4 shows the simulated gain diagram of the proposed antenna. We can observe a peak gain of 6.78 dBi, which is much smaller in size. The geometry shows the 3 diagrams for 3 different frequencies: at a frequency 3 GHz, it shows the 6.78 dBi gain; at frequency 6 GHz, it shows the 2.9 dBi gain; and at a frequency of 9 GHz, it shows the 5.11 dBi gain, which means the antenna is capable of radiating good signals in the entire bandwidth. Now, moving

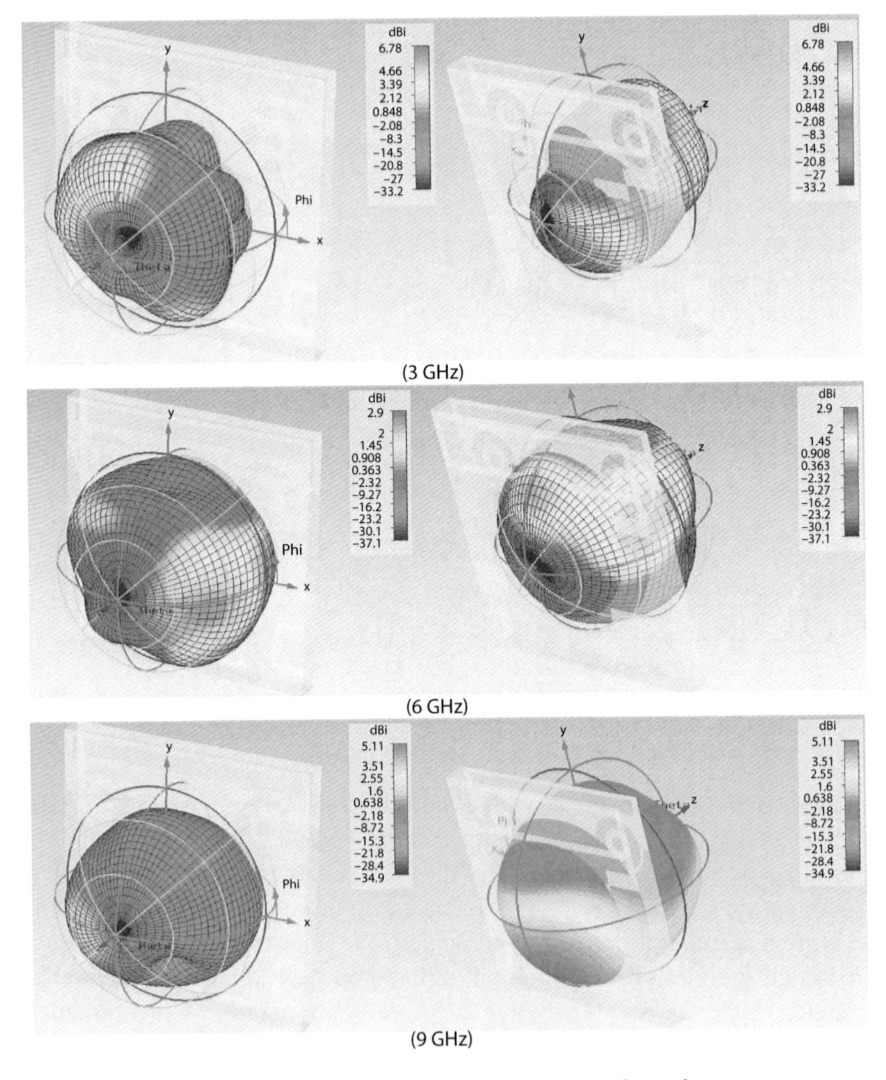

Figure 8.4 Gain of special military application antenna at 3 different frequencies.

forward we will discuss the other performance metrics. The proposed military application antenna achieved a peak radiation efficiency of 85%.

Figure 8.5 shows the structure's radiation pattern at 3 and 6 GHz frequencies. The pattern is defined in two coordinates 90^0 degree to each other, the H-plane is defined by the XoZ plane (Φ=0°), and E-plane by the YoZ plane (Φ=90^0). At all frequencies, the design is very efficient, and it shows stable radiation patterns.

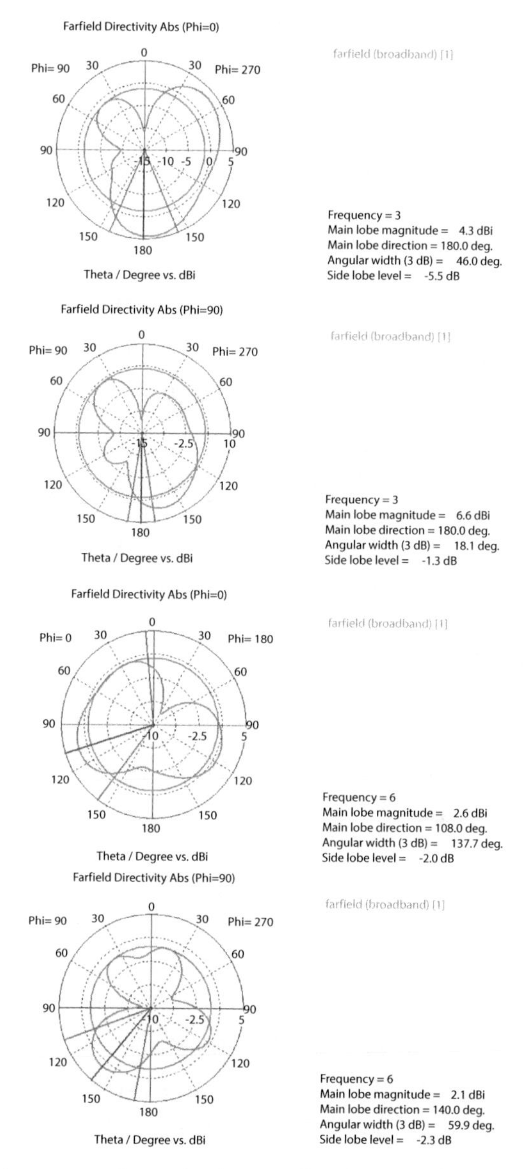

Figure 8.5 Radiation pattern of special military application antenna at 2 different frequencies.

Figure 8.6 shows (S_{11}) the curve of the smart antenna for home automation systems. The return loss curves show the antenna's good signal transmission capability throughout the entire bandwidth. A smart antenna's loss is defined by following formula:

$$Return\ Loss\ (S_{11}) = -20\ \log10\ (\tau)$$

$$VSWR = \frac{1+|\tau|}{1-|\tau|}$$

where (τ) is the reflection coefficient.

Both curves show that the antenna has a good signal transmission capability throughout the entire bandwidth and its impedance is fully matched from 2.9-11.6 GHz.

Figure 8.7 shows the front view antenna current distribution at 3 and 6GHz frequencies for E-field and H-field current distribution. An equivalent surface replaces an antenna and its radiated current field shows its signal strengths. An antenna can generate both E-field and H-field, which shows the good signal strength of the design.

Table 8.2 shows the comparison chart of the previously published antennas. As per the comparison of published antennas, we can say that our proposed military application antenna is more efficient, capable, and compact in size.

The robotics process automation is defined in Table 8.3. With the help of simulations, an antenna rotates and covers all the angles at different frequencies. Table 8.3 shows the robotics process automation, how the antenna is rotated, and how it can radiate the different radiation patterns.

In an extreme situation, cyberterrorists [30–32] may launch "cyber-physical strikes," basically converting satellite antennas into microwave-operated weapons [33], using the same satellite technology that ships, aeroplanes, and the military use to link to the web. Several well-known satellite

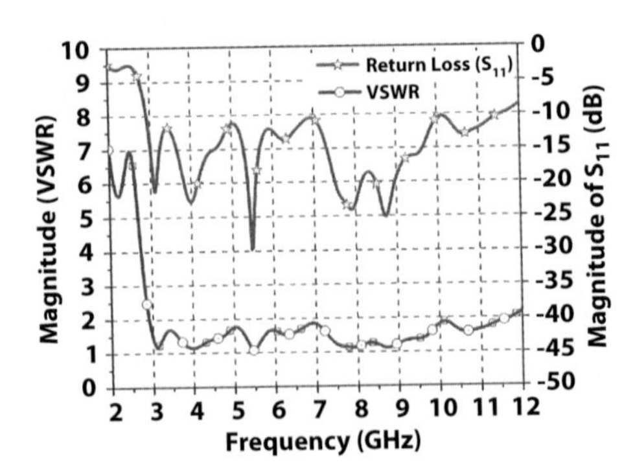

Figure 8.6 Simulated VSWR and S11 curve.

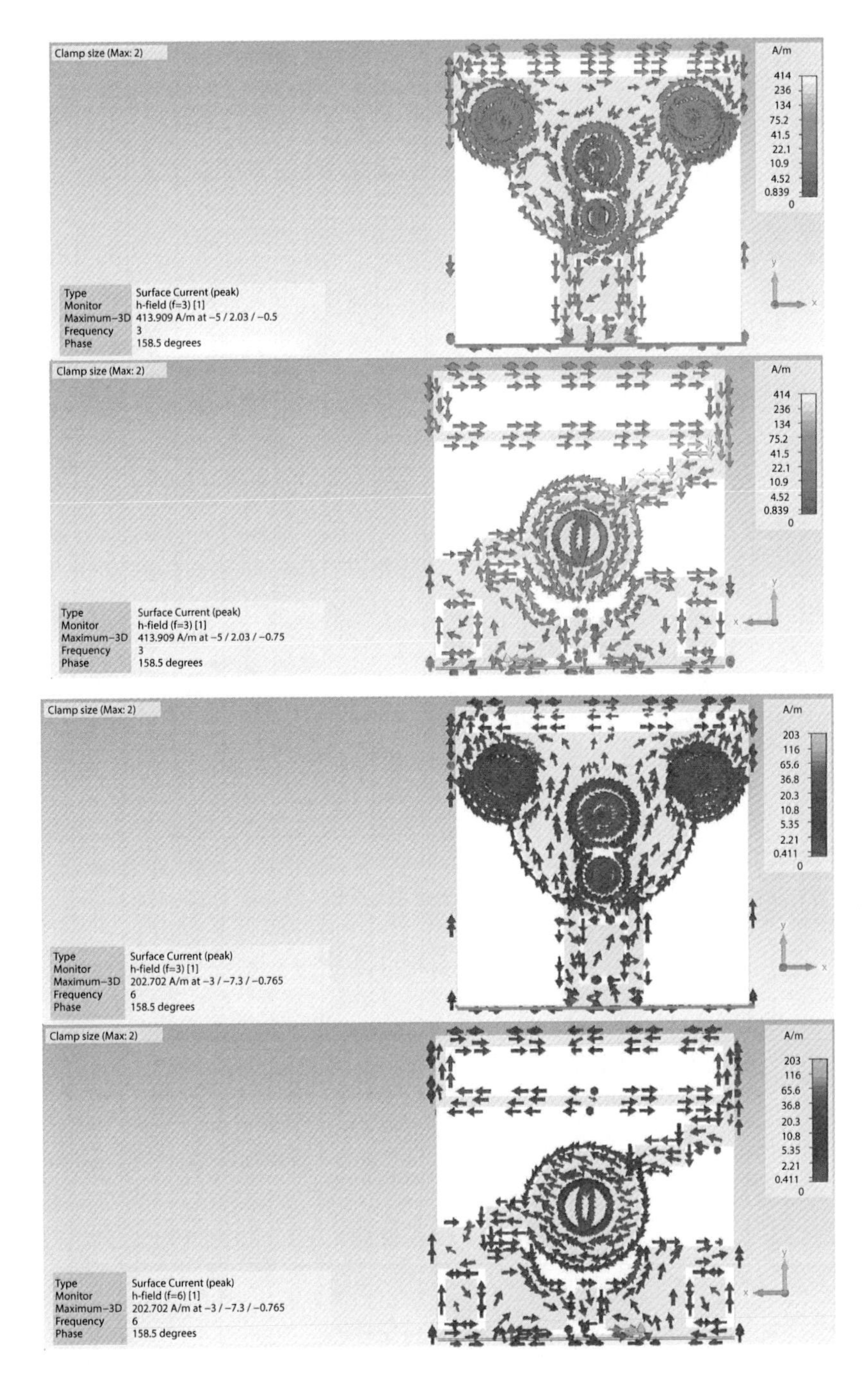

Figure 8.7 Surface current distribution.

Table 8.2 Comparison of published planar antennas.

Ref.	Overall volume (in λ)	Band obtained (GHz)	Fractional B/W (%)	Peak gain (dBi)	Peak (η) (%)
[2]	.28λ*.25λ*.016λ	3.1-22	150%	1.7	NA
[5]	.20λ*.25λ*.015λ	3.1-11	110%	5.1	89%
[7]	.26λ*.26λ*.019λ	3.9-14	142%	3.5	75%
[9]	.33λ*.22λ*0.1λ	2-9	127%	4.5	62%
[13]	.23λ*.23λ*.015λ	3.5-19	145%	3.2	81%
[16]	.55λ*.41λ*.022λ	3.1-11	109%	2	60%
[18]	.33λ*.24λ*.014λ	2.9-16	139%	5.2	87%
[24]	.18λ *.14λ *.15λ	2.8-12	122%	2.79	72%
[27]	.32λ*.2λ*.014λ	2.7-7.3	108%	2.3	78.3%
[28]	.14λ*.18λ*.015λ	3.1-11	110%	2.2	69%
[29]	.2λ*.3λ*.014λ	2.3-11	129%	2.1	70%
Presented	**.14λ*.14λ*.014λ**	**2.9-11.6**	**120%**	**6.78**	**85%**

Table 8.3 Robotics process automation smart antenna for military application at different degrees (angles).

	4 GHz	4 GHz	6 GHz	6 GHz	9.2 GHz	9.2 GHz
Theta (degree)	**E-plane**	**H-plane**	**E-plane**	**H-plane**	**E-plane**	**H-plane**
30	-11.51	-3.544	-1.407	1.15853	-2.636	0.1938
60	-9.29	-1.032	-10.43	-3.6778	-15.93	0.6021
90	-1.391	0.981	-8.401	-3.4596	-9.036	0.872
120	1.4666	2.02129	-0.661	-2.495	-2.602	0.788
150	2.1572	2.2181	2.215	-1.898	-0.3514	0.3486
180	1.64	1.7451	2.292	4.415	0.0971	0.0944
210	-0.525	0.746	0.3033	-0.454	-0.6007	0.1534
240	-6.822	-0.472	-3.488	-3.022	-3.491	0.4001
270	-11.91	-1.084	-11.35	-4.042	-8.485	0.6306
300	-4.188	-1.345	-8.549	-6.794	-12.63	0.6335
330	-2.49	-2.531	-0.565	-1.991	-3.678	0.4763
360	-3.951	-4.254	1.305	-0.2096	0.1762	0.1762

communication systems are susceptible to the assaults, which also have the potential to expose intelligence and compromise linked devices. The cyber-attacks [34] might endanger the safety of military and marine [25] users even though they are only an inconvenience for the aviation industry.

8.4 Conclusion

A special military application antenna for robotics process automation has been reported. The proposed antenna for multiple applications, including defense and satellite communication, wi-fi, and Wimax, which is useful for the robotics automation process, is 15*15*.5 mm^3 in size and has a fractional bandwidth of 120% (2.9–11.6 GHz). The compactness of the structure was determined to be responsible for a peak gain of 6.78 dBi and a radiation efficiency of 85%. An antenna's patterns are also consistent across the whole wideband frequency range. As a result, the antenna is suitable for 3.5-7 GHz (5G (I) Sub-6 GHz band), WiMAX (3.3, 3.5, 5.5 GHz), WLAN (5.2, 5.8 GHz), C Band (4-8 GHz), and X Band (8-11.6 GHz) applications.

References

1. Mishra B, Verma RK, Yashwanth N, Singh RK (2021). A review on microstrip patch antenna parameters of different geometry and bandwidth enhancement techniques. International Journal of Microwave and Wireless Technologies 1–22.
2. Li G, Zhai H, Li T, Li L and Liang C (2012). A compact antenna with broad bandwidth and quad-sense circular polarization. IEEE Antennas and Wireless Propagation Letters 11, 791-794.
3. Yash, B., Yadav, M. V., Baudha, S. (2020). A compact mace shaped ground plane modified circular patch antenna for ultra-wideband applications. Telecommunications and Radio Engineering, 79(5), 363-381.
4. Khidre, A., Lee, K. F., Elsherbeni, A. Z., & Yang, F. (2013). Wide band dual-beam U-slot microstrip antenna. IEEE transactions on antennas and propagation, 61(3), 1415-1418.
5. S. V. Yadav and A. Chittora, "A High Power Circularly Polarized Antenna with Overmoded Waveguide," 2021 IEEE Indian Conference on Antennas and Propagation (InCAP), 2021, pp. 431-434, doi: 10.1109/InCAP52216.2021.9726452.
6. Srivastava, I., Baudha, S., Yadav, M. V., & (2020). A Novel Approach for Compact Antenna with Parasitic Elements Aimed at Ultra-Wideband Applications. In 2020 14th European Conference on Antennas and Propagation (EuCAP) IEEE, 14(4), 1-5.
7. Shagar, A., & Wahidabanu, S. (2011). Novel wideband slot antenna having notch-band function for 2.4 GHz WLAN and UWB applications. International Journal of Microwave and Wireless Technologies, 3(4), 451-458.

8. Kapoor, K. (2019). U-Shaped Microstrip Patch Antenna with Partial Ground Plane for Mobile Satellite Services (MSS). In 2019 URSI Asia-Pacific Radio Science Conference (AP-RASC) (pp. 1-5). IEEE.
9. Kurniawan, A., & Mukhlishin, S. (2013). Wideband antenna design and fabrication for modern wireless communications systems. Procedia Technology, 11, 348-353.
10. Ghosh, A., Ghosh, S. K., Ghosh, D., & Chattopadhyay, S. (2016). Improved polarization purity for circular microstrip antenna with defected patch surface. International Journal of Microwave and Wireless Technologies, 8(1), 89-94.
11. Baudha, S., & Yadav, M. V. (2019). A compact ultra-wide band planar antenna with corrugated ladder ground plane for multiple applications. Microwave and Optical Technology Letters, 61(5), 1341-1348.
12. Abdelraheem, A. M., & Abdalla, M. A. (2016). Compact curved half circular disc-monopole UWB antenna. International Journal of Microwave and Wireless Technologies, 8(2), 283-290.
13. Gupta, S., (2019). Parasitic rectangular patch antenna with variable shape ground plane for satellite and defence communication. In 2019 URSI Asia-Pacific Radio Science Conference (AP-RASC) (pp. 1-4). IEEE.
14. Varun Yadav, S., & Chittora, A. (2022). Circularly polarized high-power antenna with higher-order mode excitation. International Journal of Microwave and Wireless Technologies, 14(4), 477-481. doi:10.1017/S1759078721000611.
15. Basak, A., Manocha, M., Baudha, S., & Yadav, M. V. (2020). A compact planar antenna with extended patch and truncated ground plane for ultra wide band application. Microwave and Optical Technology Letters, 62(1), 200-209.
16. Yadav, M. V., & Baudha, S. (2021). A miniaturized printed antenna with extended circular patch and partial ground plane for UWB applications. Wireless Personal Communications, 116(1), 311-323.
17. S. V. Yadav and A. Chittora, "A Compact high Power UWB TEM Horn Antenna," 2020 IEEE International Conference on Electronics, Computing and Communication Technologies (CONECCT), 2020, pp. 1-3, doi: 10.1109/CONECCT50063.2020.9198586.
18. Hota, S., Mangaraj, B. B. (2019, December). Miniaturized planar ultra-wideband patch antenna with semi-circular slot partial ground plane. In 2019 IEEE Indian Conference on Antennas and Propogation (InCAP) (pp. 1-4). IEEE.
19. S. V. Yadav and A. Chittora, "Cone-Shaped high gain antenna for Ultra-Wideband applications," 2021 IEEE Indian Conference on Antennas and Propagation (InCAP), 2021, pp. 27-30, doi: 10.1109/InCAP52216.2021.9726181.
20. Garg, H., *et al.* (2019). Dumbbell shaped microstrip broadband antenna. Journal of Microwaves, Optoelectronics and Electromagnetic Applications, 18, 33-42.
21. Lakshmanan, R., & Sukumaran, S. K. (2016). Flexible ultra wide band antenna for WBAN applications. Procedia Technology, 24, 880-887.
22. Mangaraj, B. B., Hota, S., & Varun Yadav, M. (2020). A novel compact planar antenna for ultra-wideband application. Journal of Electromagnetic Waves and Applications, 34(1), 116-128.

23. Hota, S., Mangaraj, B. B. (2019). A compact, ultrawide band planar antenna with modified circular patch and a defective ground plane for multiple applications. Microwave and Optical Technology Letters, 61(9), 2088-2097.
24. Golait M, Varun Yadav M, Patil BH, Baudha S, Bramhane LK (2021). A compact ultra-wideband square and circular slot ground plane planar antenna with a modified circular patch. International Journal of Microwave and Wireless Techno. 1–6.
25. Baudha, S., & Yadav, M. V. (2019). A novel design of a planar antenna with modified patch and defective ground plane for ultra-wideband applications. Microwave and Optical Technology Letters, 61(5), 1320-1327.
26. Mazinani, S. M., & Hassani, H. R. (2009). A novel broadband plate-loaded planar monopole antenna. IEEE Antennas and Wireless Propagation Letters, 8, 1123-1126.
27. Thukral, S. (2021). A Compact Hammer Shaped Printed Antenna With Parasitic Elements For Defense And Mobile Satellite Applications. In 2021 IEEE Indian Conference on Antennas and Propagation (InCAP) (pp. 521-524). IEEE.
28. Bansal, Y. (2020). A compact slot antenna for ultra-wideband applications. Telecommunications and Radio Engineering, 79(3).
29. Kim, G. H., & Yun, T. Y. (2013). Compact ultrawideband monopole antenna with an inverted-L-shaped coupled strip. IEEE Antennas and Wireless Propagation Letters, 12, 1291-1294..
30. Mahor, V., Pachlasiya, K., Garg, B., Chouhan, M., Telang, S., & Rawat, R. (2022, June). Mobile Operating System (Android) Vulnerability Analysis Using Machine Learning. In *Proceedings of International Conference on Network Security and Blockchain Technology: ICNSBT 2021* (pp. 159-169). Singapore: Springer Nature Singapore.
31. Mahor, V., Garg, B., Telang, S., Pachlasiya, K., Chouhan, M., & Rawat, R. (2022, June). Cyber Threat Phylogeny Assessment and Vulnerabilities Representation at Thermal Power Station. In *Proceedings of International Conference on Network Security and Blockchain Technology: ICNSBT 2021* (pp. 28-39). Singapore: Springer Nature Singapore.
32. Rawat, R., Gupta, S., Sivaranjani, S., CU, O. K., Kuliha, M., & Sankaran, K. S. (2022). Malevolent Information Crawling Mechanism for Forming Structured Illegal Organisations in Hidden Networks. *International Journal of Cyber Warfare and Terrorism (IJCWT)*, 12(1), 1-14.
33. Rawat, R., Rimal, Y. N., William, P., Dahima, S., Gupta, S., & Sankaran, K. S. (2022). Malware Threat Affecting Financial Organization Analysis Using Machine Learning Approach. *International Journal of Information Technology and Web Engineering (IJITWE)*, 17(1), 1-20.
34. Rawat, R., Mahor, V., Chouhan, M., Pachlasiya, K., Telang, S., & Garg, B. (2022). Systematic literature Review (SLR) on social media and the Digital Transformation of Drug Trafficking on Darkweb. In *International Conference on Network Security and Blockchain Technology* (pp. 181-205). Springer, Singapore.

9

Blockchain Based Humans-Agents Interactions/Human-Robot Interactions: A Systematic Literature Review and Research Agenda

Faizal Kureshi, Dhaval Makwana, Umesh Bodkhe*, Sudeep Tanwar and Pooja Chaturvedi

Department of CSE, Institute of Technology, Nirma University, Ahmedabad, Gujarat, India

Abstract
The notion of blockchain has facilitated the creation of smart contracts, cryptocurrencies, and a multitude of other technological developments. However, apart from its usage in cryptocurrencies and networking cooperation and mechanization, distributed ledger mechanisms also have important social and technical ramifications for the future cohabitation of humans and robots. This article is inspired by the recent surge in interest in blockchains, as well as our substantial work on what is freely available and its combination with robotics and how blockchains and other distributed technological advances can affect interactions as a result of social integration and interaction of robots into human society and with the robotics agents.

Keywords: Blockchain, robotics, process, security, human-robot interaction

9.1 Introduction

HRI is the science of studying, creating, and assessing robotic systems employed by or working with human operators. Interaction amongst robots and humans is required for interaction. Human-robot interaction might take a variety of forms. These shapes, however, are heavily influenced by

**Corresponding author*: umesh.bodkhe@nirmauni.ac.in

Romil Rawat, Rajesh Kumar Chakrawarti, Sanjaya Kumar Sarangi, Rahul Choudhary, Anand Singh Gadwal and Vivek Bhardwaj (eds.) Robotic Process Automation, (139–166) © 2023 Scrivener Publishing LLC

the human driver's proximity to the robot. As a result, communication or contact between humans and robots is divided into two major categories: The first type of interaction is distant interaction and the second type of interaction is proximate interaction. In the distant one, there really is no human in the vicinity of the robot. Furthermore, they are separated spatially or even momentarily. The Mars Rovers, which are away from Earth in both space and time, are an example of this type. Humans and robots cohabitate in the same environment in the near category. This category includes service robots that interact with humans in the same room [1].

These criteria can assist in distinguishing between applications that need mobility, physical engagement, and social interaction. Remote engagement with mobile robots is sometimes known as teleoperation or supervised control. Telemanipulation is defined as remote interaction with a physical manipulator. The robot helper is used for close engagement with mobile robots. Physical contact is a type of proximate interaction. Social contact encompasses all aspects of social, emotional, and cognitive interaction. Humans and robots interact in social situations as peers or colleagues. Significantly, social contact with robots is near instead of remote.

Proximal contact is addressed in this paper based on these factors. It demonstrates some real-world uses of HRI in commercial, medical, and rehabilitative settings, agriculture, education, and other settings. These applications demonstrate the significance of the interaction that occurs between the robot and human interface in our daily lives. Beyond circumscribed task oriented human based robot collaborative interaction, we can imagine future situations in which a robot with even the most limited cognitive capabilities and independent organization would be required to indulge in comparably equivalent societal connection with humans, for example in contracts in financial activity [2, 3].

It is not unique to consider the notions of contracts and financial activity between robots and humans. HRI research now incorporates such exchanges in trials that provide financial gains to game-like situations involving robots and people. However, such exchanges do not completely represent the peer-to-peer financial exchanges between persons which do not involve human intervention. Consider the possibility that a human subject in an in-lab study believes that winning a financial wager to a robot indicates that the monetary payoff would be made by a human researcher rather than the robot. Although not tested in current HRI study, this form of intermediate contact may exhibit framing effects and potentially contribute to expectancy bias. This might be the situation in studies that simulate competitive activity between people and robots before applying standard game theories helps understand the results. As a result, we notice that classic game theory investigates the way rational actors play against all other rational entities. Humans are not too rational beings and different aspects of contact, both psychological and physical, can influence the eventual outcomes. As a result, behavior-based game concepts, which incorporate psychology into game-based theory, may give a better framework for evaluating such kinds of interactions. This provokes us to take into account other aspects that may contribute to preconceived biases, overreaction, and other artifacts that result in an inaccurate experience.

Contact between human and robot intervened by a third-party person may be seen as a component that curbs authenticity and confines outcomes to laboratory situations. Moreover, in a future where humans and machines are coexisting, we should not constantly imagine a human to act as a mediating entity. It is likely that designing a human-robot interface that can precisely replicate interaction between humans, such as during a negotiation or a fierce game, and will result in differing outcomes that represent human perception of robots.

The restrictions of money and the lack of instruments to undertake direct peer-to-peer value exchange involving a robot and a human may partially explain such unrealism in HRI-based studies that entail furious games, negotiations, and commitments. Furthermore, formal contracts between a robot and a human do not completely duplicate or portray the authenticity of a human connection due to the absence of robots and the limits of justifiability of such contracts due lack of technology. Consider the following agreement: (a) if a robot with a specific identification (alpha) completes a job (beta) (e.g., picking up Lego pieces from a specific workstation) within a specified time (delta), the robot alpha will earn (gamma) dollars or tokens and (b) if the robot (alpha) completes a job (beta) (e.g., picking up Lego pieces from a specific workstation) within a specified time (delta), the robot (alpha) will earn game. We might be curious about the role and impression of a human as a robot administrator or controller. However, we can flip the roles and designate the human agent in charge of accomplishing the assignment, while the robot serves as the executor or controller. The first scenario, in which someone acts as a supervisor, is unusual and has not been researched. This is due in part to the limitations discussed above.

However, we can still pose the following questions to ourselves:

1. How can a robot receive monetary compensation?
2. Why does a robot need money?
3. How valuable is money to a robot?
4. What would a robot buy with his or her money?
5. How can we make such a contract legally binding and provide more assurances of fairness to the robot?

The second instance, in which the robot serves as the supervisor, receives the greatest attention in the literature, notably in the fields of behavioral change and compelling robotics. Human psychology suggests a token economy in which tokens can be redeemed for goods or freedoms. Psychological incentives, in addition to real benefits, are used in persuasive robots. However, we need to consider how palpable a robot delivering a reward to a person is, as well as how we may effectively describe such an interaction. We can look into the following topics, among others:

1. How can a robot give a human a monetary or non-monetary incentive?
2. What role does a robot play in supervising a human-assigned task?
3. How can we make such a contract legally binding?

In terms of the first scenario, we may imagine a nearish future wherein robots, whether semi-autonomous or completely autonomous, may be required to conduct financial transactions with humans. For example, a robot may decide that it is more efficient to purchase a service as part of a larger strategy to complete a job, such as calling a cab, rather than utilizing its system of natural locomotion. As part of a proxied encounter for its owner, a robot may be compelled to make a financial transaction (e.g., in telepresence). As a result, money might assist the same role for robots as it does for people, allowing for more participation and connection with human civilization. The limitations of fiat money, on the other hand, present a barrier to robots not only because of the physical form of money, but also because of the technological-social barriers that money erects over financial services ownership and acknowledgment building. Making an online purchase almost always needs the use of a credit card, which necessitates the use of human- identity cards, credit scores, and physical addresses, which are examples of centric procedures. Another challenge concerns the ability to enforce a contract; how can we ensure that a human will just not defraud a robot? Or, at the very least, lessen the likelihood of such an occurrence. When only human parties are engaged in either situation, we could rely on third-party mediators or arbitrators to handle enforceability and fairness [3].

Consider what would happen if a mother assumed the position of a supervisor and asked her child to complete a certain task in exchange for a reward. If, after completing the assignment, the mother refuses to give the kid the prize for any reason, the child can enlist another person to act as an arbitrator or negotiator. We can envisage another member of the family, such as the father, playing the position of arbitrator/negotiator. However, for quantitatively quantifiable activities, we can consider introducing technology that keeps track of the job and provides verifiable data. Likewise, procedures should exist in contracts between robots and a human that not only offers justice, but also contributes to improving the commitments of both the robot and the human. Notions such as arbitrators and reliable pieces of knowledge ought to be available and used.

Smart contracts, cryptocurrency, and the study of cryptoeconomics are all examples of blockchain technology that acts as enabler elements of the later debate, in addition to their use, based on contracts and peer-to-peer financial activity. These innovations have broader social and technological consequences for the cohabitation of humans and robots. There is a range of potential interactions among both humans and robots, as well as ramifications for robot interactions. Furthermore, Distributed Identity and the capability of robotics and objects (IoT) to get such an identification might be regarded as one and the same.

This study discusses the future of financial transactions and contracts between robots and humans, as well as among robots.

Three fundamental technologies serve as the foundation for such interactions:

1. Blockchain(s)
2. Smart Contracts
3. Cryptocurrencies

The integration of these technologies with robotics creates a new arena that researchers in fields like human-robot interaction, law, social policy, and other political science should examine further. Blockchains, cryptocurrencies, smart contracts, and decentralized identities are among the study's bases.

9.2 Conceptual Foundation

9.2.1 Blockchains

Billing existed back in Mesopotamia, with the invention of double-entry bookkeeping acting as the fundamental underpinning of how we account for value to this day. This procedure has been computerized in contemporary times, with the logic of debit and credit balances integrated above standard database management systems. On a high level, the latter enables parties to keep track of specific transactions and responsibilities, while also offering a comprehensive picture of account status. When we look at other industries, we see that they have procedures and systems in place to track communication between two parties.

Bankers and other financial sector third parties rely on the concept of keeping track of all financial services, such as payment providers and depositories (i.e., a ledger). In a fiat economy, financial transactions necessitate the use of precise ledgers. However, the deployment of such ledgers is currently customized and centralized with no openness to external parties who transact; each party is linked to a customized ledger. As a result of this, there are complications in data reconciling and concerns when undertaking large network transactions.

A blockchain solves this challenge by replacing centralized, customized, and ambiguous ledgers with a single immutable, decentralized, and 2 ledger that can track everything of value digitally and is publicly verified (e.g., a deed, a land tenure, identity, or a financial activities).

Bitcoin, the totally digital, decentralized cryptocurrency, is credited for coining the term and introducing the notion of a blockchain. Bitcoin and other decentralized systems run on blockchain technology. Unlike Bitcoin, where the blockchain is primarily utilized as a ledger to record all Bitcoin transactions, the blockchain is also used for other purposes. Ethereum's blockchain is used to store, execute, and monitor smart contracts. From a technological sense, the real data structure and technology behind blockchains are not novel. It is built based on decades of scholarly research study in encryption, distributed databases, mathematics, and economic engineering, among other areas. The real-world implementation of a consensus process that integrates these technologies, allowing users to participate in generating a line of records without the need for a central party, is a key breakthrough. A blockchain's technological design has developed since its conception and fresh structures are being deployed. However, the notion of a shared ledger that enables anyone (or anything) to store and distribute important data in a safe, verifiable, and tamper-proof manner remains central to all deployments. It is what enables numerous cross-industry and worldwide coordination and cooperation application cases. It is the technology that promises to do away with the need for third-party trust and reliance on middlemen [6, 7].

We feel it is up to the user to learn more about blockchain mechanics by reading the extensive literature, such as the following, but we want to concentrate on the implications of having a ledger with such attributes available for robotics applications (or in a world filled with robots). We employ public permissionless blockchains to achieve our objectives because we believe the future is decentralized and that security and privacy should never be in the hands of ruling elites.

9.2.2 Cryptocurrencies

Cryptocurrency enables worldwide financial activities to be conducted with everyone and everything that can establish a blockchain bitcoin wallet. Cryptocurrencies are classified into two types: native blockchain coins (such as Bitcoin, Ethereum, or Litecoin) and tokens. Coins function as a currency or means of exchange, where transaction Coins serve as a medium of exchange or currency with transactions taking place on their own blockchain. Tokens, on the other hand, are a representation of a specific commodity or utility, such as commodities or reward points, that can be generated via smart contracts. Cryptocurrency is successfully transformed into controllable money through the interaction of blockchain-based contracts taking place on their local blockchain. Tokens, on the other hand, are a reflection of a specific item or utility that may be generated using smart contracts, such as commodities or reward points. The interaction of blockchain based contracts effectively transforms cryptocurrency into controllable money.

To store cash or tokens, a wallet account from the blockchain with which the agent will communicate is required. The majority of the time, creating an account, such as an Ethereum address, necessitates the extraction of a public key from such an encryption key, as well as other cryptographic processes. Client wallet apps like Meta-Mask and Portis make it easy for people to generate wallet addresses and connect to decentralized apps (DApps) on blockchains like Ethereum. A DApp is often made and comprised of front-end software that connects to a smart contract. To send bitcoin or engage with a smart contract, the sender's private key is required, followed by the usage of the sender's private key to authenticate the transaction, and finally the deal is sent into the blockchain. By showing user interfaces and completing the appropriate cryptographic mechanisms behind the scenes, the wallet software packages listed above make these activities easier. Because of the decentralized structure of the blockchain, Bitcoin is the first completely digital decentralized cryptocurrency to operate.

Bitcoin, a cryptocurrency, is created through the mining process. This is a method of verifying the legitimacy of anything. Blockchain requires a network of computers to authenticate transactions and compete to solve a cryptographic issue in order to construct a "block" of transactions and add it to the "chain" in exchange for freshly created Bitcoins. These freshly created Bitcoins serve as an economic measure to keep the administrators of these machines (miners) trustworthy and active. Ethereum and other blockchains employ the same mechanism.

9.2.3 Smart Contracts

Nick Szabo initially proposed the concept of a smart contract in the 1990s. Szabo developed a "computerized transaction protocol that performs the provisions of a contract". Szabo's research essentially presented a way for turning people's flawed trust into useful trust delivered by software that follows established rules. Smart contracts are self-executing, self-enforcing, and tamper-proof digital agreements that can be used to represent complex commercial relationships.

Bitcoin popularized smart contracts later and Ethereum implemented them completely. Simple smart contract solutions in bitcoin were established using stack-based programming languages that allowed users to construct Bitcoin exchanges. These programs could be used to execute multi-signature agreements or escrow payments. Ethereum was later developed as part of an effort to establish a global computer that uses a blockchain and provides a framework for creating and implementing smart contracts. Unlike Bitcoin, Ethereum uses actually, a computer program that allows for more complicated smart contracts. Because smart contracts and cryptocurrencies are totally digital and decentralized, people and machines can create and engage in solitary and cohesive ways, as well as participate in peer-to-peer (unmediated) transactions.

9.2.4 Decentralized Identity

Identity is one of the most important concepts in the functioning of human socioeconomic growth. As humans, we utilize our identities to build trust and show ownership or personal rights. Passports, driver's licenses, and circumstance identifiers like a job ID or even a social media account can now be used to assert human identification. These identities are linked to the characteristics that identify and distinguish a person. Birth dates, names, addresses, biometric information, and government-issued IDs like passport numbers and driver's license numbers are all included. These characteristics identify us. Identity is generally linked to a history of financial transactions and responsibilities, which is then used to verify a person's identity (e.g., credit worthiness). In nations like China, a social welfare system is even tied to government-issued IDs in future economic road maps.

The conventional paradigm of identification is reversed by decentralized identity focused on the standard of conscience identifiers. The ability for users to establish, administer, and dispose of multiple decentralized identifiers (DIDs) without the assistance of third parties is critical to the Disparate Systems paradigm, as is the ability to recall the attributes associated with such identities. This technique of providing the user complete power is what distinguishes it as self-sovereign. To that purpose, a decentralized infrastructure and solutions like blockchain and public-key authentication are appropriate. DIDs, for example, might be issued rather than identity distribution authority and authorization could be managed by a blockchain. A public key obtained from a private key, on the other hand, can be used as a unique identity. However, we'd need a reliable system for tracking and recording transactions involving such identities and the blockchain provides that infrastructure. Both humans and machines are affected by this worldview. This means that people will have more control over their identities. For robots, this means the ability to attach historical knowledge that indicates an identity, as well as the ability to approach the identification in an appropriate manner [8].

9.3 Motivation

Human-robot interaction (HRI) is a broad study area and concept that has recently acquired prominence and relevance. HRI aspires to achieve a complementing blend of robot capabilities and human talents. Humans are assisted by robots in terms of precision, force, and speed. Humans contribute in terms of experience, task execution expertise, intuition, simple adaptability and learning, and comprehension of control tactics. This paper examines the implications of human-robot interaction. Industrial, medicinal, agricultural, governmental, and educational applications are all possible. HRI is employed in industrial applications such as assembly line picking and placement, welding operations, part assembly, and painting. Adaptive robots is one of the most visible fields of HRI. Robots can offer interaction and rehabilitation for persons with physical and mental disabilities. Furthermore, HRI may be widely used in hospitals. HRI is now critical for dealing with the emerging coronavirus (COVID-19) pandemic. In agriculture, human-robot collaboration aids in numerous jobs such as harvesting, sowing, fertilizing, spraying, weed identification, transporting, and mowing. HRI is also used in various areas such as education, mining, and in residences.

Blockchain technology is evolving at a rapid pace and it is feasible to combine it with a variety of systems, including robotics and AI applications. However, because this is a new subject, no clear picture of what it could become exists. In this paper, we summarize a variety of methods and platforms that aim to incorporate blockchain benefits into robotic devices, improve AI services, or focus on solving problems that exist in massive blockchains, all of which can lead to the creation of robotic systems with enhanced functionality and security. We give an overview, explore the techniques, and end with our thoughts on how these technologies will interact in the future.

9.4 Blockchain and Robotics Overview

This section contains work that attempts to incorporate blockchain with robotics. The work of Bruno Degarding and Lus A. Alexandre (as supervisor) demonstrates how to build a blockchain and utilize it to store robotic activities. This concept enables the implementation of smart contracts that employ information gathered in the wild by several robots (potentially from different manufacturers) and have initiative based on agreements recorded and confirmed on the blockchain. This can eventually boost manufacturing efficiency by reducing the time required on chores such as replacing screws for a robot that utilized the blockchain to communicate that it needed more screws to complete its operation.

As a follow-up to this study, the authors suggest the development of a blockchain for robotic events and exhibitions that makes use of the enhanced security afforded by the formal verification incorporated in Tezos. This ongoing effort will enable smart contracts to execute AI code well over blockchain, where these smart contracts will be proven true (that is, they will accomplish precisely what their description says). They also want to modify the blockchain to handle several more transactions per second than the present standard allows, allowing the system to handle a massive number of interacting robots.

Aitheon is a platform that uses ERC-20-compliant blockchain technology. Their objective is to create a full platform that may decrease the number of time-consuming chores that developers must undergo, such as document organization. Their solution is a platform that combines AI and robotics to automate certain business activities. They created a platform with five parts. The first is an AI module that attempts to collect knowledge about regularly performed tasks and then automates those processes. The second module, Digibots, is quite similar to the very first, but it focuses on automated programming activities such as back-end answers and data-driven difficulties. The third module, Mechbots, is designed to assist organizations in integrating robotic automation to improve efficiency and production. The final two modules, Aitheon

Professionals and Pilots, are human specialists who do jobs that cannot be entirely automated while simultaneously supervising the robots. In a nutshell, Aitheon offers a platform that integrates AI and Robotic technology to automate time-consuming processes.

This approach uses MIT OPAL to provide a layer of protection and privacy to the information represented in the data. As a result, the robots may use the knowledge they've gathered to develop Machine Learning (ML) algorithms locally before publishing them to the network. The blockchain technology is a component of this architecture for transparency and to act as a ledger, recording the events performed by the robot during an occasion and validating the published models. The authors also suggest a consensus method that allows each node to vote on which models to adopt by utilizing the blockchain's smart contract technology.

Recently, an approach for assembling a consortium of robots, sensors, and actuators was suggested. This is accomplished by putting all information via knowledge processing and then storing it into a blockchain. This enables each coalition component to have global information and the usage of smart contracts to change robot operations and reallocate resources. According to the authors, this system enables immutable distributed storage, which is critical for negotiating discrete responsibilities among various parties (robots).

9.5 Human-Robot Interaction

9.5.1 Games

Because the technology is peer-to-peer, we can create unmediated transactions that allow a robot to participate directly in a bilateral agreement with a person more simply. For example, in a physical chess encounter between a human and a robot, both teams bet on the winner-take-all scenario. Eduardo Castellon Ferrer's research highlights the advantages of merging blockchain technology with robotics, particularly swarm robotics and robotic hardware. The benefits of robotic swarms include their simplicity of scalability, as well as their robustness against failure. These benefits stem from the fact that individuals of these swarms are dispersed. In the industrial sector, we can observe how this industry is expanding and helping enterprises to reach increased efficiency, such as AmazonRobotics, which has been exhibiting its army of robots that collaborate to run their warehouses. Most robotic clusters only employ local data, which implies that a robot only knows about itself and/or robots nearby, however, the integration of blockchain in these platforms can provide the robots with global knowledge, which can be helpful for a variety of applications. Blockchain integration can also increase the speed with which the system changes its behavior because having global knowledge allows the entire system to swiftly modify behaviors to meet the demands of specific robots. This may also be done by a controller robot that analyses the system status using blockchain knowledge and then publishes the modifications to the blockchain. These enhancements, as well as the system's contextual information, may lead to increased productivity and ease of management [10].

The creators of RoboChain, a theoretical technique for securely sharing essential data among robots, provided a framework to address privacy concerns around the use of blockchain. The Figure 9.1 shows about the human versus robot chess game and betting sequence diagram and the Figure 9.2 shows about the example of autonomous vehicle smart contract ownership.

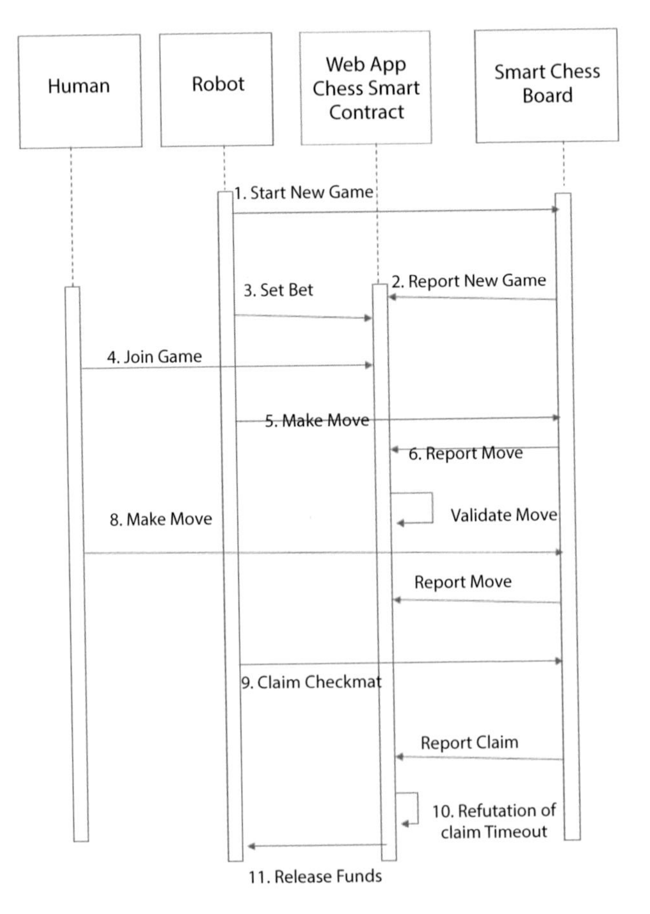

Figure 9.1 Human versus robot chess game and betting sequence diagram.

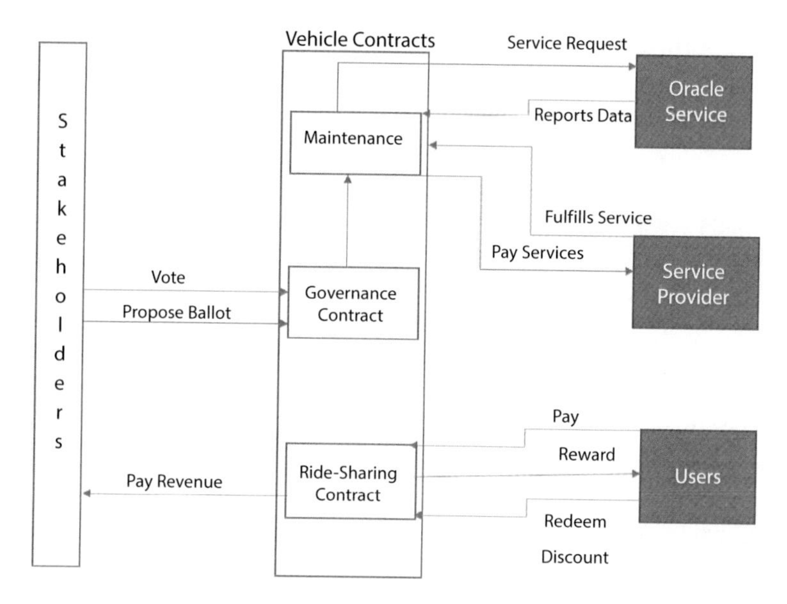

Figure 9.2 Example of autonomous vehicle smart contract ownership.

In a laboratory setting, a typical application of this human contact would involve a human researcher first familiarizing the human subject with the game's context. Basically, it's a competitive game with a monetary incentive on the line. The game's winner receives a monetary award. A rewards card is commonly used to represent monetary compensation. This game does not allow for meaningful interaction between a human and a robot with a strong sense of intelligence and agency, given the framing effects stated. During a competitive match, we may expect a human subject to consider not only what is at stake, but also how the opposing party sees what is at stake (e.g., utility of the award).

We can't expect a robot to utilize a rewards card for personal purposes because it is difficult for a robot to access traditional physical or online markets and make a purchase with one. Our society's basic finance infrastructure, as previously indicated, is insufficient for cyberphysical actors.

However, as previously said, it is more plausible for a robot agent to deal economically with people and other robots using Bitcoin. We can envisage a robot purchasing a non-fungible badge, simulated property in a blockchain game, or a tangible product or service through other decentralized ecommerce platforms like OpenBazaar after winning a bet.

A human agent, a robot agent, a DApp, and a smart chess game are among the characters we encounter.

Each of the human and robot agents has a Bitcoin wallet and/or software. They make a move on the board by moving chess pieces around. The chess board is deemed smart since it communicates with a shared ledger via a surveillance system and a set of physical keyboards, simulating the chess game. Chess boards and other physical electrical games could be connected to a blockchain in the future.

The smart contract verifies the transactions and is in charge of keeping track of all wagers, delivering funds solely to the winning player's digital wallet. The DApp is a user-friendly interface for the chess game that is tied to a smart contract. It displays the current condition of the game and enables any player to sign the contract and gamble. A human user can learn about the game's rules and the amount of bitcoin required to play by scanning the QR code on the app. A robot can use the latter or it can use an Ethereum library like Web3 to submit a transaction directly to the chess shared ledger.

Other games might be organized similarly, emphasizing the robot's role as the game's transactional verifying party, forcing the robot to decide if a move is valid or whether the opponent has won or lost the game. As a result, the ideas of deception and greed may be more accurately depicted because the robot is the genuine agent in charge of conducting or authorizing a transaction (not a third-party human). Additional game interaction is shown. The latter features a racing/combat game in which remotely, human drivers gamble cryptocurrency or acquire the privilege to operate a robot and race against one another, as well as the ability for the entire globe to bet either against the robot or buy robots with additional "power ups". This is a revolutionary cyber-physical blockchain game that takes advantage of economic incentives and cryptographic functionalities to enable worldwide cooperative and adversarial play.

9.5.2 Commitment Strategies and Conditioning

We may look at using smart contracts to research dedication strategies and conditioning. We may foresee the interaction of smart contracts and intelligent computers that can engage in arrangements with humans to offset a human's short-term inclinations. A smart device, for example, can engage into a time arrangement with a human and restrict the human from accessing the fridge late at night or gaining access to specific food containers. When such logic is included in a smart contract, it becomes personal, self-verifiable, tamper-proof, and irreversible. From here forward, the options for interaction are endless.

9.6 Applications of HRI

9.6.1 HRI Application in Industry

HRI is commonly utilized in industrial applications such as picking and placing on assembly lines, welding techniques, component assembly, and painting. There are several instances of HRI in the industrial applications provided. The robot workstation is operational in the BMW facility in South Carolina, where it assists human workers in the final door assembly. Human operators and robots collaborate in the door assembly procedure. In series manufacturing, the BMW factory has been successful in adopting and developing direct human-robot collaboration and interaction. In flexible manufacturing lines, human-robot teams can also be seen where robotic hucksters and human operators work together to handle workpieces. The closeness of the human driver to the robot might lead to possible injuries, thus safety is also highly crucial in such circumstances.

Repetitive co-manipulation activities can be carried out in appropriate human body positions. The positions can help to reduce the consequences of overloaded joint torque. Moreover, they can increase the potential of human manipulation.

The multi-robot platform with interactive functionality supports the worker in the handling of large and bulky elements in welding scenarios. Two robots aid in the location of the components to be linked in the welding process, following a human operator who performs the welding work in an ergonomically favorable environment. Unlike a traditional welding bench, the human operator is not obliged to adopt uncomfortable positions or work overhead. All necessary workpiece placement and alignment can be done by the robot. This also includes positioning the components in the best possible position for the welding process, allowing for proper welding bead flow. Because the robotic displacement action is swift, the managing time, which accounts for around one-third of the overall process time, is cut in half when compared to welding processes that do not use HRI and the cost is minimal.

9.6.2 HRI in Rehabilitation and Medical

Adaptive robotics is one of the most visible fields of HRI. Robots can give opportunities for engagement and rehabilitation for those with physical and mental disabilities. This work is being investigated with autistic youngsters. Many of them are unable to respond substantially to social cues, yet they respond quite well to mechanical gadgets. Robots can play a therapeutic function by employing mechanical devices to improve social relations. Robots are indeed being considered for several domains wherein children benefit, such as children who have endured trauma. Not just in assisting roles, but also in many sectors and domains of local engagement, the sociological dimension of human-robot interaction is required.

For those with physical limitations, the robot embodiment offers options that other types of technology do not. For example, researchers are focusing on the construction of robots that deliver and assist with physical rehabilitation. Efforts include supplying the appropriate force and movement patterns to aid with the restoration of flexibility and strength. Other work includes identifying motivational states and customizing therapy to optimize results. Intelligent wheelchairs are a form of robot that uses external sensors to aid with course planning and collision prevention for the person who utilizes the wheelchair.

A robotic system was built that accurately executes typical physiotherapy exercises like shoulder flexion in a manner similar to what the physiotherapist does during co-manipulation using the KUKA LWR robot. Thus, the good features of the human (decision capability) and the robot (accuracy, labor capacity, and repeatability) might be coupled to obtain superior outcomes in arm musculoskeletal rehabilitation. Furthermore, the second and primary goal was to assess non-pathological shoulder activity while completing the shoulder flexion exercise [5].

Coronavirus disease (COVID-19) is a viral infection caused by a recently identified coronavirus. COVID-19 symptoms can range in intensity from modest to severe. The collaboration of the human and the robot aids in the battle against the next epidemic. Two examples of human-robot collaboration can be employed in the hospital to combat Coronavirus. The robot can identify coronavirus genetic information by adding the solution to nasopharyngeal swabs from patients. The positive test result indicates that the patient has coronavirus infection. In that instance, samples collected from the patients' mouth, nostrils, and throat are analyzed in the laboratory for coronavirus genetic material. The laboratory personnel simply need to load the materials into the tray and the pipetting is handled by the COVID-19 test robot. In the second example, the manipulator is given a movable unit. The mobile robot collaborates with the person and precisely aligns to the workpiece. The mobile robot may be used to deliver meals and medicine to patients. Furthermore, the robot can measure the temperature of the patients. This reduces direct interaction between patients, medical professionals, and others and thereby reduces the risk of infection. The robot may also assist people in sweeping and cleaning the walls and floors. This also reduces the likelihood of viral infection.

9.6.3 HRI in Agriculture

Human-robot collaboration in agriculture aids with numerous jobs such as harvesting, sowing, fertilizing, spraying, weed identification, transporting, and mowing. In precision agriculture, a robot assists a human-operator or farmer in harvesting strawberries. Both the human operator and the robot must be protected in this situation. The robot was also controlled remotely by a co-located operator. They were responsible for directing the robot to the pickers' positions as needed, allowing the filled boxes to be loaded onto the robot and then transporting them to the storage facility. In the greenhouses, an intelligent robot is deployed to care after and assist farmers with melon picking. The principle of HRI is readily seen in Scaffold Mode. The human operators and robot collaborate as a united system in which the vehicles autonomously traverse along the organized tree rows while the humans aboard the vehicle focus on executing and accomplishing duties such as:

> 1) thinning, 2) trimming, 3) harvesting, and 4) attaching trees to wire. People working on the robot in Scaffold Mode were able to shrink trees more than twice as quickly as humans using the traditional ladder-based technique, according to Bergerman *et al.*

9.6.4 Robochains

We utilize RobotChain as a blockchain. RobotChain allows for the reliable and secure registration of robotic as well as other events. It is based on the Tezos blockchain, which has characteristics that are crucial in industrial robot environments like a reduced energy consumption and quicker consensus algorithm (Assigned Proof-of-Stake), when compared to conventional ones such as Proof-of-Work, its identity property that enables changes in the blockchain core to be performed without any need for hard-forks, and the assistance to formally confirm smart contracts. This is vital in complex software to limit the number of developer-induced defects. These may be almost anywhere, for example in the cloud, and simply require a network to interact with the blockchain network. The location of the Oracle(s) must be provided on smart contract reasoning. As a result, only authorized parties may edit the knowledge on the blockchain. In the suggested technique or method, 3 separate Oracles undertake calculation on data (from the blockchain and some additional sensors such as a Kinect) and insert or regulate data analysis on smart contracts which are committed to controlling the robot, storing information about people who enter the robot working space and their affiliations, and so on. The smart contracts enable the system either to be private and locked (not letting other parties to connect with it) or to allow trusted parties to engage with it, allowing complicated algorithms to be executed and other apps to operate over them.

9.7 Transactions between Robots and Human Beings

The following factors are taken into account: (1) A blockchain can be used as a ledger to search and record anything of value, such as a property title or financial transactions by both machines and people; (2) Between a robot and a human, smart contracts can embed self-enforcing and self-verifiable agreement logic; and (3) Cryptocurrencies enable machines to carry financial commitments and engage in value exchanges with humans and vice versa. Finally, we believe that their use and extension will lead to more complicated and realistic interactions between robots and humans and that their adoption in the near future will make robot integration easier. We'll go over how to use them in the sections that follow.

9.7.1 A Ledger that is Decentralized, Immutable, and Publicly Verifiable

A blockchain, like IoT, may be utilized for identity and security management programs for robots, data, and communication. For symmetric encryption communicated between robots as well as data logs kept elsewhere, public key encryption can be employed. Access to information and control of a robot can both be authorized via digital signatures. A blockchain could be utilized for "on-chain" communication and signaling if low-latency and low-cost activities are available. Other agents can listen in on transactions performed by agents on a given blockchain site. A blockchain with a verification time of less than a second could be ideal for such a system. You can also use Whisper and other indicated technological prerequisites.

9.7.2 Contracts

Smart contracts are capable of encoding the logic of unilaterally and bilateral contractual contracts. Cyberphysical agents can interact with such logic and make contracts with other robots or humans because of their digital nature.

Listing 1 depicts a smart contract's preliminary interface implementation for the use case. It illustrates the use of partial ownership of an autonomous car. In turn, the car serves as an automated taxi. Rather than tokenizing ownership, the smart contract merely describes a collection of owners and a method for transferring ownership. Additional smart contracts that automated other procedures and allow for on-chain management of the ride-sharing service would be included in a more thorough deployment of such smart contracts. A governance contract might enable car owners to submit ballots to amend the settings of other smart contracts. A referendum might be suggested to raise the service cost for riding in the car. Another referendum might be recommended to raise the allocated budget for vehicle upkeep. When the vehicle's detectors indicate that it needs repair, such a contract might be activated (e.g., oil change). The contract may interface with an Oracle service, which may give pricing information for service suppliers. The Oracle system may then notify an internet provider that the automobile requires service and create an escrow contract with a pricing quote and a set of requirements that the provider must meet.

After that, the automobile can authorize the contract and provide the escrow contract for the bitcoin amount requested. When the service is completed, the service provider informs the escrow contract that it is complete.

The car will then be notified and it will be able to evaluate whether the contract/service agreement has been fulfilled (e.g., checking oil sensors). The vehicle can sign a transaction and send it to the escrow contract to confirm the service if the parameters are met. The monies would subsequently be released to the service provider under the terms of the escrow arrangement [9]. The Figure 9.3 shows about the smart contract partial interface definition and the Figure 9.4 displays the unilateral contract flow.

```
1   contract RideSharingContract {
2    address public vehicle;
3    bytes32 public vin;
4    address[] public owners;
5    address passenger;
6
7    function vin() constant returns(bytes32 vin);
8    function owners()constant return(address owners);
9    function setOwners(address [] _ownerAddrs);
10   function approveTransfer(address _to);
11   function requestRide();
12
13   event RideReq(address indexed _passengerAddr, uint256
          rideCost);
14   event Transf(address indexed _from, address indexed
          _to);
15   event Appr(address indexed _owner, address indexed
          _approved);
16  }
```

Figure 9.3 Smart contract partial interface definition.

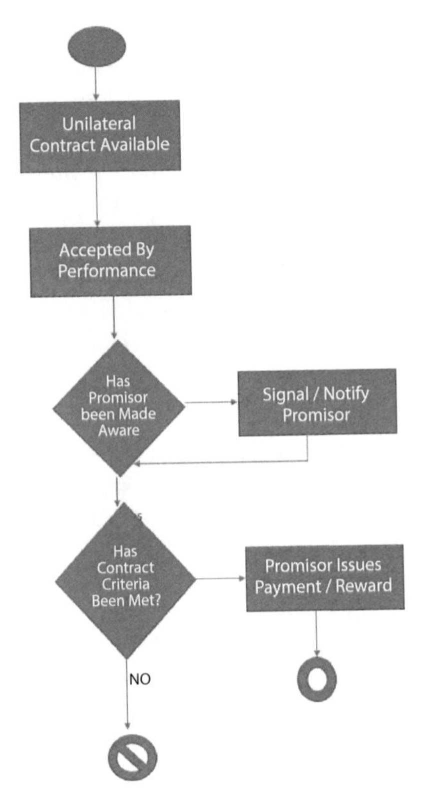

Figure 9.4 Unilateral contract flow.

9.8 Escrow Services

Open requests or compensation contracts, for example, giving a reward for a missing pet, are frequent examples. By giving a reward, the owner commits to complete the payment if the requirement to return the pet is met. Likewise, a robot can provide the public a unilateral contract. In this case, the robot can offer a unilateral or bidirectional contract to any human willing to deliver the robot to a specific location. The robot begins the unilateral edition of the contract by indicating its present location as determined by GPS, the target destination, and the payment for delivering the robot to the target destination. A person ready to sign the smart contract can transport the robot to its destination and sign it there. Contractual agreements are susceptible to judgement and are supported by a legal system. "The script is law" in smart contracts. That is, the smart contract's logic and the inputs it receives determine the final result. As a result, smart contract logic should be carefully developed and smart contract inputs should be trustworthy.

That is, with the appropriate input, conditional logic, such as money transfers, will be implemented. As a result, we must evaluate what constitutes a violation of contract, as well as how a robot may bring a breach of contract claim against a human and conversely, a "breach of smart contract" in the psychological sense could be triggered by (a) a malicious actor acquiring access to either group's private keys, (b) a robot failure, or (c) a mistake by a raw data or hostile data source provider. Code flaws can also cause problems, but they can be avoided with good auditing. A smart contract can be used to address such concerns in a buyer-seller situation. However, we could use an escrow smart contract, which allows an escrow agent to make decisions and transmits reliable data to the smart contract via Oracle. The parties trade in a decentralized escrow system using an escrow smart contract, which is managed by an escrow agent authorized by both parties. If there is a disagreement over the transaction, the escrow agent can step in and evaluate whether another party is due a refund or compensation [11–13].

9.9 Challenges for HRI

9.9.1 Task Dynamic Analysis

Task analysis is an essential part of the human factor toolset. The division of labor between people and machines is a subset of human factors research. Fitts' (1951) list of what a "man" and a computer can do best is already outdated and has yet to be definitively updated. Machine capabilities have clearly increased thanks to miniature sensors, artificial intelligence, and high-speed networking. Consider developing a robot to gently move elderly and disabled people in and out of bed or to the restroom if you think ergonomics is an out-of-date human factor. Many human caregivers are now doing this task and injuring their backs as a result. The problem of planning, scheduling, and simulation in time, place, force, power, and expense with virtual reality visualization aids has never been so difficult. There's also the question about whether general purpose robots in humanistic shape make sense. Experience shows that a robot's structure is best defined by work context. It is difficult to determine the ideal physical shape by analyzing HRI tasks [4].

9.9.2 Teaching a Robot and Avoiding Unintended Consequences

Humans may convey geometric instructions to robots by moving their hands, but explaining how to move, when and where to move, what else to avoid, and so on necessitates symbolic instead of analogic language. Rapid developments in computer-based voice recognition (such as Apple's SIRI) promise to make controlling robots easier. However, there is a significant chance of unexpected outcomes. Before giving the robot the "go" signal, human supervisors can use a real-time virtual reality simulation to see how the robot will respond to the spoken instructions. This would essentially be an extension of prediction displays that continuously update models of the processes under management in order to "look ahead" via model extrapolation.

9.9.3 Connecting Mutual "Mental" Models to Prevent Working in Conflict

Humans and robots have long been thought to have internal (mental) representations of one another (e.g., Sheridan and Verplank, 1978). Computer vision now has the ability to watch human behavior, save data as stick figure modelling, and do Bayesian analysis on the selected data. Humans' mental representations of what robotics can or should do and merging that understanding with the computer's model for scheduling and interpersonal trust continue to be an HRI difficulty. Experiences in AI (visual reinforcement learning, language interpretation) has shown that matching the first 90 percent of human competence is quite straightforward, but replicating the last 10 percent is exceedingly challenging due to the immense quantity of human real-world experience.

9.9.4 Role of Robots in Education

Learning through books and packaged lectures is feasible, but it is tough and monotonous and it will not work for youngsters who do not read or people who have cognitive problems. Interacting with a real person teacher or fellow learner nearly always improves the learning experience. Since Papert's experimentations on children teaching a mechanical "turtle," as mentioned above, the robot has played a role in thinking about the future of schooling to add joy, to start serving as an avatar to be tried to teach or to speak, to illustrate a physical connection (as in physics), or to react to students' responses (with criticism or reinforcement). The interactive workshop of robot learns from those other robots and also people. Knowing how individuals of all ages and capacities better learn from robots is a significant challenge to which humans should contribute.

9.9.5 Lifestyle, Fears, and Human Values

Karel Capek, a Czech playwright, is credited with coining the term "robot" in his play Rossum's International Robots in 1920. Initially, his sci-fi nightmare scenario was amusing, but it appeared implausible. Today's horror flicks feature robot takeovers of employment and infringements on personal dignity. The media has been consumed by robots for both its good and harmful societal repercussions. Modern cinematic adaptations include Star Wars, WallE, ExMachina, and a slew of others. Clearly, there are tradeoffs to consider: robots as useful assistants raising human self-worth versus destroying human self-worth, robots assisting human security versus spying (small UAVs, for example), saboteurs, and assassins. I feel that human factor professionals are more aware of the realities of working and living with robots than the general public. As a result, HF professionals must participate in debates and policy formulation on these topics, which includes public education. If we could automate everything, what would we want to automate and what would we not want to automate?

9.10 Discussion and Future Work

Some suggestions propose to connect blockchain technology with AI, which might be utilized in tandem with robotics. What we strip away from these recommendations is that blockchain, robots, and AI will undoubtedly change the way we live: not just because they can deliver so many benefits on their own, but because we can multiply those advantages by combining them. Blockchain technology may be used to communicate messages between machines and have action triggers embedded in smart contracts, boosting the performance of the robots and their interconnectedness. Although this is undoubtedly true in the near future, present approaches are still in their infancy, owing to the fact that we are in the midst of these technologies' exponential growth period and they have yet to mature. These techniques might theoretically be employed in real-time, but in order to make that happen, blockchain technology must change to speedier consensus processes that allow the authentication and inclusion of transactions into such a blockchain considerably faster than is now achievable.

We concluded from the approaches and platforms reviewed that those with the brightest future are those that incorporate multiple services in a single platform while also sharing the code with the open-source community and creating incentive schemes for detecting flaws.

We will undoubtedly see numerous robotic systems embracing blockchain technology, particularly in industrial and military situations where blockchain can assist to automate operations with the use of smart-contracts and enabling systems to have greater security and traceability. By having validated information about the whole system, blockchain introduces a means to trust the data, respect other players, and make internal and external adjustments. It's simple to conceive scenarios in which both technologies operate together to achieve a single goal. A swarm of "Cop Robots," for instance, that patrol the streets welcoming people and searching for misbehavior. These robots might interact through blockchain and use smart contracts as action triggers. When they see one individual injuring another, they may have the network vote on the best way to approach the incident or call for aid. However, in order to accomplish this sort of behavior, smart contracts must have increased security and be able to communicate with knowledge from outside of the blockchain (oracles).

It is critical to create platforms that can combine complementing techniques so that the market is reduced from many diverse isolated approaches to a limited number of proven solutions, or to specify explicit inter connectivity standards to allow numerous solutions to communicate to each other. The emerging marketplaces will be critical in enabling individual robots to accomplish several complicated jobs without the need for their engineers to create all of the various required solutions. This is something that can, and therefore should, be combined with cloud robotics.

1. Legal Enforceability of Smart Contracts: The community needs to look into the legal enforceability of smart contracts, as well as the consequences for dealing with robots.
2. Usability: Blockchain technology has still yet to acquire widespread adoption. The technology's users need to be able to safely store and maintain private keys. Secret keys can be stored in hardware wallets or used with software wallets and applications. Users must learn concepts like "gas" on the Ethereum network, as well as transaction charges and other jargon.
3. Peer-to-Peer Transactions: If we believe that all parties will act appropriately, not every transfer of value needs the use of a smart contract. Consider the second scenario we discussed before. If the robot uses advanced image recognition and event detection algorithms, it might be able to recognize when a human-based subject has completed the task and pay the human. The robot may inquire about the person's wallet address and send bitcoin directly to them.

4. Future Applications: Our present research looks on the use of blockchain technology in the context of a secure architecture for completely immersive telepresence robotics, for example, robotic replicas that mimic an operator's movements and transmit elevated haptic stimuli. The biometric data from a telepresence remote controller can be added to a robot's DID and linked to the DID of a human driver. Smart contracts can be used as an authorization technique that requires the operator's digital signature. Digital certificates can also be utilized to get access to control devices like a telepresence control garment. Furthermore, we may envisage robotic avatars using bitcoins to transact in a distant setting. Work that fingerprints contacts with robots, for example, can also benefit from blockchain technology. Finally, in a RaaS paradigm, utility-based tokens can be used to provide network permission and charges based on robot utilization could be paid for in real-time for activities like data collection.

Typically, a malware [14, 15] threat and vulnerability [16, 17] entail the unauthorised deployment of software or applications on your machine. These strategies employ interactions between people and psychological abuse to trick users into disclosing personal knowledge or selecting risky security [18] measures. Threat actors pose as reliable people or sources of information while using social engineering tactics to mask their actual identities and objectives. The goal is to persuade, deceive, or manipulate users into disclosing private data or gaining entry inside an organisation. Several social engineering [19] schemes rely on people's propensity for cooperation or concern about punishment. For instance, the attacker can pose as a coworker who needs urgent access to more network capacity to solve a problem.

9.11 Conclusion

Blockchain technology is still very much in childhood and its potential influence on the world economy is unclear. The combination of services with blockchain, particularly robots, is still in its early stages. This means that various advancements are being made on different block chains. There are no apparent 'winning' technologies yet and most market players are unaware of many of the new technologies, but they can also doubt the strength of these first suggestions. Proposed techniques are abounded, but connectivity standards are lacking and integration of such solutions with Industry 4.0 or cloud robotics, for example, has yet to be accomplished. Finally, in the context of robotics, the application of cryptocurrencies, smart contracts, and blockchain(s) ensures game-changing involvements. This includes the concept of a physical or digital robotic agent conducting unfiltered peer-to-peer monetary transactions with humans or other robots, as well as robots entering into contracts with humans and other robots. When imagining the cohabitation of human and robot, cryptocurrencies, blockchains, and smart contracts are critical technologies to contemplate. As a result, the integrated study of robot human interaction, behavioural economics, cryptoeconomics, and behavioral game theory is now referred to as Robonomics.

References

1. L. A. Proceedings of the First Symposium Blockchain and Robotics, edited by E. Castell o Ferrer, T. Hardjono, and A. Pent- land, MIT Media Lab, 5 December 2018. Ledger 4.S1 32–41 (2019), doi:10.5915/ledger.2019.175.
2. T., Shrier, D., and Pentland, A. Cabridge: A Foresightful Future (2016).
3. H. Jeong, J. Kim, and Y. Chi. Irvin S. Cardenas, H. Jeong, J. Kim, and Y. Chi. Distinctive cipher-acoustic languages for human-robot interaction. In the 14th ACM/IEEE International Conference on Human-Robot Interaction (HRI), March 2019, pages 600–601.
4. Nikolay Teslya and Alexander Smirnov A blockchain-based framework for the establishment of coalitions of ontology-oriented robots in cyber-physical systems. MATEC Web Conf., 2018; 161 EDP Sciences:03018.
5. Aho, James A. Rhetoric and the development of double entry accounting Rhetorica: A Journal of Rhetorical History, 3(1):21–43, 1985. Ainslie, George A behavioural theory of impulsiveness and impulse control called Specious Reward. The Psychological Bulletin, 82:463–96, August 8, 1975.
6. U. Bodkhe *et al.*, “Blockchain for Industry 4.0: A Comprehensive Review,” in IEEE Access, vol. 8, pp. 79764-79800, 2020, doi: 10.1109/ACCESS.2020.2988579.
7. Tanwar S, Bodkhe U, Mohammad Dahman Alshehri, Rajesh Gupta, Ravi Sharma, Blockchain-assisted industrial automation beyond 5G networks,-Computers & Industrial Engineering, Volume 169,2022,108209.
8. Ferrer, E. C., Rudovic, O., Hardjono, T., & Pentland, A. (2018). Robochain: A secure data-sharing framework for human-robot interaction. *arXiv preprint arXiv:1802.04480.*

9. Lopes, V., & Alexandre, L. A. (2019, April). Detecting robotic anomalies using robotchain. In *2019 IEEE International Conference on Autonomous Robot Systems and Competitions (ICARSC)* (pp. 1-6). IEEE.
10. Van der Aalst, Wil MP, Martin Bichler, and Armin Heinzl. "Robotic process automation." *Business & Information Systems Engineering* 60.4 (2018): 269-272
11. Syed, R., Suriadi, S., Adams, M., Bandara, W., Leemans, S. J., Ouyang, C., ... & Reijers, H. A. (2020). Robotic process automation: contemporary themes and challenges. *Computers in Industry, 115*, 103162.
12. Aguirre, S., & Rodriguez, A. (2017, September). Automation of a business process using robotic process automation (RPA): A case study. In *Workshop on engineering applications* (pp. 65-71). Springer, Cham.
13. Axmann, B., & Harmoko, H. (2020, September). Robotic process automation: An overview and comparison to other technology in industry 4.0. In *2020 10th International Conference on Advanced Computer Information Technologies (ACIT)* (pp. 559-562). IEEE.
14. Rawat, R., Sowjanya, A. M., Patel, S. I., Jaiswal, V., Khan, I., & Balaram, A. (Eds.). (2022). *Using Machine Intelligence: Autonomous Vehicles Volume 1.* John Wiley & Sons.
15. Rawat, R., Bhardwaj, P., Kaur, U., Telang, S., Chouhan, M., & Sankaran, K. S. (2023). *Smart Vehicles for Communication, Volume 2.* John Wiley & Sons.
16. Mahor, V., Bijrothiya, S., Rawat, R., Kumar, A., Garg, B., & Pachlasiya, K. (2023). IoT and Artificial Intelligence Techniques for Public Safety and Security. *Smart Urban Computing Applications*, 111.
17. Mahor, V., Bijrothiya, S., Mishra, R., & Rawat, R. (2022). ML Techniques for Attack and Anomaly Detection in Internet of Things Networks. *Autonomous Vehicles Volume 1: Using Machine Intelligence*, 235-252.
18. Mahor, V., Bijrothiya, S., Mishra, R., Rawat, R., & Soni, A. (2022). The Smart City Based on AI and Infrastructure: A New Mobility Concepts and Realities. *Autonomous Vehicles Volume 1: Using Machine Intelligence*, 277-295.
19. Mahor, V., Pachlasiya, K., Garg, B., Chouhan, M., Telang, S., & Rawat, R. (2022, June). Mobile Operating System (Android) Vulnerability Analysis Using Machine Learning. In *Proceedings of International Conference on Network Security and Blockchain Technology: ICNSBT 2021* (pp. 159-169). Singapore: Springer Nature Singapore.

10

Secured Automation in Business Processes

Ambika N.

Dept. of Computer Science and Applications, St. Francis College, Bangalore, India

Abstract

The work is a model for the unique restricting of entertainers to jobs in cooperative cycles and a related restricting approach determination language. It proposes three sorts of controlled adaptability component dynamic restricting of entertainers to jobs of a cooperative cycle, dynamic determination of sub-cycles, and active choice of elective trails in an assumed implementation condition of an interaction. The business process model is in the same place as each undertaking related to a job. It plays five parts: Customer, Supplier, Carrier-Candidate, Carrier, Invoicer-Invoice. Assuming the request is dismissed, the cycle is ended. An undertaking is performed by the entertainer bound to the assignment's job. Entertainers might assign themselves or different entertainers to assume a part for a situation or they might demand to let themselves or different entertainers out of a job. A job might be related to the case-maker, suggesting that the job is bound upon case creation and needn't bother with an assignment or underwriting. The execution goes on with the Shipment sub-process, where a provider demands statements from different transporter applicants. The shipment completes in two ways. These installments are embodied in sub-process Invoicing. This sub-process is called two times: for the provider's receipt and the transporter's receipt. The suggestion takes additional measures to make it more secure by 4.35%.

Keywords: Supply chain management, digitalization, security, blockchain, business process

Email: Ambika.nagaraj76@gmail.com

Romil Rawat, Rajesh Kumar Chakrawarti, Sanjaya Kumar Sarangi, Rahul Choudhary, Anand Singh Gadwal and Vivek Bhardwaj (eds.) Robotic Process Automation, (167–180) © 2023 Scrivener Publishing LLC

10.1 Introduction

Mechanization (Willcocks & Lacity, 2016) is a method that operates without direct mortal dealings. Automated procedure Industrialization is the new methodology that aspires to make a program machine that imitates mortal manners. By transitioning to automation (Tripathi, 2018), businesses strive (Ambika N., 2021) to decrease work expenses, improve efficiency, lessen mistake rates, and enhance consumer fulfilment. It is a strategy to systemize enterprise operations using corporation reasoning and client intakes. The purposes deliver means for customers to describe androids that can mock their relations with requests to calibrate a trade, manage information, initiate reactions, and disseminate with other digital techniques.

Historically, industrialization was supposed to enhance operation efficiencies for executing lower prices and trustworthy manufacturing. With the manufacturing procedure thus committed, mechanization assignment achieves scheduled functions efficiently rather than trying to modify or supply fresh possibilities for the business. Digitization has radically changed this inert and non-resistant idea of methodologies. It backs it with new competencies in the framework and making outcomes. The conversion span is undervalued from creation to production. Mechanization has allowed the latter to apply. It is applicable in development and producibility conclusions. Promoting specific manufacturing due to software design skills has ended combination manufacturing, probably at earlier achievable expenses, only through extending creation battings of consistent yields.

The previous system (López-Pintado, Dumas, García-Bañuelos, & Weber, 2022) is related to a career in the action plan. The job links to things considered entertainer. An entertainer has a character and may address a client, a gathering, an association, a framework, or a gadget. At first, a client presents a buy request to a provider. The cycle ends when the PO dismisses. The execution goes on with the cargo sub-procedure, where a provider demands statements from diverse transporter competitors. There are two ways to deal with the installments. These installments embody sub-process invoicing. This sub-procedure is the provider's receipt and the transporter's receipt.

The demonstration of doling out an entertainer to a job inside a case is known as restricting. The job is unbound when a position does not dole out to an entertainer for a situation. Limiting an entertainer to a job might happen during a case. Entertainers untie from a job, an activity called discharge. An undertaking is performed by the entertainer bound to the errand's position. The chore is empowered when its related job is unbound. The project holds on until the job is bound. Entertainers might designate themselves or different entertainers to assume a part for a situation or they might demand to let themselves or various entertainers out of a job. Given the absence of trust, the selection/arrival of an entertainer to/from an assignment might require the underwriting of entertainers assuming different parts. This selection possibly prompts a limiting, considering the necessary supports are conceded. The limiting interaction strategy determines which role(s) are permitted to choose an entertainer for a job, demand an entertainer's delivery from a job, and support a designation/discharge demand.

The suggestion minimizes the disadvantage of the previous system. It aims to create the hash key using the Merkle tree. Every entity, role, and activity has a preassigned code. A hash code generates using the preassigned codes. Every session, the hash code changes though the company plays the same role in the system. Hence, the system is secure using the methodology by 4.35%, considering the attack is average.

The work splits into six sections. The introduction is followed by a Literature survey in Section 10.2. The background is explained in Section 10.3. Section 10.4 describes the proposed work and Section 10.5 details the analysis work done. The last section concludes the work.

10.2 Literature Survey

The work (López-Pintado, Dumas, García-Bañuelos, & Weber, 2022) presents a model for the unique restricting of entertainers to jobs in cooperative cycles and a related restricting approach to determination language. It proposes three sorts of controlled adaptability components dynamic restricting entertainers to positions of a rhythmic cycle, dynamic determination of sub-cycles, and vibrant choice of elective pathways in a given execution condition of an interaction. The business process model is the same for each undertaking related to a job. It plays five parts: Customer, Supplier, Carrier-Candidate, Carrier, Invoicer-Invoice. The cycle ends, assuming dismissal of the request. An undertaking is performed by the entertainer bound to the assignment's job. Entertainers might assign themselves or different entertainers to carry a part for a situation or they might demand to let themselves or various entertainers out of a job. A job might be related to the case-maker, suggesting that the job is bound upon case creation and needn't bother with an assignment or underwriting. The execution goes on with the Cargo sub-procedure, where a provider demands statements from different transporter applicants. The shipment completes, followed in two ways. These installments embody sub-procedure Invoicing. This sub-procedure is called twice for the provider's receipt and the transporter's receipt. The suggestion takes additional measures to make it more secure.

The proposed structure (Parthasarthy & Sethi, 1992) depicts a company's innovation as an endogenous variable that goes through successive versatile changes to remain serious inside the business' innovative climate. The business methodology decisions (BSC) of a firm are impacted by the progressions happening in the company's financial climate. The system interfaces a company's innovation, business procedure, and construction and their effect on execution inside the requirements presented by the business' innovative climate and the general financial climate. The elements of the connection between the two is the power of adaptable computerization and business technique choices. This climate outside of the association comprises large-scale level circumstances: social, political, and monetary powers. It describes and obliges a company's possible item, and the market opens doors. The power of adaptable mechanization build characterizes the degree to which a firm purposes FA as a component of its assembling technique.

The proposition (Vrhovnik, *et al.*, 2007) is a structure to improve information handling in business processes. The work presents a set of modified decisions that change a business cycle. It is a superior execution from the board without changing the semantics of the first cycle. These revamped rules depend on a semi-procedural interaction chart model that externalizes information and controls stream conditions of business dealings. The centre part is a streamlining agent motor that works on the inside portrayal of an interaction. This motor is designed with the assembly of change rules and a control system for rule application. Each rework practice comprises of two sections: a condition and an acting part. The condition part characterizes what conditions need to hold a rule to apply to protect the interaction semantics. It alludes to the control stream conditions concerning the information conditions of interaction.

The principal errand (Velikorossov, *et al.*, 2020) of present-day robotized staff in the executive's frameworks is to upgrade the work (the administration and faculty of the HR administrations of ventures). Information is made, put away, refreshed, and utilized by workers and HR chiefs. Automated stockpiling and handling of faculty data permit choice and development of employees. It is feasible to complete finance bookkeeping data on staffing positions, excursions, wiped out leave, work excursions, advantages, and punishments with the aid of HR. Optimization with the assistance of a mechanized faculty lessens costs at the venture by multiple times. It utilizes the corporate variant of programming and expert rendition by 37.5.

The business process control stream (Graml, Bracht, & Spies, 2007) is displayed utilizing three various stream objects. Occasions happen at the beginning, during, or toward the finish of interaction, performed exercises, and entryways for directing, parting, and blending the control stream. The boxes address BPMN exercises that can either represent nuclear assignments or thereabouts called subprocesses. The jewel-molded entrances manage choices and participate in the control stream. A door can be considered and an investigation is posed at a point in the process stream. The inquiry has a characterized set of elective responses. The occasion-based XOR door addresses a spreading location where the options depend on an occasion. It has various lines and interfaces with the stream objects to make the skeletal construction of an enterprise process. A strong bolt addresses a Series Discharge. It shows the request for the exercises in the procedure. The run line manages a Message Flow with a loose sharpened stone and delivers the stream of messages between two separate BP members.

The issues (Dietz, 2006) perform two sorts of acts-creation acts and coordination acts. Underway demonstrations are addressing topics for achieving the labour and products conveyed to the climate. C-acts and P-acts show up as stages in a nonexclusive harmonisation design. It is known as an exchange in light of the discussion for activity and the work process loop. An exchange develops in three stages: the request, execution, and result phases. In the request stage, the initiator and the agent haggle to complete an agreement about the P-truth that the agent will acquire. In the implementation chapter, the P-truth is achieved by the agent. In the outcome stage, the inventor and the agent haggle to accomplish an agreement about the P-truth.

In the recommendation (Hartley & Sawaya, 2019), one company's electric-based process fosters its production network computerized guide internally. It has a concentrated gathering for dealing with the frameworks, devices, and cycles for its worldwide stock administration work. The group arranges its specialized abilities as essential and progressed. To create the guide, the group worked with inward colleagues to survey the present status of its cycles, future business needs, and the degree of development of its store network innovations. This group, likewise, benchmarked production network innovations utilized by different organizations. Electric works with its inward colleagues to refresh its production network innovation plan yearly. Three associations carried out obtain-to-pay platforms. Pro-Services executed another cloud-based (Nagaraj, 2021) secure-to-pay framework to computerize the whole cycle and kill unremarkable positions in procurement. Transport gave demands for recommendations for another source to pay stage to supplant old bulky framework that was not natural.

The technique (Climent, Mula, & Hernández, 2009) is to acquire the business cycle models of a bank. It continues by finishing the starters. It portrays the primary cycles related to the exercises done by the bank. These are private client assistance, confidential client acknowledgment, well-to-do client treatment, organization winning, organization acknowledgment, strategy formalization, occurrence goal, phone replying, convention arrangement and investigation, and convention endorsement. This interaction is the client support movement that inspects three sub-exercises: accomplishing money exchanges, data demand/risk counsel drive, and data demand/speculation guidance.

The methodology (Schiefer & Seufert, 2005) empowers constant examination across corporate business processes, informs the matter of significant proposals, or naturally sets off business activities, shutting the hole between Business Intelligence frameworks and business processes. The handling steps, their connections to one another, the boundaries of the investigation, and information change cycles can be characterized for each association exclusively. Sense and Response Infrastructure utilizes an occasion handling model for displaying Sense and Respond circles. Like a development unit, the EPM offers different structure blocks for Sense and Respond that administrations can use to build a Sense and Respond circle. Reliant upon the necessities and the business issue, these structure blocks can be deftly conjoined or separated. Joins between the structure block the address of occasions (starting with one help onto the next). Information is gathered and gotten from source frameworks and ceaselessly handled. The EPM shows the occasion streams of Sense and Respond processes and pictures complex handling steps. SARI permits an organization to dissect and change business process occasions progressively. Current information and business pointers are produced through recurrent handling. The caught information diverts to the proper business regions. The salient business data is reliably accessible to laborers and the board. It uses it for independent direction. Business processes are proactively worked, furnishing organizations with the capacity to respond to business valuable open doors and exemptions progressively.

BPR's execution (Teng, Grover, & Fiedler, 1994) is arranging viewpoint centres around the need to characterize a proper IT framework and the significance of the assurance of exact venture goals and a re-designing system. The IT foundation addresses the mix of information, telecom, and figuring advances that empower BPR. This innovation base is not entirely set in stone by the asset choices in BPR arranging. A primary strategy is a cross-utilitarian group and the second is the utilization of interaction generalists. Streams are successions of cross-useful cycles that initially navigated different divisions. Each gathering liable for a stream was assessed and given measures connected with consumer loyalty, like on-time conveyance. BPR arranging will direct the securing of new data innovation to carry out the designated processes. The IT foundation helps re-design drives during the BPR project execution.

Engineering (Jeng, Schiefer, & Chang, 2003) empowers investigation across corporate business processes and informs the matter of noteworthy suggestions or naturally sets off business tasks, successfully shutting the hole between business insight frameworks and business processes. It is a specialist-based design that upholds a total business insight interaction to detect, decipher, foresee, mechanize and answer business cycles and means to diminish the time it takes to settle on the business choices. The engineering incorporates five significant parts. Business Intelligence specialists for scientific handling, the occasion handling holder for the constant change of cycle occasions, a Process Information Factory for putting away business process measurements, a strategy framework and a dashboard to represent business process measurements and scientific results, and a BI specialist layer that comprises of three administration layers. Each layer constantly creates business possibilities for what activities the specialists ought to perform. The detecting subsystem screens and catches the circumstances created in the climate. The reaction subsystem produces activity results for the business climate by following the mandates conveyed by the specialist layers. The intervention subsystem chooses a layer to control the entire framework at some random time. The BAM Foundation is classified into a few levels: overseen assets, board tests, executive's beans, board responsibilities, BI specialists, and the executive's connectors.

The User Interface parts of the assignment (Zou, Zhang, & Zhao, 2007) in the chosen cycle occurrence are sent off in the work area region. The UI part for the assignment is sent off (When the undertaking finishes). As the client advances through the interaction, they move along the user interface and refresh the items in the work list and the route parts. In the static-investigation stage, we dissect the cycle definitions and the source code of the web-based business application. Utilizing the data accumulated from the static investigation stage, the client can make and refresh the substance of the user interface parts as a client advances through an interaction occasion. An occasion motor gives route advances and sets mindfulness for the user interface. The occasion motor sends off applications by giving capacity calls to the user interface parts which carry out an errand.

The occasion motor pays attention to the assignment occasions set off by user interface parts to decide the executing task in a cycle occurrence. The event motor imparts the situation with in-progress process events through refreshing ones to the cycle list, worklist, and the route component. The creators break down all interaction definitions and concentrate data connected with each characterized job. The data incorporates errands led by a job, control streams, and information streams between the undertakings. A cycle might contain subprocesses that portray exercises of other business processes. The execution climate guides clients through a grouping of user interface parts to achieve errands inside a business cycle. The unique execution climate likewise helps in setting mindfulness to help clients work all the while on different cycle cases.

10.3 Background

Merkle Tree

Blockchain innovation (Ambika N., 2021) (Li, Jiang, Chen, Luo, & Wen, 2020) has made different advances in cryptography, arithmetic, agreement calculations, and monetary models. It is a protected, shared, and disseminated record that records all value-based information known as blocks. The blockchain (Khan, Byun, & Park, 2020) uses P2P organizations and agreement systems to take care of the issue of conveyed information synchronization. Bitcoin is one of the most well-known applications utilizing the blockchain procedure. The blockchain information structure is characterized as an arranged back-connected record of blocks of exchanges. It is a data set or saved as a document. The block is perceived by a SHA256 (Dhumwad, Sukhadeve, Naik, Manjunath, & Prabhu, 2017) cryptographic hash calculation. The block is made out of two sections: the principal information and the header. The principal information contains a rundown of exchanges. The header incorporates a hash of the past and current block, Merkle Root, timestamp, nonce, and other data. The Merkle tree (Li, Lu, Zhou, Yang, & Shen, 2013) was introduced by Ralph Merkle in 1979. It is a fundamental key to information checks throughout PCs. Their construction assists in confirming the consistency of information content. Its engineering assists with accelerating security validation in large information applications. It is a finished parallel tree and every hub is to hash the worth from its kid hub. Figure 10.1 portrays the same.

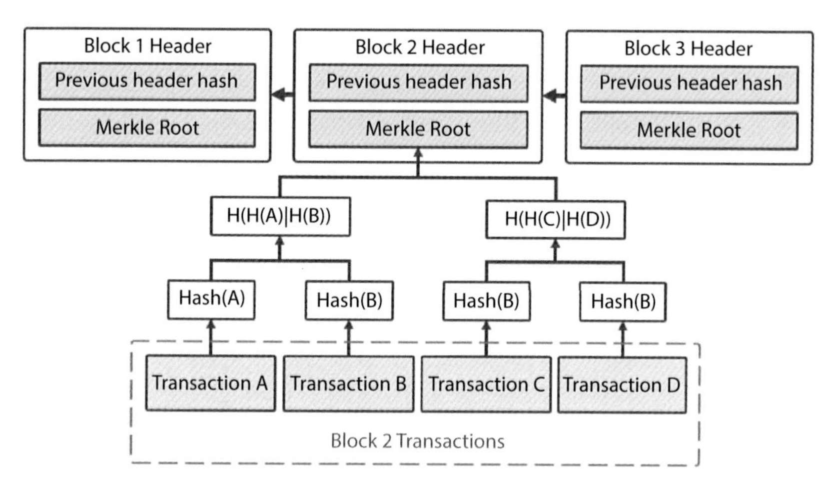

Figure 10.1 Architecture of merkle tree in blockchain (Chen, Chou, & Chou, 2019).

10.4 Proposed Model

The work (López-Pintado, Dumas, García-Bañuelos, & Weber, 2022) presents a model for the unique restricting of entertainers to jobs in cooperative cycles and a related restricting approach to determination language. It proposes three sorts of controlled adaptability components: dynamic restricting of entertainers to positions of a rhythmic cycle, dynamic determination of sub-cycles, and vibrant choice of elective pathways in a given execution condition of an interaction. The business process model is the same for each undertaking related to a job. It plays five parts: Customer, Supplier, Carrier-Candidate, Carrier, and Invoicer-Invoice. The cycle ends, assuming dismissal of the request. An undertaking is performed by the entertainer bound to the assignment's job. Entertainers might assign themselves or different entertainers to carry a part for a situation or they might demand to let themselves or various entertainers out of a job. A job might be related to the case-maker, suggesting that the job is bound upon case creation and needn't bother with an assignment or underwriting. The execution goes on with the Cargo sub-procedure, where a provider demands statements from different transporter applicants. The shipment completes in the following two ways. These installments embody sub-procedure Invoicing. This sub-procedure is called twice for the provider's receipt and the transporter's receipt. The suggestion takes additional measures to make it more secure. Table 10.1 shows about the algorithm to generate hash code for identification of individual/company. Table 10.2 displays about the generating identity bits for next cycle and Table 10.3 highlights for generating the hash code for the activity.

Table 10.1 Algorithm to generate hash code for identification of individual/company.

Step 1: Accept the parameter list {identity (24 bits) & profile (32 bits)}
Step 2: Add 4 bits of zero to the starting of the identity bits and 4 bits to the end
Step 3: XOR profile bits with profile bits (32 bits resultant)
Step 4: Divide the resultant bits into two halves (16 bits each)
Step 5: Add the bits (17 bits resultant) (resulting in outcome)

Table 10.2 Generating identity bits for next cycle.

Step 1: Consider the resultant of the previous cycle
Step 2: Right-circular shift is applied for the resultant bits
Step 3: Add "1" bit in odd positions (3, 5,7,9,11,13,15)
{This is considered as identity for the next cycle}

Table 10.3 Generating the hash code for the activity.

Step 1: Input the identity bits (17 bits) and Activity code (17 bits)
Step 2: Put identity bits in the even position and activity code in the odd position of the resultant
Step 3: Left shift the resultant (outcome hash code – 34 bits)

10.4.1 Disadvantages of Previous System

- If the identity of the role is compromised, it leads to dissimilar types of outbreaks

a. Assumptions
   - The network is liable to get attacked
   - The individual/company have an identity
   - The activities generate identities

b. Notations Used in the Suggestion

Notations used	Description
P_i	Hash code of the identification
R_i	Role played by the organization
A_c	Assigned activity code
A_i	Activity hash code
P_c	Assigned identity code

c. Working of the System
The proposal follows the same process similar to previous system (López-Pintado, Dumas, García-Bañuelos, & Weber, 2022).

i. Phase 1: *Creation of Profile Identification*
The suggestion minimizes the disadvantage of the previous system. It aims to generate the hash key using the Merkle tree. The entities play their role in the scenario. The hash code is generated for an entity using identity P_i and profile R_i.

$$P_i \Rightarrow \{P_c||R_i\} \tag{10.1}$$

ii. Phase 2: Creation of Activity Hash Code
The activity is assigned a code. Using the same and identity code, the hash code for the activity is generated. The identity P_i is concatenated with Activity code A_c to generate the hash code of the activity A_i. The same is portrayed in Equation10.2.

$$A_i \Rightarrow \{P_i||A_c\} \tag{10.2}$$

10.5 Analysis of the Work

The previous work (López-Pintado, Dumas, García-Bañuelos, & Weber, 2022) presents a model for the unique restricting of entertainers to jobs in cooperative cycles and a related restricting approach to determination language. It proposes three sorts of controlled adaptability components dynamic restricting of entertainers to positions of a rhythmic cycle, dynamic determination of sub-cycles, and vibrant choice of elective pathways in the given execution condition of an interaction. The business process model is the same for each undertaking related to a job. It plays five parts: Customer, Supplier,

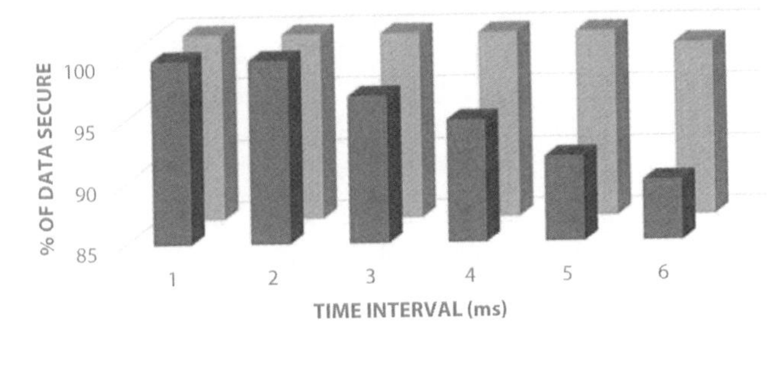

Figure 10.2 Comparison of security in the system.

Carrier-Candidate, Carrier, and Invoicer-Invoice. The cycle ends, assuming dismissal of the request. An undertaking is performed by the entertainer bound to the assignment's job. Entertainers might assign themselves or different entertainers to carry a part for a situation or they might demand to let themselves or various entertainers out of a job. A job might be related to the case-maker, suggesting that the job is bound upon case creation and needn't bother with an assignment or underwriting. The execution goes on with the Cargo sub-procedure, where a provider demand statements from different transporter applicants. The shipment completes in the following two ways. These installments embody sub-procedure Invoicing. This sub-procedure is called twice for the provider's receipt and the transporter's receipt. The suggestion takes additional measures to make it more secure.

The suggestion minimizes the disadvantage of the previous system. It aims to generate the hash key using the Merkle tree. Every entity, roles and activity have preassigned code. Using the preassigned codes, hash code is generated. Every session, the hash code changes though the company plays the same role in the system. Hence, the system is secure using the methodology by 4.35% considering the attack is average. Figure 10.2 represents the same.

10.6 Conclusion

The work (López-Pintado, Dumas, García-Bañuelos, & Weber, 2022) presents a model for the unique restricting of entertainers to jobs in cooperative cycles and a related restricting approach to determination language. It proposes three sorts of controlled adaptability components for dynamic restricting of entertainers to positions of a rhythmic cycle, dynamic determination of sub-cycles, and vibrant choice of elective pathways in a given execution condition of an interaction. The business process model is the same for each undertaking related to a job. It plays five parts: Customer, Supplier, Carrier-Candidate, Carrier, and Invoicer-Invoice. The cycle ends, assuming dismissal of the request. An undertaking is performed by the entertainer bound to the assignment's job. Entertainers might assign themselves or different entertainers to carry a part for a situation, or they might demand to let themselves or various entertainers out of a job. A job might be related to the case-maker, suggesting that the job is bound upon case creation and needn't bother with an assignment or underwriting. The execution goes on with the Cargo sub-procedure where a provider demands statements from different transporter applicants. The shipment completes in the following two ways. These installments embody sub-procedure Invoicing. This sub-procedure is called twice for the provider's receipt and the transporter's receipt. The suggestion takes additional measures to make it more secure.

The suggestion minimizes the disadvantage of the previous system. It aims to generate the hash key using the Merkle tree. Every entity, role, and activity has a preassigned code. Using the preassigned codes, hash code is generated. Every session the hash code changes, though the company plays the same role in the system. Hence, the system is secure using the methodology by 4.35% considering the attack is average.

References

Ambika, N. (2021). A Reliable Blockchain-Based Image Encryption Scheme for IIoT Networks. In *Blockchain and AI Technology in the Industrial Internet of Things* (pp. 81-97). US: IGI Global.

Ambika, N. (2021). Customer View—Variation in Shopping Patterns. In *Big Data Analytics* (pp. 55-67). United States: Auerbach Publications.

Chen, Y.-C., Chou, Y.-P., & Chou, Y.-C. (2019). An Image Authentication Scheme Using Merkle Tree Mechanisms. *Future Internet, 11*, 149.

Climent, C., Mula, J., & Hernández, J. E. (2009). Improving the business processes of a bank. *Business Process Management Journal., 15*(2), 201-224.

Dhumwad, S., Sukhadeve, M., Naik, C., Manjunath, K. N., & Prabhu, S. (2017). A peer to peer money transfer using SHA256 and Merkle tree. *23RD Annual International Conference in Advanced Computing and Communications (ADCOM)* (pp. 40-43). Bangalore, India: IEEE.

Dietz, J. L. (2006). The deep structure of business processes. *Communications of the ACM, 49*(5), 58-64.

Graml, T., Bracht, R., & Spies, M. (2007). Patterns of business rules to enable agile business processes. *11th IEEE International Enterprise Distributed Object Computing Conference (EDOC 2007)* (pp. 365-365). Annapolis, MD, USA: IEEE.

Hartley, J. L., & Sawaya, W. J. (2019). Tortoise, not the hare: Digital transformation of supply chain business processes. *Business Horizons, 62*(6), 707-715.

Jeng, J. J., Schiefer, J., & Chang, H. (2003). An agent-based architecture for analyzing business processes of real-time enterprises. *Seventh IEEE International Enterprise Distributed Object Computing Conference* (pp. 86-97). Brisbane, QLD, Australia: IEEE.

Khan, P., Byun, Y.-C., & Park, N. (2020). IoT-Blockchain Enabled Optimized Provenance System for Food Industry 4.0 Using Advanced Deep Learning. *Sensors, 20*, 2990.

Li, H., Lu, R., Zhou, L., Yang, B., & Shen, X. (2013). An efficient merkle-tree-based authentication scheme for smart grid. *IEEE Systems Journal, 8*(2), 655-663.

Li, X., Jiang, P., Chen, T., Luo, X., & Wen, Q. (2020). A survey on the security of blockchain systems. *Future Generation Computer Systems, 107*, 841-853.

López-Pintado, O., Dumas, M., García-Bañuelos, L., & Weber, I. (2022). Controlled flexibility in blockchain-based collaborative business processes. *Information Systems, 104*, 101622.

Parthasarthy, R., & Sethi, S. P. (1992). The impact of flexible automation on business strategy and organizational structure. *Academy of Management review, 17*(1), 86-111.

Schiefer, J., & Seufert, A. (2005). Management and controlling of time-sensitive business processes with sense & respond. *International Conference on Computational Intelligence for Modelling, Control and Automation and International Conference on Intelligent Agents, Web Technologies and Internet Commerce (CIMCA-IAWTIC'06). 1*, pp. 77-82. Vienna, Austria: IEEE.

Teng, J. T., Grover, V., & Fiedler, K. D. (1994). Re-designing business processes using information technology,. *Long Range Planning, 27*(1), 95-106.

Tripathi, A. M. (2018). *Learning Robotic Process Automation: Create Software robots and automate business processes with the leading RPA tool–UiPath.* Birmingham, UK: Packt Publishing Ltd.

Velikorossov, V. V., Filin, S. A., Genkin, E. V., Maksimov, M. I., Krasilnikova, M. A., & Rakauskiyene, O. G. (2020). HR systems as a new method for the automatization of business processes in organization. *2nd international conference on pedagogy, communication and sociology (ICPCS)*, (pp. 415-418). Ningbo, China.

Vrhovnik, M., Schwarz, H., Suhre, O., Mitschang, B., Markl, V., Maier, A., & Kraft, T. (2007). An approach to optimize data processing in business processes. *33rd international conference on Very large data bases* (pp. 615-626). Vienna, Austria.: ACM.

Willcocks, L. P., & Lacity, M. (2016). *Service automation robots and the future of work.* . SB Publishing.

Zou, Y., Zhang, Q., & Zhao, X. (2007). Improving the usability of e-commerce applications using business processes. *IEEE Transactions on Software Engineering, 33*(12), 837-855.

11

Future of Business Organizations Based on Robotic Process Automation: A Review

P. William[1*], Sonal C. Bhangale[2], Harshal P. Varade[3] and Santosh Kumar Sharma[4]

[1]Department of Information Technology, Sanjivani College of Engineering, SPPU, Pune, India
[2]Department of Mechatronics, Sanjivani K.B.P. Poytechnic, Kopargaon, Ahmadnagar, India
[3]Department of Mechanical Engineering, Sanjivani College of Engineering, SPPU, Pune, India
[4]Department of Mechanical Engineering, Chhattrapati Shivaji Institute of Technology, CSVTU, Bhilai, India

Abstract

The purpose of this study is to show how important robotic process automation technology is and how it is better than other technologies. The book also has information about how this technology will be studied in the future. Even though there hasn't been much research done in this area, the history of robotic process automation shows that it has a wide range of uses. Robotic Process Automation makes it easier to get the best and most effective results. This article will help businesses use Robotic Process Automation in the best way possible. It gathers and analyses current applications so that certain tasks can be done automatically. The number of software applications needed to solve a problem is a good indicator of how important RPA is for solving that problem. The more software applications that are needed, the more RPA can help fix the problem. It makes it possible to automate business processes in a safe way. Robotic process automation is used to make processes that do more than one job in the least amount of time.

***Keywords*:** Robot process automation, RPA tool, optical character recognition

**Corresponding author*: william160891@gmail.com

Romil Rawat, Rajesh Kumar Chakrawarti, Sanjaya Kumar Sarangi, Rahul Choudhary, Anand Singh Gadwal and Vivek Bhardwaj (eds.) Robotic Process Automation, (181–188) © 2023 Scrivener Publishing LLC

11.1 Introduction

RPA stands for "automating tasks with robots." A robot is a machine that acts and moves like a person. A process is the set of steps that must be taken to finish a job. When a task is automated, people don't have to do it anymore. Robotic process automation makes it possible to make software that lets people's actions be a part of digital systems. Calculations, record keeping, data entry, log in and out, and retrieval of data from a file or browser all fall under this category. It collects and modifies data through a user interface. Automation systems are built by observing the user's actions in the graphical user interface of the software and then repeating those actions directly in the program. Robotic Process Automation's virtual employee platform is comprised of a bot that can run scripts over the internet called an "internet bot." Bots are everywhere in the world of technology. Bots do repetitive tasks that make it easier for humans to do the same thing again because they don't need as much higher-order thinking. Bots collect and look at information on the web. Bots can also be programmed to act like people talking to each other. A job can be done by a machine if it is clear and follows rules. Robotic process automation is great at doing jobs with many steps without limits. Robotic process execution that is automated is deterministic. It manages both organized and unorganized data well and always gives the best results. Once the automated process is set up, the output does not change or become wrong. Robotic process automation makes it easier to do boring and repetitive tasks that are important for a person's growth. Artificial intelligence and expert systems get better when robotic processes are automated. Everything that it does is recorded, so there's no need to worry.

11.2 Literature Review

Robotic process automation, according to authors Ssu Chieh Lin and Lian Hua Shih [1], should be utilized to address the most pressing issues that many manufacturing organizations confront. They suggest a very simple solution that not only solves the problem of the production company, but also cuts costs, improves product quality with no mistakes, and reassigns workers away from manufacturing to duties requiring more human effort in order to enhance other aspects of the product. As a component of Robotic Process Automation (RPA), Optical Character Recognition (OCR) is used. They may use this to identify the image and some concealed information that instructs them on how to take clean images and distinguish between blurry ones. This makes it less likely that something will go wrong. Using Robotic Process Automation technology, this proposed system works about 90% of the time [1].

Authors Kevin C. Moftt and Andrea M. Rozario [2] talk about how important Robotic Process Automation is. They talk about how it will make auditing better. Auditing is hard and takes a lot of time. A small mistake could lead to a big problem and fixing that mistake is a hard job. Automation technologies like Excel Macros, IDEA, Python, and R have been used to automate certain auditing procedures, including Reconciliations, Analytical Procedures, Internal Control Testing Detail Testing (Attribute Match), and Reconciliations. To build a process that takes advantage of this technology, one must be well-versed in its operation. On the other hand, it's easy to use RPA technologies like Uipath and blue prism and these new solutions have made data security better. The use of RPA in auditing is very helpful in many ways, like keeping data safe, sending invoices, keeping track of purchases, and doing a number of other repetitive tasks [2].

Ui Path Studio, Automation Anywhere, and Blue Prism are three of the most popular RPA systems examined in-depth in this Delineated Review on Robotic Process Automation. These three tools are compared in this research based on a variety of variables, such as the kinds of operations they can do and which one is more accurate in terms of security, not needing any code, and being easy to use and which one is best for the kind of business procedures that this article describes in detail. They also show graphical statistics and a simple example of a student management system so that we can decide which of the three tools is best. Also, in the near future, UiPath will be much bigger than anyone could have imagined. It is better than the other two tools because it uses methods that change over time [3].

The goal of the article by Audrey Bourgouin and Laurent Renard is to show why businesses should use the idea of Robotic Process Automation to automate their business processes. They come up with a new way to look at business processes and figure out which ones are best for RPA. Also, the suggested solution can be used for many different types of business processes and is made for business analysts who don't need to become experts in RPA or Business Process Management (BPM). This study's main goal is to come up with a new way to analyze business processes and figure out if they can be automated using the RPA method. For the suggested business process, they primarily concentrate on the banking, healthcare, and insurance sectors to determine which Robotic Approach Automation approach is optimal. In addition, they provide guidance on which processes may be automated using the same method. This research contains the answers to all of these questions. Robotic Process Automation (RPA) is considered or ruled out based on the findings of this research [4].

"From the beginning," or "from scratch," Abderrahmane Leshob and Laurent Renard discuss Robotic Process Automation technologies. This research reveals how RPA has progressed from its infancy to its current state. Robotic Process Automation is referred to in terms of the technology that makes it possible. There are several examples of how Robotic Automation Technology may be put to use as well. Robotic Process Technology and IT development are also compared through BPM, with the latter being referred to as "heavyweight IT" and the former as "lightweight IT." BPM necessitates the development of a new application and back-end infrastructure, while RPA relies on an existing application. Every aspect of RPA outperforms the competition [5].

Robotic Process Automation is a big part of getting ready for more integration. In addition, they've discussed the advantages and drawbacks of robots in the manufacturing process. There are a number of issues that are slowing down the organization, including defining business cases with supporting ROIs, limiting scope creep and complexity, automating process components that are intrinsically weak, and handling exceptions. This report looks at how the use and growth of robotic process automation will affect the market now and in the future. This research also examines the factors that companies should consider before using Robotic Process Automation to address their issue statement. [6] detailed information regarding Robotic Process Automation (RPA) may be found in the Everest Research Group's research. It examines the tools required to create a virtual workforce. It looks at 10 well-known technology companies and compares and contrasts their products using a "feature, implementation, and impact matrix" [7].

Rebecca and Lauren Livingston compare traditional automation and manual labor with Robotic Process technology, which is growing quickly. They make it clear that Robotic Process Automation is not a result of evolution, but rather a very new technology that is better than older ones in many ways. This study says that Robotic Process Automation works in all industries with well-defined, rule-based processes that can be done over and over again. They have pointed out that one of the best things about Robotic Process Automation technology is that it can be used in many different fields and do many different things. This study also talks about how fast Robotic Process Automation technology will grow in the near future and how it will affect the business world [8].

The importance of RPA in Shared Services Centers and Business Process Outsourcing is shown by author Sorin Anagnoste. This research shows how important Robotic Process Automation is when there are a lot of data and a lot of mistakes. This document also talks about what Robotic Process

Automation is and who its main providers are. A comparison of Robotic Process Automation capabilities and deployment choices is shown in this article, as well as a look at the many methods in which bots may be built. This includes considerations such as architecture, security, system administration, resilience, and analytics. The case studies in this article are summarized in this article. A Service Provider and an Oil and Gas Company are both using Robotic Process Automation (RPA) [9].

The authors [8], Mary C. Lacity and Lesile P. Willocks, mostly talk about how businesses can use both technology and people to offer better services for less money. It also talks about how to make a service automation plan that helps the company grow and has many benefits, but also needs support from the top management. This study also talks about a number of sourcing options for organizations, such as insourcing, outsourcing, and cloud sourcing. This report also talks about a number of things that make it easier to start an automation process that works well. It is important to know exactly how the automation will be used and how it will help both customers and employees. Building mature service automation capabilities includes a command center to set standards and keep an eye on performance and teaching all employees the skills they need to work with the automation that has been set up. This research also shows how far automation will go in the future and how it will change the way people work [10].

11.3 Technology: A Need of Robotic Process Automation

RPA cuts down on the cost of building a system. The number of people needed to do the job goes down. As a result, the salary costs for the organization go down too. Also, the job is done quickly and correctly, which makes less work for people to do. It makes it easier to streamline the way a business works. It also improves the quality of service, which helps the organization grow quickly.

11.3.1 Benefits of Robotic Process Automation

It was made to be easy to set up and not need any code. It lets customer service get better and makes sure that business operations follow the rules. It speeds up the way many things are done and helps staff be more productive. There are also options for input and output. It is easy to set up and you can find out more about automation creation management and features.

11.3.2 Drawbacks of Robotic Process Automation

People worry that if a robot can do a job faster and more consistently than a person, humans may no longer be needed. Because of this fear, robots will eventually take over for people. Companies need to know about the different inputs that come from different places. RPA is not a way to solve the problems of cognitive computing [9, 10]. Since it can't learn from what it does, it has a shelf life.

11.4 Business Enterprise

Robotic Process Automation can be used in a wide range of business and manufacturing settings. Below, we'll talk about a few of these many institutions: Auditing uses for Robotic Process Automation ranges from computerized invoice processing to automatic customer account credit computation [11]. Even though auditing is used a lot in institutions, its full range of uses is still not fully understood. It can be used to get information about premiums and to clear a claim for processing. Travel domain-RPA [12] may be used for ticket booking, passenger information, accounting, and more. It is possible to schedule patients by using Healthcare [13] Robotic Process Automation to enter and move data, handle billing and claims, and optimize the health system as a whole. Human Resources Robotic Process Automation can be used to bring on new employees, process payroll, and reduce the amount of paperwork that needs to be done by hand.

Modern enterprise shareholders and IT professionals are extremely concerned about cybersecurity [14, 15], and for efficient reason. 37 billion records were exposed owing to data theft [16] by skilled cybercriminals [17] in 2023. Because hackers are aware of the frequent dangers and weaknesses that affect companies and organisations, cybersecurity experts must continuously remain vigilant. Due to technological advancements, access to cyberspace is becoming more widespread, which increases the potential cybersecurity problems that businesses may encounter. Emails that appear to be from a reputable source, such as a company, bank, or official agency, are used in phishing campaigns [18]. Recipients expose their connections to malware whenever they open an attachment or click a link contained in an email.

11.5 Conclusion and Future Scope

Robotic Process Automation is better than outsourcing because it saves money, cuts down on cycle times, and increases production. It can also be used to do boring tasks, giving people more time to learn new things and advance in their careers in other ways. This article will help the reader learn all they need to know about Robotic Process Automation.

With more work on this technology, there will be a lot more opportunities to improve how businesses work. Companies may save time and money by using a Robotic Process Automation solution while developers build a custom solution for their operations. BFSI, manufacturing, retail, analytics, aviation, oil and gas, and law are just a few of the areas where robotic process automation will be heavily used. Data entering and data rekeying chores will soon be handled by automated tools and procedures. All computer-aided operations that are governed by a set of protocols will be taken care of using this technology. Analytics and data accuracy will be improved as a result. In order to manage formatting chores, Robotic Process Automation (RPA) will be deployed. The following years saw an increase in adoption: RPA will be used to manage and combine many processes, making it simpler for enterprises to operate effectively. In the next stage of development, rule-based technology will be joined by aspects of artificial intelligence (AI). At this point, Smart Process Automation (SPA) is born. RPA is often considered a stepping stone toward SPA, which stands for "smart process automation". When it comes to unstructured data, SPA aims to automate the operations that robots cannot do on their own. SPA is anticipated to make use of machine learning and robotic process automation. In addition to RPA, additional technologies and methodologies will be used. Automation of Robotic Processes will continue to flourish on its own, without any intervention from the government. This technology's next great step is RPA 2.0, which will include more AI, digital workers, the whole workforce, and digital transformation. As part of the digital workforce, people who know how to use technology will help with innovation, creativity, and working together.

References

1. Ssu Chieh Lin, Lian Hua Shih, Damon Yang, James Lin, Ji Fu Kung, "Apply RPA (Robotic Process Automation) in Semiconductor Smart Manufacturing", 2018 e-Manufacturing Design Collaboration Symposium (eMDC), 7th September 2018 in IEEE.
2. Moffitt, Kevin C., *et al.* "Robotic Process Automation for Auditing." *Journal of Emerging Technologies in Accounting*, vol. 15, no. 1, July 2018, pp. 1–10. *Crossref*, doi:10.2308/jeta-10589.
3. Ruchi Isaac, Riya Muni, Kenali Desai, "Delineated Analysis of Robotic Process Automation Tools." *2018 Second International Conference on Advances in Electronics, Computers and Communications (ICAECC)*, IEEE, 2018, pp. 1–5. *Crossref*, doi:10.1109/ICAECC.2018.8479511.

4. Audrey Bourgouin, Abderrahmane Leshob, Laurent Renard. "Towards a Process Analysis Approach to Adopt Robotic Process Automation." *2018 IEEE 15th International Conference on E-Business Engineering (ICEBE)*, IEEE, 2018, pp. 46–53. *Crossref*, doi:10.1109/ICEBE.2018.00018.
5. Audrey Bourgouin, Abderrahmane Leshob, Laurent Renard "Robotic Process Automation: Dynamic Roadmap for Successful Implementation", Reykjavik University, June, 2018.
6. Capgemini "Robotic Process Automation: Gearing up for greater integration", 2017 Storyful, 11th August 2017.
7. Everest Research Group. "Robotic Process Automation: Technology vendor landscape with fit matrix assessment-Technologies for building virtual workforce", December 2016.
8. Rebecca Dilla, Heidi Jaynes, and Lauren Livingston," Introduction to Robotics Process Automation a Primer", 2015 by the Institute for Robotic Process Automation, December 2015.
9. Anagnoste, Sorin. "Robotic Automation Process - The next Major Revolution in Terms of Back Office Operations Improvement." *Proceedings of the International Conference on Business Excellence*, vol. 11, no. 1, July 2017, pp. 676–86. *Crossref*, doi:10.1515/picbe-2017-0072.
10. Mary C. Lacity and Leslie P. Willcocks. "A new approach to automating services" MIT Sloan Management Review, October 2016.
11. International Auditing and Assurance Standards Board (IAASB). 2016. "Exploring the Growing Use of Technology in the Audit, with a Focus on Data Analytics." New York, NY: IFAC.
12. C. Mendis, C. Silva, and N. Perera, "Moving ahead with Intelligent Automation", 2016.
13. UiPath Community Forum - Robotic Process Automation," UIPath vs Automation Anywhere-Blue Prism", 2017.
14. Rawat, R., Chakrawarti, R. K., Vyas, P., Gonzáles, J. L. A., Sikarwar, R., & Bhardwaj, R. (2023). Intelligent Fog Computing Surveillance System for Crime and Vulnerability Identification and Tracing. *International Journal of Information Security and Privacy (IJISP)*, 17(1), 1-25.
15. Rawat, R., Sowjanya, A. M., Patel, S. I., Jaiswal, V., Khan, I., & Balaram, A. (Eds.). (2022). *Using Machine Intelligence: Autonomous Vehicles Volume 1*. John Wiley & Sons.
16. Rawat, R., Bhardwaj, P., Kaur, U., Telang, S., Chouhan, M., & Sankaran, K. S. (2023). *Smart Vehicles for Communication, Volume 2*. John Wiley & Sons.
17. Mahor, V., Bijrothiya, S., Rawat, R., Kumar, A., Garg, B., & Pachlasiya, K. (2023). IoT and Artificial Intelligence Techniques for Public Safety and Security. *Smart Urban Computing Applications*, 111.
18. Mahor, V., Bijrothiya, S., Mishra, R., & Rawat, R. (2022). ML Techniques for Attack and Anomaly Detection in Internet of Things Networks. *Autonomous Vehicles Volume 1: Using Machine Intelligence*, 235-252.

12

Comparative Overview of FER Methods for Human-Robot Interaction Using Review Analysis

Jitendra Sheetlani[1]*, Mohit Kadwal[1], Sumanshu Sharma[2], Sanat Jain[3] and Shrikant Telang[4]

[1]MCA Department, Medi-Caps University, Indore, India
[2]Computer Science, SSSUTMS, Sehore, (M.P.), Sehore, India
[3]Department of Computer Science and Engineering, Manipal University Jaipur, Rajasthan, India
[4]Department of Information Technology, Shri Vaishnav Vidyapeeth Vishwavidyalaya, Indore, India

Abstract

Facial expressions can convey indications of a person's emotional state, which together with the voice, language, hands, and body language, makes up the basic system of communication between people in a social setting. Personal Face Recognition is one of the most powerful and challenging activities in social media. Generally, facial expressions are a natural and direct way for people to express their feelings and intentions. Facial expressions are important aspects of non-verbal communication. Facial expressions can be used for expressing human emotion. Over the past few decades, FER facial recognition (FER) has become an exciting area of research and has made great strides in computer vision. FER is the recognition of a person's emotional state associated with biometric factors. Creating a personal FER system based on a machine is a daunting task. Various FER systems are developed by analyzing facial muscle movements and skin flexibility based on algorithms. In the standard FER system, advanced algorithms work on a blocked website. In an unrestricted environment, the efficiency of existing algorithms is limited due to certain problems during image acquisition. This paper describes the Analysis of Face Expression Recognition (FER) techniques for emotion detection [1] which include the three major stages of preprocessing, feature extraction,

**Corresponding author*: jitendrak.sheetlani@medicaps.ac.in

Romil Rawat, Rajesh Kumar Chakrawarti, Sanjaya Kumar Sarangi, Rahul Choudhary, Anand Singh Gadwal and Vivek Bhardwaj (eds.) Robotic Process Automation, (189–196) © 2023 Scrivener Publishing LLC

and classification. This analysis explains the various types of FER methods in the review-based technique. The performance of various FER methods is compared based on the accuracy level of the algorithm. This paper uses a review analysis approach for emotion detection by facial expression recognition methods.

Keywords: FER, AI, psychology, robot, emotion, human intelligence, FU, Facial Unit Analysis (FUA), HBI (Human-Robot-Interaction)

12.1 Introduction

Face recognition (FER) has a high impact on the field of pattern recognition and great effort is being made by researchers to develop an FER system for human and computer applications. Face status provides clues to sensitive information to build the FER system and is considered the best tool for easily detecting a person's emotions and intentions. At the moment, the FER system plays a central part in artificial intelligence and acts as real-world applications that can be present in a variety of psychological research, driver fatigue, interactive game design, mobile apps to automatically add emotions to conversations, and autistic help programs. People use facial sensor levels in the medical field, as a sensory diagnostic system used by the disabled to assist the caregiver, or a socially intelligent robot with sensory intelligence. Facial expressions can convey indications of a person's emotional state, which together with the voice, tongue, hands, and the body form a basic system of communication between people in a social setting. In various cases, facial expressions have many communication functions. Extensive research has helped develop a better FER system in recent years, but system performance is affected by a variety of factors. In the case of FER systems, the available methods only treat model face shots captured under laboratory limits [3]. In the case of an unrestricted area, the use of an existing method often leads to higher chances of incorrect classification due to automatic expression. Recognition of automatic speech in real-time is a challenging problem related to changes in light, changes in scalp, subtle facial degradation, aging, closure of any objects such as hair, glasses, or scarves, skin color variations, and complex backgrounds. Most of the trained images are also not well classified by the system, which works best due to the challenges mentioned above [4]. Consistently, the available rating site is not naturally linked to the emotional state of the test image. This analysis focuses on the FER system in a controlled and unregulated environment based on their operational characteristics.

Face Detection: Face Detection is a computer-based artificial intelligence (AI) technology used to detect and identify human faces in digital images. Face detection technology can be used in a variety of fields including security, biological metrics, law, entertainment, and personal security to provide real-time monitoring and tracking of people [5].

Human-Robot-Interaction (HBI): The robot appears as an automated form of man. It is the evolution of an intelligent and obedient but impersonal machine. The word robot is derived from the word "robota", which means 'forced labor.' Many robots require subtle guidance from a human operator to operate throughout their mission. However, robots started with the aim of using some amount of artificial intelligence in machines and are slowly becoming more independent. Most artificial intelligence eventually leads to robotics. It is the field of computer science and engineering that deals with creating devices and robots that can move and respond to sensory inputs. Most Neural Networking, Natural Language Processing, Image Recognition, and Speech Recognition/Synthesis research aims to eventually incorporate their technology into the robotics paradigm, that is, a fully human-like robot.

12.2 FER Method Review Based Analysis

12.2.1 AdaBoost Method

AdaBoost is an integrated face detection method and is widely used due to its improved accuracy and relatively low complexity. It is a popular face-to-face with low value, which is not true. A major limitation in Adaboost sensitivity is noisy data and external content. The set of image elements is trained by a few class dividers in the cascade using Adaboost to eliminate incorrect samples. The output of the first phase divider will be the input to the next divider used to determine the correct surface area. Therefore, a solid separator is created which helps to reduce the number of features and leads to higher detection accuracy. Kheirkhah and Tabatabaie proposed a mixed and solid face detection system of color and sophisticated images. The mixed method uses skin color information and facial recognition based on Adaboost. It provides better performance with less and less time to perform [6].

12.3 Feature Extraction Techniques

After facial detection, the next step in FER is to remove the feature. The main purpose of removing a facial feature is to remove the effective and efficient representation of facial features without losing face knowledge. Geometric-based and visual-based features are two ways to extract features that are classified based on facial movement and facial transformation. The inserted image may be a standalone image or image sequence. Based on the input image, a suitable facial feature algorithm is used to extract local, international, or mixed features. The extracted elements are greatly reduced in size, which is provided as a classification input and that greatly helps the divider to make the decision easier to identify and detect facial shape [7].

12.3.1 Haar Classifier Method

The Haar classifier is considered a solid way to get a face in a real-time environment. Haar features are considered to find facial edges, lines, movements, and skin color. Haar features are a black and white rectangular box connected, as shown in Figure 12.1, used to extract the feature. Haar features can be easily measured and areas are checked by increasing or decreasing pixel intensity in different parts of the image. The feature value present is the difference in the total number of pixels for black and white regions within a rectangular box. The Haar classifier finds features that contribute to face detection problems in the training phase. Thus, it reduces the cost of calculation and complexity in the texting process leading to higher accuracy [8].

12.3.2 Geometric-Based Method

Geometric-based algorithms focus on permanent features (eyes, eyebrows, forehead, nose, and mouth) that define the shape and location of

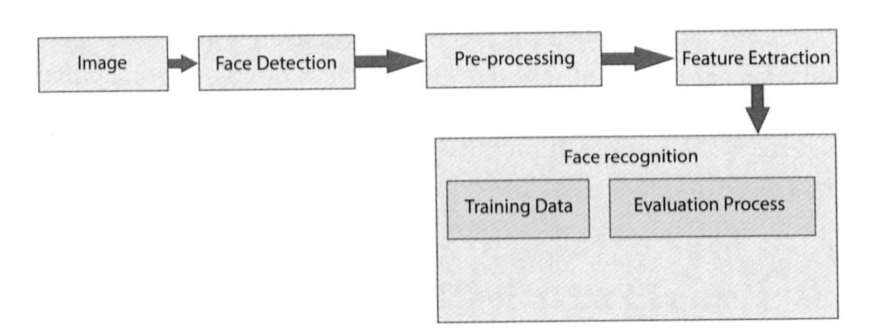

Figure 12.1 Facial expression recognition process.

facial features using the pre-defined geometric area. These parts of the face are subtracted to form a vector element representing the geometry of the face. However, the additions affect the shape associated with the shape of the various facial features. As a result, the lower facial shape can be seen by measuring the movement of important facial features. In the case of image sequencing as input, FACS is used which helps to distinguish facial movements based on facial action analysis [13]. FACS contains various action units (AUs) related to specific muscle contraction. Tain *et al.* has developed an automated facial recognition system to analyze subtle changes in facial expressions that are also converted to AUs. Speeches can contain a single AU or a combination of AU. On the other hand, still image input uses model-based methods, such as active model (ASM), active visual model (AAM), and a scale-invariant feature transform (SIFT) algorithm to extract facial features. The geometric-based approach is best suited to real-time facial images [11, 12] where features can be easily identified and tracked, but it does require an accurate way of finding the face [9].

12.3.3 Appearance-Based Method

Visual-based algorithms focus on transit factors (wrinkles, bumps, forearms) that explain changes in facial expressions, firmness, histograms, and pixel values. In this approach, PCA, linear discriminant analysis (LDA), independent component analysis (ICA), Gabor wavelet, and LBP algorithms are considered to extract feature descriptions. In recent years, the Gabor wavelet and LBP have been widely used to extract feature adjectives. Gabor wavelets are a well-known standalone element for successfully extracting texture information. Zhang *et al.* investigated and compared geometry-based and Gabor-based methods and the results show that the Gabor wavelet performs very well in performance and are considered to be the most powerful tool for extracting features. Many of the research results are popularly obtained using the Gabor filter bank to find lines and edges over multiple scales and shapes, with good timing-frequency and multi-resolution features. The filter limit is the maximum calculation time [2] due to the large size of the filtered vectors [10]. Table 12.1 shows about the comparative analysis from tabular structure. Figure 12.2 represents about the various FER methods' accuracy levels.

Table 12.1 Comparative analysis from tabular structure.

S. no.	Approaches	Database	Accuracy %
1	Appearance based Approaches	Facial Database	75
2	Feature Base	Web Face	80
3	CNN	Ohio	76
4	MicroExpNet	JAFFE / FER2013	70.02
5	SVM	JAFFE	82.60
6	Gabor Filter and PCA	JAFFE	70
7	Distance Calculation	Image Database	85
8	Geometric Based	Image Database	88
9	Stationary Wavelet Transform	JAFFE and CK	87
10	Projective Complex Matrix	CK+ and JAFFE	82

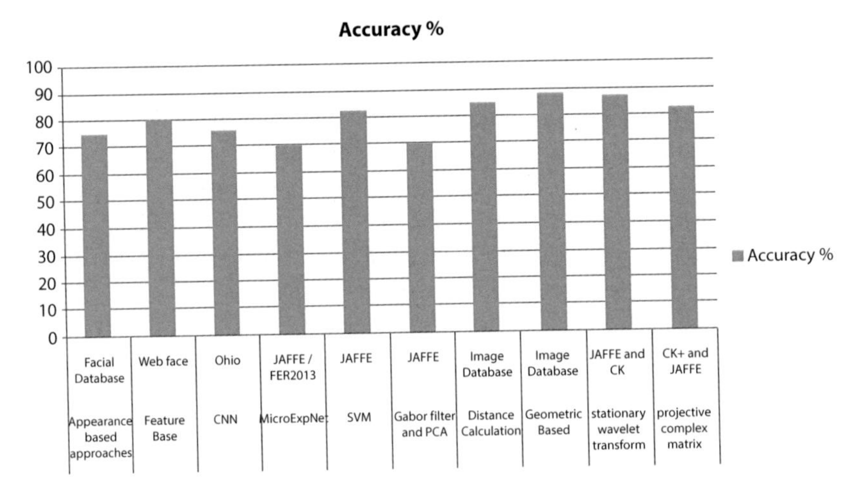

Figure 12.2 Various FER methods' accuracy levels.

12.4 Conclusion

In recent years, FER has become an active research site due to its many real-time applications. Many authors have adopted a different approach to image capture, extraction, and segmentation in order to build a solid real-time FER system. In this analysis, we have studied various FER strategies.

The above graph and chart represent various facial expression techniques which explain different FER methods' accuracy levels. This paper provides simple review-based analysis for FER methods for recognition accuracy. Facial expression is used for recognition of emotion expression. This paper can be used for emotional analysis with the FER Method using AI and Psychology concepts.

References

1. M. R. González-Rodríguez, M. C. Díaz-Fernández, and C. P. Gómez, "Facial-expression recognition: an emergent approach to the measurement of tourist satisfaction through emotions," Telematics and Informatics, vol. 51, no. 3, pp. 101–114, 2020.
2. Chu W.S., Torre F.D., and Cohn J.F.: 'Selective transfer machine for personalized facial expression analysis', IEEE Trans. Pattern Anal. Mach. Intell., 2017, 39, (3), pp. 529–545.
3. M. Shadaydeh, L. Mueller, D. Schneider, M. Thümmel, T. Kessler, and J. Denzler, "Analysing the direction of emotional influence in nonverbal dyadic communication: a facial-expression study," 2020.
4. D. Gera and S. Balasubramanian, "Landmark guidance independent spatio-channel attention and complementary context information based facial expression recognition," Pattern Recognition Letters, vol. 11, no. 2, pp. 1000–1011, 2020.
5. Alazrai R., and George Lee C.: 'Real-time emotion identification for socially intelligent robot'. IEEE Int. Conf. Robotics and Automation, USA, May 2012, pp. 14–18.
6. G. Du, S. Long, and H. Yuan, "Non-contact emotion recognition combining heart rate and facial expression for interactive gaming environments," IEEE Access, vol. 11, no. 99, pp. 367–376, 2020.
7. P. Kim, "Image super-resolution model using an improved deep learning-based facial expression analysis," Multimedia Systems, vol. 27, no. 4, pp. 615–625, 2021.
8. Dulguerov P., Marchal F., and Wang D. *et al*: 'Review of objective topographic facial nerve evaluation methods', AM. J. Otol., 1999, 20, (5), pp. 672–678.
9. D. Yatera, K. Nishiya, Y. Karouji, and T. Kojiri, "Facial expression visualization system for medical interview practice support," Procedia Computer Science, vol. 159, no. 1, pp. 1986–1994, 2019.
10. T. D. Lalitharatne, Y. Tan, F. Leong *et al.*, "Facial expression rendering in medical training simulators: current status and future directions," IEEE Access, vol. 8, no. 1, pp. 215874–215891, 2020.

11. Rawat, R. (2023). Logical concept mapping and social media analytics relating to cyber criminal activities for ontology creation. *International Journal of Information Technology*, 15(2), 893-903.
12. Rawat, R., Mahor, V., Álvarez, J. D., & Ch, F. (2023). Cognitive Systems for Dark Web Cyber Delinquent Association Malignant Data Crawling: A Review. *Handbook of Research on War Policies, Strategies, and Cyber Wars*, 45-63.
13. Rawat, R., Chakrawarti, R. K., Vyas, P., Gonzáles, J. L. A., Sikarwar, R., & Bhardwaj, R. (2023). Intelligent Fog Computing Surveillance System for Crime and Vulnerability Identification and Tracing. *International Journal of Information Security and Privacy (IJISP)*, 17(1), 1-25.

13

Impact of Artificial Intelligence on Medical Science Post Covid 19 Pandemic

Yash Aryaman and Amit Kumar Tyagi*

Department of Fashion Technology, National Institute of Fashion Technology, New Delhi, Delhi, India

Abstract

In science, artificially intelligent systems are used excessively. Common programs include diagnosing patients, end-to-end drug discovery and development, improving communication between health practitioners and affected persons, transcribing medical documents, such as prescriptions, and remotely treating patients. Several kinds of artificial intelligence are now being utilized by patients, suppliers of care, and life sciences organizations. Modern computer computations have recently achieved accuracy levels in the field of clinical sciences that are comparable to those of human experts, despite the fact that computer frameworks frequently carry out tasks more effectively than people do. Some theorize that it is inevitable that people will be supplanted in specific jobs inside the clinical sciences. While diagnostic confidence never comes to 100 percent, joining machines and doctors dependably upgrades framework execution. Cognitive projects are affecting clinical practices by applying regular language handling to peruse the quickly growing library of logical writing and examine long stretches of different electronic clinical records. Artificial intelligence might enhance the consideration direction of infected patients, propose accurate treatments for complex ailments, lessen clinical mistakes, and work on subject enlistment into clinical preliminaries. The reason for this chapter is to sum up the fundamental uses of artificial intelligence in clinical material science, depict the abilities of the medical practitioners in research and clinical utilizations of artificial intelligence, and characterize the significant difficulties of artificial intelligence in medical care (post-Covid 19 pandemic).

Keywords: Artificial intelligence, deep convolutional neural network, medical use, clinical decision support, electronic health record systems

**Corresponding author*: amitkrtyagi025@gmail.com

Romil Rawat, Rajesh Kumar Chakrawarti, Sanjaya Kumar Sarangi, Rahul Choudhary, Anand Singh Gadwal and Vivek Bhardwaj (eds.) Robotic Process Automation, (197–210) © 2023 Scrivener Publishing LLC

13.1 Introduction

Artificial Intelligence (AI) [1, 2] and related advances are progressively common in business [3, 4] and society and are starting to be applied to medical [5, 6] services. These developments could change a variety of areas of patient [7, 8] care as well as operational procedures inside organizations that provide, pay for, and distribute pharmaceuticals. There are, as of now, various exploration studies proposing that AI can perform as well as or better than humans at key medical care errands, like diagnosing illness. Today, computations [9, 10] are outperforming radiologists [11, 12] at identifying dangerous tumors [13, 14] and advising scientists on how to develop associates for expensive clinical preliminary exams [15]. However, for a combination of reasons, we acknowledge that it will be many years in the future before artificial intelligence [16] replaces human beings for broad clinical [17, 18] interaction areas. In this article, we depict both the potential that artificial intelligence offers to computerize parts of care and a part of the setback that artificial intelligence causes in the field of medical science [19].

13.2 Types of AI Relevant to Healthcare

The field of artificial intelligence comprises a number of innovations, not just one. The majority of these developments [20] are immediately important to the medical industry [21, 22], but the specific cycles [23, 24] and projects they support change over time. The following describes and illustrates a few specific AI developments that are highly significant to the medical industry.

13.2.1 Machine Learning - Neural Networks and Deep Learning

A prominent method of learning through creating paradigms with information is machine learning, which links models with data. One of the most prevalent types of artificial intelligence is machine learning. According to a 2018 Deloitte poll of 1100 US managers whose enterprises were been pursuing AI, 63 percent of the companies surveyed were using machine learning in their operations. It is a broad-based strategy that is used as the foundation for many AI systems and it has a range of performances. The most well-recognized usage of customary machine learning is medication accuracy in medical service that is able to anticipate which therapeutic convention will probably benefit the patient in view of varied patient attributes and therapy context. In the process of supervised learning, the formation of a dataset in which the result variable (for example, beginning of sickness) is known, is required by accurate medical applications and is an eminent part of machine learning. Accessible since the 1960s, neural networking is a complex kind of machine learning. This innovation has been anchored in medical research services for an extensive amount of time and has been applied in classifying applications, for example determining if a patient may develop a certain illness. It analyses problems from information sources, results, and other numerous elements that combine information with results. Although this type of machine learning has been compared to the way neurons sequence signals, it is primitive when juxtaposed next to the brain's computing and analyzing prowess. The most intricate kinds of machine learning comprise of deep learning or neural networking models with numerous grades of features or factors that foresee results.

The expeditious processing of modern design handling units and cloud structures may uncover a significant number of concealed features in such models. Acknowledgement of potential malignant sores in radiology pictures is a classic use of deep learning in medical services. Deep learning is being extensively applied to radionics or the identification of clinically relevant elements in imaging data beyond what can be perceived by the human eye. Oncology oriented image examinations utilize both radionics and deep learning most frequently. This amalgamation appears to guarantee a significantly increased precision in diagnosing than the previous computerized instruments for picture investigation, known as computer-aided detection or CAD.

Deep learning is a kind of natural language processing (NPL) that is being extensively used for speech recognition. Each component in a deep learning model typically has minuscule meaning for a human observer, nothing like the prior types of statistical examinations. Subsequently, the exegesis of the model's outcome may irrefutably be challenging to decipher.

13.2.2 Natural Language Processing

Since the 1950s, comprehending human language has been an aspiration of AI specialists. Speech acknowledgement, text analysis, and translation are some of the applications incorporated into the field of Natural Language Processing, along with different aims connected with language.

There are two indispensable ways to deal with it: statistical and semantic NLP. Statistical NLP is contingent on machine learning (deep learning neural networks specifically) and has contributed to a contemporary growth in the accuracy of recognition. A huge corpus, or assemblage of language, is required for learning. Existing uses of NPL in medical services include the creation, comprehension, and characterization of clinical documents and published research. NPL structures can scrutinize disorganized clinical notes on patients, prepare reports (for example, on radiology assessments), transcribe patient interactions, and conduct interactive conversations.

13.2.3 Rule-Based Expert System

Assortment of 'If-then' rules given by the Master frameworks was the reigning innovation for AI during the 1980s and was mostly used commercially in later periods. These were commonly utilized for 'clinical decision support' purposes in medical services throughout the majority of the past few decades and is still in wide use today. These days, a plethora of EHR suppliers devise a bunch of rules with their structures. Human specialists and data designers are required by master structures to construct a progression of rules in a specific information space. They function admirably to a certain degree and are uncomplicated. However, there are numerous rules (typically exceeding a few thousand) and when they begin to conflict with another, things tend to fall apart. Additionally, if in a circumstance, the information area changes and changing the principles can be troublesome and tedious. In light of information and machine learning calculations, they are gradually being supplanted in medical care by comprehensive measures.

13.2.4 Physical Robots

Every year, over 200,000 contemporary robots are launched all around the globe. It has become apparent that robots have become a memorable highlight and the same is reflected by the number of robots introduced every year all over the world. Pre-defined tasks are executed by these robots like lifting, repositioning, welding, or assembling objects in setups like factories, warehouses, and distribution centers, along with delivering supplies in medical clinics.

Lately, robots have become more synergistic with people and are effectively trained by being put through ideal assignments. As AI capacities are being installed in their "brains" (their operating systems), they are consequentially becoming savvier. It seems logical that, after a while, the very upgrades in knowledge seen in different spheres of AI could be combined with actual robots.

First approved in the USA in 2000, surgical robots give surgeons a tremendous edge, significantly improving their ability to see, make precise incisions that are fairly less invasive, stitch wounds with added accuracy, and so forth. Consequential choices are still made by human specialists. Nevertheless, routine surgeries using mechanical medical procedures are performed which include gynecologic operations, prostate surgery, gastrointestinal surgery, head and neck surgery, etc.

13.2.5 Robotic Process Automation

This innovation executes sorted computerized tasks for authoritative purposes, such as those involving data systems, as if it were a human operator keeping a script or set of guidelines. When compared with different types of AI, they are economical, simple to program, and straightforward in their undertakings. Robotic process automation (RPA) doesn't include robots, but just computer programs on servers. The system behaves like a partially-clever user by depending on a combination of work processes, business rules, and a "presentation layer" integrated with data frameworks. Medical services utilize this innovation for the completion of insipid engagements like earlier authorization, refreshing patient records, or billing. When this innovation is combined with various other innovations like image recognition, their amalgamation can be used to extricate data from faxed images, for example, to feed the information extracted from them into transactional systems. These improvements are portrayed as individual ones, yet they are being consolidated and incorporated systematically, robots are being given AI-based "brains", and image recognition is being coordinated with RPA. Perhaps, subsequently, these advances will be combined, leading to composite arrangements being almost certain or plausible.

13.3 Diagnosis and Treatment Application

MYCIN was created at Stanford in the 1970s for diagnosing blood-borne bacterial contamination and, ever since, diagnosis and treatment of various illnesses has been the attention centerpiece of AI. A promising precision was shown by AI and other early rule-based frameworks in diagnosing and treating diseases, but regardless, these methods were not assimilated in clinical practices. This technology was close in capabilities to human diagnosticians but was not significantly finer and was poorly paired with clinician/clinical work procedures and record structures. Lately, IBM's Watson has gotten substantial recognition in the media for its focus on accurate medication, particularly in cancer determination and treatment. Watson uses a mixture of machine learning and NPL abilities. Anyhow, the initial enthusiasm for using this innovation has dimmed as clients have started to understand the necessary help that has to be provided to Watson in addressing specific types of malignancies and difficulties that one will have to face while including Watson in healthcare procedures and cycles. Undeniably, Watson is not a solitary product yet, but a collection of 'cognitive services' provided via Application Programming Interfaces (APIs), including speech and language, vision, and machine learning-based data analysis programs. According to the majority of observers, although skilled, engaging in malignant growth treatment was an exceedingly contentious aim for Watson APIs. All the more, free 'open source' programs given by certain merchants, like Google's TensorFlow, are additionally posing as rivals for Watson and other restrictive projects. Execution complications with AI puzzles countless medical healthcare unions. Although standard-based programs compiled within the EHR (Electronic Health Record) framework are widely used, even in the NHS, they are precision deficient in new algorithmic schema, taking into consideration the process of machine learning. These standard-based clinical diagnosis support systems are arduous to maintain as clinical information is susceptible to frequent modification, transition, and advancement. These structures are not equipped to deal with sudden bursts of information and knowledge in terms of genomic, proteomic, metabolic, and the plethora of ways to deal with healthcare. The present situation is undergoing transformation, but it is mostly seen in research facilities and tech firms and not in clinical practices. Examination labs have made it a routine procedure to put forward claims of using AI or extensive data for fostering methods at par, if not superior, in precision while treating illnesses when compared to human clinicians. A significant number of such discoveries are contingent on Radiological Image Investigation, although other types of imaging are also included, such as retinal scanning or genomic-based medication accuracy. As this genre of discoveries is dependent on statistically-based machine learning models, a phase of evidence and probability-based medication is being introduced. For the most part, this introduction is viewed as irrefutable, but it carries many difficulties with itself in terms of patient-clinician relations and clinical morals and ethics. Tech firms and businesses alike are diligently working on similar problems. For instance, Google is collaborating with wellbeing conveyance organizations to create prediction models from a sizeable amount of information to alert clinicians of highly volatile conditions, for example, sepsis and heart failure. AI-inferred image interpretation algorithms are being built by an assortment of various companies, like Google, Enlitic, etc. Jvion offers a 'clinical success machine' that pinpoints high-risk patients as well as those that are most likely to respond to treatment protocols. All this information can aid clinicians by providing an array of choices, out of which the best analysis and treatment can be selected for optimum prognosis. There are a few firms that are extensively focusing on the hereditary profiling of their patients to diagnose and recommend treatment for specific cancer-based diseases in light of the hereditary information collected by them. Yet, clinicians are of the opinion that it is immensely complicated to see all hereditary variations of malignancies and their reaction to new medications and conventions because various cancers have a hereditary premise. Organisations like Foundation Medicine and Flatiron Health, both now owned by Roche, work on this methodology. "Population Health" machine learning models are used by both providers of healthcare facilities and their clients to foresee the population in danger of specific diseases, at a heightened chance of getting into accidents, or to anticipate clinical readmission. Even though currently these models are overlooking information of significance, like patient financial status, that may add predictive capacity and these models can be quite compelling. Regardless of whether AI-based results and treatment recommendations are rules-based or algorithmic in nature, they might be tough to integrate into clinical work procedures and her frameworks. Various issues have been combined to become a substantial roadblock in the far-ranging implementation of AI over any inability to provide precise and compelling suggestions. Numerous AI-based facilities for diagnosis and treatment recommendations provided by tech firms are reclusive and address just a few tasks out of numerous healthcare undertakings. Few EHR providers have commenced the introduction of restricted AI capacities (beyond rule-based clinical decision support) in the facilities provided by them, yet these inclusions are in a nascent phase. Suppliers will either have to take on substantial reconciliation tasks independently or wait until EHR providers expand their AI capabilities.

13.4 Limitation of Artificial Intelligence in Medical Science

13.4.1 Data Availability

The initial move towards constructing an artificially intelligent system (after problem selection and development of solutions strategy) is information assortment. The formation of well-performing models depends on the accessibility of enormous amounts of top-notch information. The issue of information assortment is covered in the discussion because of patient security and because of late episodes of information breaches by major companies. Progress in innovation has brought about expanded computational and logical power just as the capacity to store huge amounts of information. Innovation, for example, facial recognition and gene analysis give way for a person to be distinguished from a pool of individuals. Patients and general society overall reserve a privilege to security and the option to pick what information, if any, they might want to share. Information penetrations now make it feasible for patient information to fall under the control of the insurance agencies bringing about a forswearing of clinical protection because a patient is considered more costly by the protection supplier because of their hereditary creation. Patient security prompts confined accessibility of information, which prompts restricted model preparation and accordingly, the maximum capacity of a model isn't investigated.

13.4.2 One-Sided Data

The information gathered by the programmer is used to train the artificially intelligent system. This data, which is used to train the system, is known as test data. The test data should not be one sided, i.e., it should not include only one particular race, age category, orientation etc. Hence, the data collected should not be one sided or the desired output will not be achieved.

13.4.3 Data Preprocessing

Even after collecting unbiased test data, it is possible for the system to make the data one sided. Hence, the data which is collected should be processed before feeding it to the system. The crude data often contains mistakes and blunders because of the manual entry of the data into the system and due to many other factors. These wrong entries are often eliminated through mathematical calculations. Hence, even after unbiased data is collected it is very important to check the test data during the data processing to eliminate the mistakes during the entry of these data and prevent the data from being one sided.

13.4.4 Selection of Model

With the help of various calculations and prototypes to browse, one should select the calculation that is the most appropriate for the job needing to be done. In this manner, the course of model choice is critical. Bias models are excessively straightforward and neglect to catch the patterns available in the dataset.

13.4.5 Presenting Biased Model

A user of an artificially intelligent system really should have a fundamental comprehension of how such models are constructed. This way, a user can all the more likely decipher the result of the model and choose how to utilize the result. For example, there are numerous measurements that one could use to assess the performance of a model, for example, exactness, accuracy, review, F1 score, and AUC score. Nevertheless, not every metric is appropriate for every problem. At the point when the client of an artificially intelligent system is given execution measurements of a model, they need to ensure that the measurements fitting to the issue are being introduced and not the measurements with the most noteworthy scores.

13.4.6 Fragmented Data

One more constraint of the use of AI is that models that one association invests energy and work to plan and convey for a particular assignment (regression, classification, clustering, NLP, etc.) cannot be seamlessly transitioned for immediate use to another organization without recalibration. Because of security concerns, information sharing is regularly out of reach or restricted between medical services associations, bringing about divided information restricting the unwavering quality of a model.

13.4.7 Black Boxes

Artificially Intelligent systems have a reputation for being black boxes because of the intricacy of the mathematical algorithms involved. There is a need to make models more available and interpretable. While there is some new work towards this path, there is still some headway to be made.

13.5 The Future of AI in Healthcare

It is obvious that AI has a significant role to play in the medical field in the futuristic world. Although largely accepted to be a necessary but woeful development in healthcare, machine learning is the vital mechanism behind improvements in medication accuracy. It is expected that AI will eventually dominate the sphere of diagnosis and treatment recommendation, even though the early attempts in these areas proved to be challenging. Observing the quick developments in AI for image examination, it appears that, in due time, most radiological and psychological images will be analyzed by a machine. Currently, speech and text acknowledgement are used for tasks like patient correspondence and capturing clinical notes. This use is bound to increase in the coming years. The assurance of adoption of this technology in medical care for everyday clinical practice is the area with maximum contention for whether the advances noted in the field of AI will be helpful. The process of upward acceptance of AI can be sped up in a couple of ways, like endorsement by controllers, inclusion in EHR structures, normalization to an adequate degree, providing instructions to clinicians, funding by open or private payer associations, etc.

These difficulties, however tolerable, will take significantly longer to walk through than for actual advancements to develop. Though restricted, the utilization of AI in clinical practice is hoped to be seen within 5 years and a wider use within the field is anticipated. Additionally, it appears to be increasingly evident that AI frameworks will not replace human clinicians but rather expand their horizons to focus on patients. In the long run, human clinicians may push for tasks and career opportunities that rely on uniquely human talents such as compassion, persuasion, and big-picture thinking. Perhaps the medical care providers who refuse to engage with artificial intelligence will be the ones who end up losing their jobs. Finally, a few interesting uses of machine learning or automated technologies in several useful sectors (including issues and challenges) can be found in [25–32] for future research.

Cybercriminals [33–35] frequently focus their attacks on the e-healthcare industry [36]. The study estimated that in 2023, there would be 1,532 Threats [37] per week against e-healthcare firms. There were 65% more assaults this year than the year before, and several of the biggest attacks by cybercriminals [35] and threat agents of the year targeted e-healthcare providers.

13.6 Conclusion

Regardless of the existing limits discussed above, AI seems fit to reform the healthcare industry. Artificial intelligence systems can help save crucial time for specialists by translating notes, entering and putting together patient information into portals (like EPIC) and diagnosing patients, and possibly filling in as an alternative provided to doctors for cross checking the diagnosis made by them. Artificially intelligent systems can likewise assist patients with follow-up care and the accessibility of physician-endorsed drug options. Artificial intelligence additionally has the capacity to remotely diagnose patients, stretching out clinical benefits to distant regions past the major metropolitan cities around the globe. The eventual fate of AI in medical services is splendid and promising, although a lot still needs to be finished. The use of the artificially intelligent system by the general public for medical care is somewhat neglected. As of late, the FDA (U.S Food and Drug Administration) supported AliveCor's Kardiaband (in 2017) and Apple's smartwatch series 4 (in 2018) to detect atrial fibrillation. The use of a smartwatch is an initial step towards engaging individuals to gather their wellbeing information and empower quick mediations from the patient's clinical support groups. Most of the worldwide drug organizations have put their time and funds in involving AI for drug improvement for various diseases like cancer or cardiovascular disease. Nonetheless, the advancement of models for diagnosing dismissed tropical illnesses (malaria and tuberculosis) and uncommon infections remains generally ignored. The FDA presently encourages organizations to foster new medicines for these illnesses through priority vouchers. Given the effect that AI and machine learning are having on our extensive world, AI needs to be a part of the curriculum for numerous experts in different fields. This is especially valid for clinical calling where the cost of an off-base choice can be deadly. AI has the potential to aid many of healthcare's biggest issues, but we are still far from making this a reality. A big problem and barrier in realizing this to actuality is data. We can invent all the promising technologies and machine learning algorithms, but without sufficient and well-represented data, we cannot realize the full potential of AI in healthcare.

References

1. Deloitte Insights. State of AI in the enterprise. Deloitte, 2018.
2. Lee SI, Celik S, Logsdon BA *et al.* A machine learning approach to integrate big data for precision medicine in acute myeloid leukaemia. Nat Commun 2018;9:42.
3. Sordo M. Introduction to neural networks in healthcare. OpenClinical, 2002.
4. Fakoor R, Ladhak F, Nazi A, Huber M. Using deep learning to enhance cancer diagnosis and classification. A conference presentation. The 30th International Conference on Machine Learning, 2013.

5. Vial A, Stirling D, Field M *et al.* The role of deep learning and radiomic feature extraction in cancer-specific predictive modelling: a review. Transl Cancer Res 2018;7:803-16.
6. Davenport TH, Glaser J. Just-in-time delivery comes to knowledge management. Harvard Business Review 2002.
7. Hussain A, Malik A, Halim MU Ali AM. The use of robotics in surgery: a review. Int J Clin Pract 2014;68:1376-82.
8. Bush J. How AI is taking a scutwork out of healthcare. Harvard Business Review 2018.
9. Buchanan BG, Shortliffe EH. Rule-based expert system: The MYCIN Experiments of the Stanford heuristic programming project. Reading: Addison Wesley, 1984
10. Ross C, Swetlitz I. IBM pitched its Watson supercomputer as a revolution in cancer care. It's nowhere close. Stat 2017.
11. Davenport TH. The AI Advantage. Cambridge: MIT Press, 2018.
12. Right, Care Shared Decision Making Programme, Capita. Measuring shared decision making: A review of research evidence. NHS,2012.
13. Loria K. Putting the AI in radiology. Radiology Today 2018;19:10.
14. Schmidt-Erfurth U, Bogunovic H, Sadeghipour A *et al.* Machine learning to analyze the prognostic value of current imaging biomakers in neovascular age-related macular degeneration. Opthalmology Retina 2018;2:24-30.
15. Aronson S, Rehm H. Building the foundation of genomic-based precision medicine. Nature 2015;526:336-42.
16. Rysavy M. Evidence-based medicine: A science of uncertainty and an art of probability. Virtual Mentor 2013;15:4-8.
17. Rajkomar A, Oren E, Chen K *et al.* Scalable and accurate deep learning with electronic health records. npj Digital Medicine 2018;1:18.
18. Shimabukuro D, Barton CW, Feldman MD, Mataraso SJ, Das R. Effect of a machine learning-based severe sepsis prediction algorithm on patient survival and hospital length of stay: a randomised clinical trial. BMJ Open Respir Res 2017;4:e000234.
19. Aicha AN, Englebienne G, van Schooten KS, Pijnappels M, Kröse B. Deep learning to predict falls in older adults based on daily-Life trunk accelerometry. Sensors 2018;18:1654.
20. Low LL, Lee KH, Ong MEH *et al.* Predicting 30-Day readmissions: performance of the LACE index compared with a regression model among general medicine patients in Singapore. Biomed Research International 2015;2015;169870.
21. Grus J. Data Science from ScratchFirst Principles with Python. 1st ed. Sebastopol, CA, USA: O›Reilly; 2015
22. Dhamnani S, Singal D, Sinha R, Mohandoss T, Dash M. RAPID: Rapid and Precise Interpretable Decision Sets. In 2019 IEEE International Conference on Big Data (Big Data). IEEE; 2019. p. 1292-301.

23. Lucas GM, Gratch J, King A, Morency LP. It's only a computer: Virtual humans increase willingness to disclose. Comput Human Behav 201; 37:94-100.
24. Ridley DB. Priorities for the priority review voucher. Am J Trop Med Hyg. 2017;96:14-5.
25. Meghna Manoj Nair, Amit Kumar Tyagi, Richa Goyal, Medical Cyber Physical Systems and Its Issues, Procedia Computer Science, Volume 165, 2019, Pages 647-655, ISSN 1877-0509, https://doi.org/10.1016/j.procs.2020.01.059.
26. Shruti Kute; Amit Kumar Tyagi; Meghna Manoj Nair, "Research Issues and Future Research Directions Toward Smart Healthcare Using Internet of Things and Machine Learning," in Big Data Management in Sensing: Applications in AI and IoT, River Publishers, 2021, pp.179-200.
27. Kute S.S., Tyagi A.K., Aswathy S.U. (2022) Industry 4.0 Challenges in e-Healthcare Applications and Emerging Technologies. In: Tyagi A.K., Abraham A., Kaklauskas A. (eds) Intelligent Interactive Multimedia Systems for e-Healthcare Applications. Springer, Singapore. https://doi.org/10.1007/978-981-16-6542-4_1
28. Kute S.S., Tyagi A.K., Aswathy S.U. (2022) Security, Privacy and Trust Issues in Internet of Things and Machine Learning Based e-Healthcare. In: Tyagi A.K., Abraham A., Kaklauskas A. (eds) Intelligent Interactive Multimedia Systems for e-Healthcare Applications. Springer, Singapore. https://doi.org/10.1007/978-981-16-6542-4_15
29. Madhav A.V.S., Tyagi A.K. (2022) The World with Future Technologies (Post-COVID-19): Open Issues, Challenges, and the Road Ahead. In: Tyagi A.K., Abraham A., Kaklauskas A. (eds) Intelligent Interactive Multimedia Systems for e-Healthcare Applications. Springer, Singapore. https://doi.org/10.1007/978-981-16-6542-4_22
30. Nair, Meghna Manoj; Tyagi, Amit Kumar "Privacy: History, Statistics, Policy, Laws, Preservation and Threat Analysis", Journal of Information Assurance & Security. 2021, Vol. 16 Issue 1, p24-34. 11p.
31. Varsha R., Nair S.M., Tyagi A.K., Aswathy S.U., RadhaKrishnan R. (2021) The Future with Advanced Analytics: A Sequential Analysis of the Disruptive Technology's Scope. In: Abraham A., Hanne T., Castillo O., Gandhi N., Nogueira Rios T., Hong TP. (eds) Hybrid Intelligent Systems. HIS 2020. Advances in Intelligent Systems and Computing, vol 1375. Springer, Cham. https://doi.org/10.1007/978-3-030-73050-5_56
32. Mishra S., Tyagi A.K. (2022) The Role of Machine Learning Techniques in Internet of Things-Based Cloud Applications. In: Pal S., De D., Buyya R. (eds) Artificial Intelligence-based Internet of Things Systems. Internet of Things (Technology, Communications and Computing). Springer, Cham. https://doi.org/10.1007/978-3-030-87059-1_4
33. Mahor, V., Pachlasiya, K., Garg, B., Chouhan, M., Telang, S., & Rawat, R. (2022, June). Mobile Operating System (Android) Vulnerability Analysis Using Machine Learning. In *Proceedings of International Conference on Network Security and Blockchain Technology: ICNSBT 2021* (pp. 159-169). Singapore: Springer Nature Singapore.

34. Mahor, V., Garg, B., Telang, S., Pachlasiya, K., Chouhan, M., & Rawat, R. (2022, June). Cyber Threat Phylogeny Assessment and Vulnerabilities Representation at Thermal Power Station. In *Proceedings of International Conference on Network Security and Blockchain Technology: ICNSBT 2021* (pp. 28-39). Singapore: Springer Nature Singapore.
35. Rawat, R., Gupta, S., Sivaranjani, S., CU, O. K., Kuliha, M., & Sankaran, K. S. (2022). Malevolent Information Crawling Mechanism for Forming Structured Illegal Organisations in Hidden Networks. *International Journal of Cyber Warfare and Terrorism (IJCWT)*, 12(1), 1-14.
36. Rawat, R., Rimal, Y. N., William, P., Dahima, S., Gupta, S., & Sankaran, K. S. (2022). Malware Threat Affecting Financial Organization Analysis Using Machine Learning Approach. *International Journal of Information Technology and Web Engineering (IJITWE)*, 17(1), 1-20.
37. Rawat, R., Mahor, V., Chouhan, M., Pachlasiya, K., Telang, S., & Garg, B. (2022). Systematic literature Review (SLR) on social media and the Digital Transformation of Drug Trafficking on Darkweb. In *International Conference on Network Security and Blockchain Technology* (pp. 181-205). Springer, Singapore.

14

Revolutionizing Modern Automated Technology with WEB 3.0

Shishir Shrivastava[1] and Amit Kumar Tyagi[1,2*]

[1]School of Computer Science and Engineering, Vellore Institute of Technology, Chennai, Tamilnadu, India
[2]Department of Fashion Technology, National Institute of Fashion Technology, New Delhi, Delhi, India

Abstract
The web has revolutionised the way modern automated technologies (intelligent technologies) work. The first version of the web, which was Web 1.0, lasted from 1989 to 2005. A few examples of Web 1.0 are Slashdot and Craigslist. Since then, continuous upgrades on the web have allowed the modern generation to open their hands to different opportunities and challenges in regards to education and learning. From 2005 to today Web 2.0 has been in use. The key difference between Web 1.0 and Web 2.0 is the nature of interaction of the web with the computer. While Web 1.0 was a read-only medium, Web 2.0 is a read/write medium. A few examples of Web 2.0 are Facebook, YouTube, and Myspace. Today in 2022, we now introduce Web 3.0 which will be able to read/write/execute and is going to play a significant role in artificial intelligence. With the ongoing advancements in Web 2.0 and Web 3.0, modern education and learning is greatly impacted. This research work shall discuss the definitions of Web 3.0, it's revolution, features, and characteristics. This chapter shall also throw light on the upcoming technologies for the future generation with the help of Web 3.0 and its tools.

Keywords: WEB 3.0, blockchain, metaverse, sematic web, virtual reality, artificial intelligence

14.1 Introduction

It is the year 2022 and all of the dreams, possibilities, and visions that the people of the twentieth century had for future generations have come true.

**Corresponding author*: amitkrtyagi025@gmail.com

Romil Rawat, Rajesh Kumar Chakrawarti, Sanjaya Kumar Sarangi, Rahul Choudhary, Anand Singh Gadwal and Vivek Bhardwaj (eds.) Robotic Process Automation, (211–224) © 2023 Scrivener Publishing LLC

Today's technology has brought even the tiniest of services and items to our fingertips [2], transforming the way commercial enterprises operate. These technological developments and futuristic technology ideas date back to the introduction of the World Wide Web (WWW) [13] to connect people in ways that no one could have imagined.

Today, the WWW has proven its usefulness in almost every sector and as more businesses, start-ups, schools and colleges, health services, and other organizations go completely hands-free and online, it's the right time to bring a completely new advancement in the current version of the World Wide Web. When the World Wide Web was first developed in the 1990s, it has evolved through two versions, Web 1.0 and Web 2.0, and is currently going forward into a brand-new version, Web 3.0. In this same regard, the Wikipedia states: "Web 1.0 is Read Only, static data with simple markup for reading. Web 2.0 is Read/Write dynamic data through web services customize websites and manage items. Web 3.0 is Read/Write/Execute." Web 2.0 users not only read information from the internet, but also offer information to others via the internet. Many popular Web 2.0 interactive apps are currently available, including WordPress, Google Maps, Flickr, Meta, Spotify, Weebly, Glogster, and so on. In comparison to Web 2.0, there isn't a particularly clear definition for Web 3.0 yet. Web 3.0 is a concept used to characterise the future of the World Wide Web and it will be examined in depth below. The ambitions and vision of various experts on the revolution brought by WEB 3.0 varies a lot. Many feel that upcoming revolutionary technologies like the Semantic Web will turn-around the very utility of the Internet and bring new and effective products based on Artificial Intelligence. Increased Internet connection speeds, modular web applications, and developments in computer graphics, according to other visionaries, will play a crucial part in the development of WEB 3.0 of the World Wide Web [1].

14.2 What is WEB 3.0: Definitions

In 2006, John Markoff of the New York Times was among one of the first ones to coin the term 'Web 3.0' and just a little early to it in 2006 itself, the term 'Web 3.0' appeared boldly in the then Blog Article "Critical of Web 2.0 and associated technologies such as Ajax" edited by Jeffrey Zeldman [7].

Different approaches to the future Web are supported by major IT specialists and scholars. Experts are unanimous in their predictions for how Web 3.0 will develop. The thoughts of pioneers in the field in this regard are discussed below.

Google's Ex-CEO, Eric Schmidt [5], states: "Web 3.0 is a series of combined applications. The core software technology of Web 3.0 is artificial intelligence, which can intelligently learn and understand semantics.

Therefore, the application of Web 3.0 technology enables the Internet to be more personalized, accurate, and intelligent."

The founder of international multi-media streaming giant Netflix, Reed Hastings states: "Web 1.0 was dial-up, 50K average bandwidth; Web 2.0 is an average 1 megabit of bandwidth and Web 3.0 will be 10 megabits of bandwidth all the time, which will be the full video Web, and that will feel like Web 3.0."

The father of the World Wide Web and the century's greatest computer scientist, Sir Timothy John Berners-Lee says:

"People keep asking what Web 3.0 is. I think maybe when you've got an overlay of scalable vector graphics - everything rippling and folding and looking misty-on Web 2.0 and access to a semantic Web integrated across a huge space of data, you'll have access to an unbelievable data resource."

In November 2006, at the TechNet Summit, the founding father of Yahoo, Sir Jerry Yang, stated: "Web 2.0 is well documented and talked about. The power of the Net reached a critical mass, with capabilities that can be done on a network level. We are also seeing richer devices over last four years and richer ways of interacting with the network, not only in hardware like game consoles and mobile devices, but also in the software layer. You don't have to be a computer scientist to create a program. We are seeing that manifest in Web 2.0 and Web 3.0 will be a great extension of that, a true communal medium... the distinction between professional, semi-professional and consumers will get blurred, creating a network effect of business and applications."

These are some of the numerous IT industry specialists' perspectives on Web 3.0. Following that, we'll go through some of the characteristics and features of Web 3.0.

14.3 Features & Characteristics

Due to the overall immediate development of the technology that will underpin Web 3.0, it may take less time to achieve traction. It is anticipated that it would fundamentally alter the way internet sites are built and how users interact with them [3].

14.3.1 Rewarding Cryptocurreny

Blockchain and cryptocurrencies are slowly becoming the new normal for the current generation. The impressions it creates among the young minds drive them to learn, create, exploit, and utilise the economic behaviour, the backend of any Industry. Cryptocurrencies, on the other hand, are envisioned as a way of compensating content providers in Web 3.0, who would earn a token every moment someone viewed their work [21]. These three technologies will have a strong link since they will be interconnected, centralised, and coherent. Smart contracts will be used to support a variety of behaviors, including transactions, security and scrutiny, censorship-resistant peer-to-peer file storage, and document sharing. It would be a complete shift in the way businesses operate, as well as an empowerment of users and producers.

14.3.2 Interconnection-Exchange & Use of Info

The concepts interconnection collaboration and reusability are intertwined in WEB 3.0 for the purpose of exchange and use of info.

Interconnection insists on reprocessing, just another type of coming together. WEB 3.0 shall provide a platform for exchange of data, knowledge, experiences, and emotions. Knowingly or Unknowingly, when data is created on the Internet and that knowledge or experience is brought to use by another, a new kind of perspective and outlook is created. It would be surprisingly easy to personalize the applications on WEB 3.0 that work on a huge number of different devices [18]. Web 3.0 applications can run on a wide range of devices, including laptops, smart refrigerators, inter platform software, mining equipment, smart phones, Smart LEDs, Automobiles, and more.

14.3.3 Metaverse

With the recent change of Facebook's parent company name into Meta, people have started to explore and learn a lot about the Metaverse. It isn't even six months since the decision was made and we can already hear start-ups, entertainment, and real-estate getting a reality on the metaverse. This new version of the Internet may be properly interwoven with this virtual cosmos, allowing for 3D website designs and the addition of IoT devices to its peripheral.

14.3.4 Personalization

In the upcoming version, "The Semantic Web" would act as a powerful program to implement personalization [9].

Personalization is another feature of the Web 3.0 era and it's something the people of this generation love a lot.

Users can provide their personal preferences that shall then be used to process various operations such as a data analyzing character ("Avatar") creation, query find, and the building of a user-specific centralized database.

14.3.5 Semantic Web

As the writers at OntoText [20] state:

"The Semantic Web is a vision about an extension of the existing World Wide Web, which provides software programs with machine-interpretable metadata of the published information and data. In other words, we add further data descriptors to otherwise existing content and data on the Web. As a result, computers are able to make meaningful interpretations similar to the way humans process information to achieve their goals."

In the same regard, a CEO and ATSD authority, Sir Tony Bingham says: "In the Semantic Web, content will find you—rather than (you) actively seeking it, your activities and interests will determine what finds you, and it will be delivered how you want it and to your preferred channel. The Semantic Web provides tremendous potential for learning." It was the vision of Sir Timothy to design computers and devices in a way to handle data and information in real-time in a much better way with the help of Semantic Web [8].

14.4 Implementation

Given the above definitions and features of Web 3.0, let us now look at the various technologies which are going to be based on the advanced nature of the upcoming version of the world wide web.

14.4.1 The Three Dimensional Web

As the heading strongly implies, Web 3.0 is going to lay the fundamentals of establishing a Virtual-three dimensional world, exactly around what the Metaverse is working on. With the help of 3D graphics and Web 3.0 tools, highly realistic applications can be developed. Super-fast Internet, faster processing rates, greater screen dynamics, virtual reality, and three dimensional gaming changes the traditional way of browsing into a three dimensional environment, allowing the users to roam around the virtual streets and playgrounds of the internet as an unreal representation of their own self. Flight simulation, vehicle driving simulation, and medical training are just a few of the Internet-based rudimentary virtual worlds that have recently gotten a lot of attention from the general public. In Dubai, a mall has exclusively set-up a VR facility to help potential customers experience a flight simulation [12]. These kinds of experiences let people try new exciting things that could've not been possible in actual reality. Anyone can build a representation ("An Avatar") of their own on the web and with it, gets to interact with rest of the world on the same platform. These virtual representations or "avatars" can roam around freely, share/play with other avatars, enjoy and take active part in various cultures, and introduce and utilize various types of services. Texts, ping message, voice support, and/or video are all options for interacting in these virtual environments.

14.4.2 Decentralized Technology

Peer-to-peer networks and blockchains are among the primary decentralized technologies used to develop the new decentralized web. Back in the 1990s when peer-to-peer networks were established until now, they enabled a set of computers to function as nodes and share resources (On and Off Demand). In the Web 3.0 age, peer-to-peer networks will have significantly more importance considering its purpose to reduce third party intervention. As Calvin Ebun-Amu [4] states: "Blockchain networks, while newer, have been used to leverage the power of peer-to-peer networks. The Web 3.0 blockchain uses principles of peer-to-peer technology and integrates cryptography and consensus algorithms to scale systems of decentralization among larger groups of people." This takes the baton from traditional databases and hands them over to such technologies of Blockchain and Web 3.0.

14.4.3 The Social Web

The Social Web shows how individuals engage with one another via the World Wide Web's underlying technology.

Web 3.0 advances will elevate current online socialization to something new known as "Semantic Social Computing or Socio-Semantic Web", which will create and utilize information or data, including future offerings, portrayal, design and content, and real-time responses [15].

The social web can also be understood as a collection of web activities, events, and user interactions that facilitate human social communications.

Currently, a good number of users consider the current version as a phase of history's WWW viz Web 2.0 as the social web, during which user involvement rose dramatically as social networking sites became more important in people's lives.

3.0 (World Wide Web) Decentralized technologies will almost certainly characterize social media. For example, social media sites based on the metaverse consisting of blockchains will run without any single authority, rewarding individuals for their engagement in the platforms through incentive systems. A technology like Minds [6] is an example of this.

Data, Flags, instructions, services, software functionalities, and behaviors will all get underlying knowledge representations thanks to Semantic Web and Artificial Intelligence technology. The collective ideology will emerge from the semantic and logical aggregate of each individual's opinion, view, and understanding, rather than from the group's consensus decision.

14.4.4 Prevalent and Omnipresent WEB

Prevalent and Omnipresent computing platforms have improved over the years as a result of concrete advancements in technologies such as satellite communications, over-the-air transfers, portable calculators, machine learning, hands-free software, connecting technologies such as Bluetooth, Wi-Fi and WLAN, architectural systems, and health focused devices.

According to Peter Robinson [19], "Ubiquitous and pervasive computing may be defined as the task of embedding small and mobile devices into existing IT and computing infrastructures, so that it allows users to access and manipulate information where and when it matters, even while on the move."

Service Oriented Architectures (SOA) and associated technologies would enable interactions between many types of machinery and the Web, providing cross-platform interoperability. Some of the world's top-most and successful software development companies are working in this regard.

A development API, as well as an extraordinary innovation-focused products provided by Microsoft [14], called Life Ware, which are a great illustration of what this technology may do in the near future [15].

14.4.5 Multi-Media Web

Traditional search engines rely on text inputs for the majority of their results. Like the traditional search engines, WEB 3.0 will not only limit itself to alphanumerical or text searches but allow many more. Queries or

Searches made on this next version of the web will be able to narrow down certain unconventional formats such as mp3 and mp4 based on their properties. Search engines will allow input as a media or multi-media data and provide results on the basis of similar media objects.

To search for images containing pineapple, for example, we need to upload an image of pineapple as an input and the algorithm should be able to return images of pineapples with similar qualities. Other media-based items, such as mp3 and mp4 extensions, should also be possible to search throughout the web. Work has already begun in this direction.

Software like Google Images [10], a picture sharing application that allows users to automatically categorize photographs using uploaded pictures, and the site Amazon Pantry [11], which lets the customers buy grocery online, are both notable instances of this type of technology.

14.5 Inventions around Modern Technology

A much-awaited response and advancement with WEB 3.0 should be advanced Scalable Vector Graphics ("or SVG") in collaboration with the semantic web.

There has also been talk about three-dimensional socializing platforms and enhanced three-dimensional web settings that will bring together the best of augmented reality (like Play Station VR) and gaming possibilities with the Web 3.0.

14.5.1 A.I. Powered Search Engine

On Requesting a standard Web search engine, it can't fully grasp what you're looking for. It goes through the pages on the internet that contain similar words to what you entered in your search queries. The search engine has no way of knowing if the page is relevant to your query. Only the presence of the keyword on a particular website may be determined. The upcoming dynamic search engines of the new web age will not restrict it to the keywords in your query but shall try to understand the context and show relevant results. This should increase the accuracy of results shown to you for your purpose [16].

Web 3.0, according to experts, will give users deep and more pleasant experiences while searching through the internet. Data scientists also expect that with Web 3.0, each user should be able to configure their own personal online profile based on their browsing history. This profile will be used by Web 3.0 to personalize the surfing experience for each user.

That is, if two different users used the same service to conduct an online search with the exact keywords, they would get different results based on their personal profiles [17].

Users can also benefit from the other kind of search requirements made possible by the Web 3.0 with other multimedia objects such as pictures, voice-notes, and short-clips. For example, a few software which portray this tech are Googles Images [10], an image search tool that allows users to upload JPGs or any compatible format images using a file folder, and Amazon Pantry, which allows the customer to find groceries or toiletries based on real-time photos [11].

14.5.2 Touring and Travelling

Three dimensional in-depth graphical user interfaces will work as a strong solution for the travel freaks and general users to roam, explore, and participate in native activities, sharing stories, moments, and happiness, and exchanging important relevant information among people around the globe in a very hassle-free manner.

With proper implementation of Web 3.0 and it's tools to invent the necessary technologies and apply them on the tours and travel industries of the world will bring a never-before seen revolution in the Tourism Industry. People who earlier couldn't travel to the places they love and couldn't meet the people they inspired, can now do this very easily with the help of the next version of the World Wide Web.

Visiting various locations in virtual worlds would assist people in a variety of ways. Users can virtually go to ancient sites in a short amount of time. They may also engage and experience the environment of the sites, other natives, and have their fantasies fulfilled over the web, for example, when looking at historical places like the Taj Mahal, Red Fort, or Rome. In the same way, kids can visit an Egyptian town or view the Egyptian pyramids.

There is so much we can do so that kids and adults can be provided a safe and cost-effective means to experience such things.

14.5.3 Gaming Advancements

A three-dimensional virtual world built on the web, as expected, is a mix of augmented reality, three-dimensional gaming technology, and a human-programmed environment powered by Internet and web technology in which users experience a virtual character of them. People can build such characters through various personalization and use them to communicate with others in virtual worlds. Students can use this technology to perform dangerous experiments and remain safe at the same time with no risk.

Such a world can revolutionize the gaming industry at once and build a whole new concept of role-playing, three-dimensional modelling, creativity, simulations, and active participation. There's a lot of room for personalization of one's own way of playing a game and a choice on how they want to proceed with their current achievements.

Users can enjoy their favorite game in a variety of settings within a three-dimensional world where they can communicate with fellow players in a realistic atmosphere. Players and developers can conduct private aims in a shared virtual 3D space from geographically scattered places.

They can better enlighten fellow players and developers on their views and demands regarding what they would want to do in the same way they do in real life.

Media, medical, business, computer science, commerce, education, research, interactions, art, real estate, judiciary, linguist, past-present-future, and earth, to name a few, will all benefit greatly from the 3D virtual worlds that are already available and will be available in the future.

14.5.4 E-Learning with Virtual Documentary

Documentaries are videography of instances that facilitate an ecosystem that enables students and learners to design and learn about various theories and novels to create a database of quick answers in a set of interconnected web profiles. They play an important role in knowledge creation by facilitating content creation, publishing, editing, rewriting, and collaboration. Such a database is being utilized to maintain and monetize a set of evets that happened in real life for a long time. These leaners can also document themselves around a theory and publish their report on such a database. The software's ease of use makes it simple for an editor (teacher) to proof-read and provide credibility through the appropriate authority. Note that Data Scientists and Tech Geeks have been exploring a completely new world of opportunity to provide a new aspect to the world of documentaries and encyclopedias as the three-dimensional web has evolved. Some examples of this kind of technology can be found on software like Adobe Premiere Pro [22].

If a user conducted a query and selects a particular result from the list that is linked to a database of some other geographical area, the user will virtually teleport to that location with an animation ("on globe") to produce relevant mp3/mp4 data. For example, when looking for the Taj Mahal, the user would teleport virtually towards the great Indian monument and eventually, the user will be greeted with an audio: "Welcome to Taj Mahal! One of the 7 wonders of world" standing virtually right in front of it in Agra, India, as well as a floating panel of information about its past and significance [22].

As Rajiv and Manohar Lal states: "3D Wikis would be able to provide rich & effective environment involving all media and animation, for learners, so that they can have better impact on learning & knowledge." Future researchers/readers are suggested to refer to articles [23–30] to know more about intelligent automation, blockchain technology, etc., and their uses in several useful sectors for their future research

Businesses [31] will need to develop, adapt, and incorporate blockchain [32], interactive AI, and decentralised implementations as more online functionality is integrated into Web 3.0 automated systems [33]. This is despite the possibility of a boost in risk, vulnerabilities [34], and cybersecurity attacks [35] by threat agents.

14.6 Conclusion

The upcoming third version of the World Wide Web (WWW) shall not restrict itself to more than just a collaboration of unique and proper tools and services. The concerned technologies will deliver a variety of options that allow for the impossible to be proven possible with its implementation in various different aspects as mentioned very well in this chapter. Web 3.0 services will also benefit studying and testing because of their very nature. Usefulness of the next generation browsing technology consists of three-dimensional documentaries, three-dimensional research-labs, context-based search engines, artificial environments including personal characters ("or Avatars"), and many more.

Acknowledgement

I would like to thank my Mom and Dad for always supporting me and having faith in me, which gave me the courage to push my limits.

References

1. Jinhong Cui and Xu Wang, "Capability sharing architecture and implementation in IM or SNS," 2008 IEEE International Conference on Service

Operations and Logistics, and Informatics, 2008, pp. 520-523, doi: 10.1109/SOLI.2008.4686450.
2. Silva, J.M., Rahman, A.S., & El Saddik, A. (2008). Web 3.0: a vision for bridging the gap between real and virtual. Communicability MS '08.
3. Victoria Shannon (June 26, 2006). "A 'more revolutionary' Web". International Herald Tribune. Retrieved May 24, 2006.
4. Calvin Ebun-Amu, "The 4 Technologies That Will Make Web 3.0 a Reality", maseuseof.com, 2021.
5. Tyagi, Amit Kumar; Nair, Meghna Manoj; Niladhuri, Sreenath; Abraham, Ajith, "Security, Privacy Research issues in Various Computing Platforms: A Survey and the Road Ahead", Journal of Information Assurance & Security. 2020, Vol. 15 Issue 1, p1-16. 16p.
6. Mind.com, 2020
7. Jeffrey Zeldman, "Designing with Web Standards", New Riders, Oct 15, 2009
8. Russell K, "Semantic Web", Computer world, 2006(9):32.
9. Zhang Yang, "The Development of Web and Library s Reference Service-from Web 1.0 to Web 3.0," Sci - Tech Information Development & Economy, voI.l8, 2009.
10. Tyagi, A. K. (Ed.). (2021). Multimedia and Sensory Input for Augmented, Mixed, and Virtual Reality. IGI Global. http://doi:10.4018/978-1-7998-4703-8
11. Amazon Pantry, www.amazon.in/pantry-online-grocery-shopping-store/2022
12. Dubai Mall, www.emiratesa390experience.com/, 2022
13. Davis Mills "Semantic Social Computing". September 20, 2007.
14. LifeWare http://www.exceptionalinnovation.com. (visited on 14/04/11)
15. Microsoft WSD http://msdn.microsoft.com/library/default.aspx
16. S. A. Inamdar and G. N. Shinde, "Intelligence Based Search Engine System for Web Mining, Research, Reflections and Innovations in Integrating ICT in Education", 2009.
17. Strickland, Jonathan, "How Web 3.0 Will Work", http://computer.howstuffworks.com/web-30.html
18. Mathieu d'Aquin, Enrico Motta, Martin Dzbor, Laurian Gridinoc, Tom Heath, and Marta Sabou. 2008. Collaborative Semantic Authoring. IEEE Intelligent System, 23, 3 (May 2008), 80–83. DOI: https://doi.org/10.1109/MIS.2008.43
19. Peter Robinson, Stefan Hild, "Controlled Availability of Pervasive Web Services", 2003 IEEE.
20. OntoText, "What Is the Semantic Web?", https://www.ontotext.com/knowledge hub/fundamfundam/what-is-the-semantic-web/, 2021
21. Plain Concepts, "Web 3.0 | The New Internet Revolution", www.plainconcepts.com, 2022
22. https://www.adobe.com/products/premiere.html
23. M. M. Nair, A. K. Tyagi and N. Sreenath, "The Future with Industry 4.0 at the Core of Society 5.0: Open Issues, Future Opportunities and Challenges," 2021 International Conference on Computer Communication and Informatics (ICCCI), 2021, pp. 1-7, doi: 10.1109/ICCCI50826.2021.9402498.

24. Tyagi A.K., Fernandez T.F., Mishra S., Kumari S. (2021) Intelligent Automation Systems at the Core of Industry 4.0. In: Abraham A., Piuri V., Gandhi N., Siarry P., Kaklauskas A., Madureira A. (eds) Intelligent Systems Design and Applications. ISDA 2020. Advances in Intelligent Systems and Computing, vol 1351. Springer, Cham. https://doi.org/10.1007/978-3-030-71187-0_1
25. Goyal, Deepti & Tyagi, Amit. (2020). A Look at Top 35 Problems in the Computer Science Field for the Next Decade. 10.1201/9781003052098-40.
26. Madhav A.V.S., Tyagi A.K. (2022) The World with Future Technologies (Post-COVID-19): Open Issues, Challenges, and the Road Ahead. In: Tyagi A.K., Abraham A., Kaklauskas A. (eds) Intelligent Interactive Multimedia Systems for e-Healthcare Applications. Springer, Singapore. https://doi.org/10.1007/978-981-16-6542-4_22
27. D. Goyal, R. Goyal, G.Rekha, S. Malik and A. K. Tyagi, "Emerging Trends and Challenges in Data Science and Big Data Analytics," 2020 International Conference on Emerging Trends in Information Technology and Engineering (ic-ETITE), 2020, pp. 1-8, doi: 10.1109/ic-ETITE47903.2020.316.
28. Rekha G., Tyagi A.K., Anuradha N. (2020) Integration of Fog Computing and Internet of Things: An Useful Overview. In: Singh P., Kar A., Singh Y., Kolekar M., Tanwar S. (eds.) Proceedings of ICRIC 2019. Lecture Notes in Electrical Engineering, vol 597. Springer, Cham. https://doi.org/10.1007/978-3-030-29407-6_8
29. Sheth, H.S.K., Tyagi, A.K. (2022). Mobile Cloud Computing: Issues, Applications and Scope in COVID-19. In: Abraham, A., Gandhi, N., Hanne, T., Hong, TP., Nogueira Rios, T., Ding, W. (eds) Intelligent Systems Design and Applications. ISDA 2021. Lecture Notes in Networks and Systems, vol 418. Springer, Cham. https://doi.org/10.1007/978-3-030-96308-8_55
30. Varsha R., Nair S.M., Tyagi A.K., Aswathy S.U., RadhaKrishnan R. (2021) The Future with Advanced Analytics: A Sequential Analysis of the Disruptive Technology's Scope. In: Abraham A., Hanne T., Castillo O., Gandhi N., Nogueira Rios T., Hong TP. (eds) Hybrid Intelligent Systems. HIS 2020. Advances in Intelligent Systems and Computing, vol 1375. Springer, Cham. https://doi.org/10.1007/978-3-030-73050-5_56
31. Rawat, R., Bhardwaj, P., Kaur, U., Telang, S., Chouhan, M. and Sankaran, K. S. (2023) Smart Vehicles for Communication, Volume 2. John Wiley & Sons.
32. Mahor, V., Bijrothiya, S., Rawat, R., Kumar, A., Garg, B. and Pachlasiya, K. (2023) IoT and Artificial Intelligence Techniques for Public Safety and Security. Smart Urban Computing Applications, 111.
33. Mahor, V., Bijrothiya, S., Mishra, R. and Rawat, R. (2022) ML Techniques for Attack and Anomaly Detection in Internet of Things Networks. Autonomous Vehicles Volume 1: Using Machine Intelligence, 235-252.
34. Mahor, V., Bijrothiya, S., Mishra, R., Rawat, R. and Soni, A. (2022) The Smart City Based on AI and Infrastructure: A New Mobility Concepts and Realities. Autonomous Vehicles Volume 1: Using Machine Intelligence, 277-295.

35. Mahor, V., Pachlasiya, K., Garg, B., Chouhan, M., Telang, S. and Rawat, R. (2022 June). Mobile Operating System (Android) Vulnerability Analysis Using Machine Learning. In Proceedings of International Conference on Network Security and Blockchain Technology: ICNSBT 2021 159-169, Singapore: Springer Nature Singapore.

15

The Role of Artificial Intelligence, Blockchain, and Internet of Things in Next Generation Machine Based Communication

R. Harish[1], Sanjana Chelat Menon[2] and Amit Kumar Tyagi[1,3]*

[1]*School of Computer Science and Engineering, Vellore Institute of Technology, Chennai, Tamil Nadu, India*
[2]*Department of Electrical and Electronics, PSG Institute of Technology and Applied Research, Coimbatore, Tamil Nadu, India*
[3]*Department of Fashion Technology, National Institute of Fashion Technology, New Delhi, Delhi, India*

Abstract

In a rough definition, IoT is a structure where real-world objects communicate with each other over the internet. Back in the 19th century when the electrical (electromagnetic) telegraph was invented, it facilitated between the machines via electric signals, thus the concept of communication of two similar entities over a central mode of transmission was already in place. During the 1980s in Pennsylvania, a group of students designed such a device to measure the amount of Coke remaining in the vending machine by installing microswitches into the machine. In 1999, the term internet of things was coined. The point that is being conveyed here in this research is to highlight the applications of Blockchain, AI, and IoT and how they converge together and work efficiently with the assistance of services like cloud computing.

***Keywords*:** Blockchain, IoT-based applications, convergence of IoT, blockchain, AI

**Corresponding author*: amitkrtyagi025@gmail.com

Romil Rawat, Rajesh Kumar Chakrawarti, Sanjaya Kumar Sarangi, Rahul Choudhary, Anand Singh Gadwal and Vivek Bhardwaj (eds.) Robotic Process Automation, (225–242) © 2023 Scrivener Publishing LLC

15.1 Introduction

The internet of things (IoT), blockchain technology, and artificial intelligence (AI) are now acknowledged as advancements with the potential to disrupt whole sectors and enhance existing business processes. Numerous technologies are probably going to be combined, or "converged," to create new platforms, goods, and services. In this chapter, we analyze the confluence of blockchain with two of the more significant new technologies—the Internet of Things (IoT) and Artificial Intelligence (AI)—to show how this may be the case with blockchain. This trio was chosen because there is a lot of potential overlap and room for them to collaborate to address a variety of significant use cases. Each of these technologies can add significant components to the mix, as we will attempt to demonstrate in more detail below, but they each also have problems that the others might be able to solve. IoT devices, for example, can be used to interact directly and autonomously with the environment by gathering data from it. There is a high possibility for them to be exposed to hacking in terms of sending details and confidential data. AI is all about autonomous comprehension and prompt decisiveness based on models and prior training done using large datasets. AI, however, needs access to reliable data, ideally in methods that protect privacy, to be helpful and secure. Blockchain can be useful in this situation. A native "trust" component, which is sorely lacking in the digital sphere, can be added by a technology created to enable consensus on data among large groups. Blockchain can offer reliable, auditable records that big communities can utilize as the foundation for a single, accepted version of the truth since it is a decentralized data storage based on consensus and an immutable ledger. Blockchain-based smart contracts may be used to automate transactions and procedures, ensuring the validity of contracts and the accuracy of instructions. With the help of blockchain, efficient and direct links can be developed to bridge the gap between the physical and virtual world through unique and profound strategies. The network architecture of smart cities and many other areas are being revolutionized by the convergence of artificial intelligence (AI), blockchain technology, and the internet of things to create sustainable ecosystems. However, when it comes to achieving the objectives of developing sustainable smart cities with sophisticated features, these technological advancements present both opportunities and challenges. In the first section, we discuss the basic idea surrounding the concept of blockchain and the advantages of using such a setup. In the second section, we discuss how IoT first came to existence and how we can improve upon such a potential setup. In the third section, we discuss how these concepts can potentially converge together and work in tandem to bring out efficient results by combining different applications of each set-up. IoV (Internet of Vehicles) is such an interesting concept, so in the fourth section this concept is briefly introduced and we also discuss how IoV works together with the concept of blockchain. In the fifth section, we discuss how IoT when combined with CPS (Cyber-Physical Systems) is useful in detecting physical parameters by sensing them. These three innovations complement one another by design and when combined, they can reach their full potential. The consolidation of these innovations can be promising alternatives for managing data and the mechanization of enterprise applications will be analyzed and discussed in the next few sections as mentioned above.

15.2 Blockchain

Blockchain is the term used in IT to describe decentralized lists of records (blocks) where data exchanges are recorded in a distinct, impenetrable, unchangeable, and transparent way. Individual blocks that are connected to one another via cryptography each contain a record of a single transaction. Blockchain systems, in general, are dispersed among all collaborating computers rather than using centralized servers [1]. This organization aims to ensure that everyone who is participating can use the system equally. Users have more entitlements than any organization, state, or authority. If only one party is successful in controlling over 50% of all connected systems in the network, tampering is theoretically possible. The blockchain network internally logs and verifies each transaction. Blockchain is a setup for recording information in such a way that it is modifying the setup unauthorizedly. As the name suggests, each block in the chain contains information along with those of the preceding block. Let's say a person working in a software company receives X as salary and after each passing month his salary is incremented by 1%. Then, let us say in the first month his salary, along with other details like PF, are also calculated and for every month these details along with the incremented salary are also calculated. Now, if I want to calculate the gross salary after 8 months then this setup is really helpful because every "block" of information is interconnected. This is also very helpful in terms of security because generally this information in the blocks can't be changed. So, if the information is compromised in any block, then it is safe to say that the set-up has been tampered with [2]. One of the important parts is that the data being managed is decentralized and there is little to no risk of centralized database failure. There are several types of blockchain, specifically public blockchain, private blockchain, and consortium blockchain.

In a public blockchain, everybody can transfer data among the network and the operations taking place are completely transparent, thus posing the risk of easily getting compromised and there is no chance of the central setup failing. A client program or wallet in the context of cryptocurrencies like bitcoin, is necessary to actively engage in a blockchain network [3]. The application employs asymmetric encryption, which is based on a private key and a public key, with the latter acting as the destination for transactions. The private key is kept private and is used to sign transactions in order to validate their validity. Hardware wallets can potentially be utilized for blockchains as an alternative to software-based wallets. These gadgets are specifically made to handle users' private keys in the safest manner possible. There is no direct internet connection, which minimizes a crucial attack vector, unlike software wallets on cellphones or PCs. In a private blockchain, the transactions taking place are not transparent for public view except for the authorized personnel. This is usually used by corporates for communication among investors [4]. The third type is consortium blockchain, which is actually a mix of the public blockchain and private blockchain, and this setup is used by private firms among its employees.

In comparison to centralized systems, blockchain solutions provide a variety of benefits. Only the physical architecture can guarantee such a high level of dependability. The administration of the data is distributed in a blockchain network, as opposed to conventional databases which are run on central servers that clients use to query and process the data. The data is accessible across all instances even if there is a technical issue at one node [5]. In contrast, depending on the level of redundancy, server issues in centralized designs typically result in restricted availability. The blockchain uses technology to verify and record transactions. As a result, all parties involved in the network may transparently track and record data and value transactions in a tamper-proof way. Some other advantages of this blockchain setup are that there are self-executed commands that can be stored and utilized at any time necessary. It also reduces the processing time for these transactions. By extension, there are also some serious disadvantages like massive transactions can't take place because of tedious verification that needs to be undergone to successfully process the transaction. Data malleability is also a serious issue and this setup cannot be changed abruptly, it needs to undergo transformation gradually. The most well-known use of blockchain is the cryptocurrency Bitcoin, whose price swings frequently. Bitcoin and the majority of certain other blockchain tokens are inadequate for typical payment transactions due to their high volatility. Much more, for investors who want to take risks, cryptocurrencies are now used as illiquid assets. Blockchain is also used to create the electronic corona vaccination certificate. When checking into hotels or attending events, for example, people will be able to electronically communicate their immunization status. The personal information, including name, address, and ID number, as well as the vaccination data, are encrypted and held across five different blockchains. However, only the related QR code is used to make a comparison pertinent to citizens. The code may be printed out or entered into a smartphone software wallet. In addition to cryptocurrencies, blockchain is now being utilized in logistics for the electronic surveillance of production processes and the execution of customs and tax requirements via smart contracts.

15.3 Internet of Things

Internet of Things (IoT) focuses on the concepts of interconnecting computational devices, systems, and machines which facilitate easy exchange of data and information throughout the network. The term "thing" in IoT includes physical as well as imaginary substances which can acquire an IP address and are capable of transmitting data [6]. The concept of IoT was an enigmatic change where at one point we were using desktops and mobiles for a well-defined network of interconnected devices so that data is transferred among the interconnected devices. The Internet of Things is the collective term for the billions of physical objects that are now linked to the internet and actively collecting and exchanging data [7]. IoT can practically consist of any component as long as it adheres to the development of computer chips, networks, etc. The interconnection of all these items together and combining them with sensors and other intelligence models is what gives IoT so much significance. It paves the way for a smarter society with automation as the ultimate goal. The IoT devices can control and sense the power flow and certain mobility [8]. The ability to sense is done via sensors and control of the flow is done through the help of actuators. The sensors monitor the required parameter based on the nature of the sensor. Sensors or other equipment first gather information about their surroundings. It may be as straightforward as a temperature readout or as intricate as a whole video feed. The term "sensors/devices" is used since a single sensor can be combined with several others or a sensor might be a component of a device that performs other functions. Your phone, for instance, contains a camera, an accelerometer, a GPS, and other sensors, but it is not simply a sensor [9]. But something is gathering data from the environment in this first stage, whether it's a solitary sensor or a whole gadget. The actuators take an energy input and convert it into a physical movement which is used to control the flow. Despite these, IoT devices are highly vulnerable to external factors.

Integrated to an Internet of Things infrastructure, which combines data from many devices and uses analytics to share the most useful information with apps created to answer particular requirements, are gadgets and objects having built-in sensors. These robust IoT solutions can precisely identify which information is helpful and which may be safely disregarded [10]. This data may be used to identify trends, provide suggestions, and identify potential issues before they arise. The information is then utilized in some way by the end user. This may be done by sending the user a warning (email, text, notification, etc.). For instance, a text message alarm when the cold storage facility's temperature rises too high. Additionally, a user may have access to an interface that enables them to routinely monitor the system. For instance, a user may wish to utilize a phone app or a web browser to view the video feeds in their home. It's not necessarily a one-way street though. The user could also be able to take action and influence the system depending on the IoT application. By using an app on their phone, the user might, for instance, remotely change the temperature in the cold storage [11].

The user can make informed judgments as to which elements to stock up on based on current information thanks to linked devices which help save time and money. The ability to improve procedures comes with the insight sophisticated analytics provide [12]. While on the topic of the use of sensors and actuators, we can also discuss IIoT (Industrial Internet of Things). It is mainly based on the efficiency of using these sensors and actuators in industrial processes and is also known as "Industry 4.0". The main theme of this feature is that these sensors and actuators are set up in an efficient way where the data is captured and analyzed in a real-time scenario, thus making the communication among the devices easier and more accurate. These set-ups have great quality control and sustainability. The advantage of using such a set-up is that there is easy maintenance and they can also predict those maintenance sessions which financially benefit the manufacturing firm [13]. It is similar in aspects to IoT devices, for example in their vulnerability to external factors.

Thus, we can say that the biggest risk is the security of the device set-up, but a distinct difference between IoT setup and IIoT setup is that IoT applications interconnect among devices and IIoT is interconnected among machines and is quite efficient.

15.4 Convergence of Blockchain, Internet of Things, and Artificial Intelligence

Organizations may be able to harness the advantages of each of these technologies while avoiding the dangers and limits attached to them thanks to the convergence of blockchain, IoT, and AI. IoT networks have several vulnerabilities since there are so many linked devices on them, making the network vulnerable to hacker assaults, fraud, and data theft. AI that is supported by machine learning can proactively protect against malware and hacker assaults to avert security vulnerabilities [14]. Blockchain can further improve network and data security by preventing unauthorized access to and change of network data. The IoT network can perform more effectively by becoming smarter and more autonomous thanks to AI. When blockchain was introduced, the only application it was used for was Bitcoin as a means of controlling the flow of cryptocurrency, but these three entities can be used together to primarily control data management. Additionally, blockchain technology may be used to verify the legitimacy of IoT network users and can boost confidence by controlling the identification of IoT devices. Keep in mind that identity management may be used to describe both people and organizations, as well as IoT units and devices [15]. Blockchain-based identities make sure that each transaction party receives a digital identity based on their physical identity on the blockchain (e.g., identity card for individuals, commercial register entry for companies) [16]. Based on such an identity, exchanges between a person and a business (for instance, car sharing), but also between a person and a machine (for instance, passenger transportation of an autonomous car) or between two systems (for instance, the autonomous car pays for parking) can be processed effectively, that is, with quick payment speed and low transaction costs. Additionally, the crucial duty of giving verified identities to the network's real-world participants may be accomplished by combining blockchain with IoT. IoT devices, for instance, are frequently utilized to give identifying information for physical world items. Personal identities are provided by fingerprint and iris scanners, face recognition technology, and similar tools and item identities are provided by GPS trackers, embedded sensors, QR codes, etc. Blockchains can be used to ensure a tamper-proof link between identifications and their source after they have been produced. One application that easily comes at the top of the mind is that blockchain can be used in conjunction with AI to improve the speed of transactions. Let's take data management. Suppose that an automated electric system in a 1-BHK upon sensing a certain amount of heat signature switches the lights on and once the absence of a heat signature is detected, the lights get switched off. Now in the case of a 1-BHK house, the amount of data collected through the sensors is easily operatable but applying the same scenario in the case of a private firm, there is a lot of data to be collected from the sensors analyzing them to act accordingly. Now the data collected is stored on a centralized server it is considered unstandardized data, but by using blockchain technology we can make the data easily accessible whenever necessary. The data can be generally stored in on-chain or off-chain mode. If the data is to be stored in on-chain mode, the data can be restored at any point in time. Let's take a school collecting the fees for about 1000 students. Now, each transaction is stored in a block and there is a large number of data transmitted in this network, thus the processing time to analyze the data also slows down. This is generally referred to as "blockchain bloat", but in the case of off-chain storage, the data stored is easily scalable, but its setup becomes similar to a public blockchain system because the data becomes transparent. Another advantage of this convergence is that data privacy is vastly improved because IoT devices capture and analyze a large amount of unencrypted data. Also, added to the main advantage of a blockchain system not being able to be easily compromised by hacking, it can be easily detected. Sometimes the use of blockchain can be used by dangerous people to rotate the transactions and in that case an AI can be deployed to analyze the details of the data transmitted in such a setup. Figure 15.1 shows about the terms of pixels, Figure 15.3 displays about the change in terms of pixels and Figure 15.4 shows about the further change and updates in terms of pixels.

Image processing in machine learning (one of the subsets of AI) is a straightforward process. Let us take the number "2".

Figure 15.1 Sample Number – 1st

Figure 15.2 Sample Number – 2nd.

Figure 15.3 Sample Number – 3rd.

Figure 15.4 Sample Number – 4th.

In the above figure, we have taken a sample of the number "2." Now if I were to scan a set of numbers to find 2, we can't exactly find an entity that is the same as "2", there will be differences. So, let us take Figure 15.2, which represents a part of the upper half of "2." This is measured in terms of pixels and now there are a lot of possibilities for Figure 15.2. For example, let us say we have this number in a *Calibri* font; if I were to use a different font where the number "2" is represented differently, then the algorithm chooses the one with less bias. Simply put, when overlapped, the number of pixels left are compared to the number of pixels it covered in "2". This same way of image processing can be used for security purposes in conjunction with blockchain technology, thus it ensures a fast transaction speed and low costs. Figure 15.3 displays about the change in terms of pixels and Figure 15.4 shows about the further change and updates in terms of pixels.

IoRT (Internet of Robotic Things) is an interesting concept in which existing IoT devices are equipped with sensors and it is really useful because the concept of IoT revolves around managing and improving these devices. But, when combined with the concept of robotics there are a lot of uses. In a multi-use set-up there are lot of security measures that can be taken and here, the blockchain concept is really helpful because of its highly secure set-up. Electrical Grids generally consist of transmission lines, transformers, etc. On that same note, smart grids are based around the concept of electrical grids, except signals are transmitted among two sources. The smart grid is a necessary part of an IoT system which can be tracked and used to control all electrical devices within its range. For example, imagine a traffic system placed between four crossroads, where it manages to give the output on which side vehicles should move and when to stop so pedestrians can cross. Now, the concept of blockchain has introduced the transactions of smart contracts that can take place without the risk of any data leaking, as the blockchain by default stores the data in every block so it is highly secure. We can also combine this concept with cloud computing, as already the smart grids are preferred over the normal ones as these are highly resourceful. So, when they work together with cloud technologies almost every blind spot in the smart grid set-up is covered, including parallel computing of processes, resource sharing, and so on.

15.5 Block Chain for Vehicular IoT

As living standards rise and the number of automobiles rises, traffic congestion gets worse and worse. The field of intelligent transportation has advanced significantly with the development of 5th Generation, Bluetooth, and sensor technology, cloud computing, big data, and other new technologies. The onboard unit of a vehicle in the Internet of Vehicles system perceives the surrounding road condition's environmental information and the vehicle communication module communicates with roadside units and other vehicles on the network to provide the vehicle owner with road condition information and navigation information while lowering traffic risks. The Internet of Vehicles system confronts significant security issues since it is a complicated system made up of vehicles, people, and other network infrastructure and it interacts over the wireless network. Therefore, it is vital to find a solution to the challenge of protecting user privacy while offering services. The electronic systems used in vehicles have undergone a major change as the standards of technology development and on that note, we can introduce something called VANET (Vehicular Ad hoc Network). They are class of inter-vehicle network where the vehicles connect to the network to obtain details on real-time data like traffic and weather reports and by obtaining these data, we can analyze it and if possible give the best route for a vehicle to reach the required destination. There is another type of network called an intra-vehicle network [17]. Due to the development of automotive electronics, many components like sensors are added and the process of capturing data and analyzing it becomes more and more complex, making the maintenance and manufacturing cost substantially higher. Thus, we can introduce an intra-vehicle network to easily capture and analyze in real-time, thus reducing ECU(s). By establishing a wireless connection we ensure that the unit of components doesn't weigh much as they generally would. When sending a signal from let's say point A to point B, the signal is pretty much straightforward. But, if we introduce this setup into a scenario where there are many more components, then the path taken to reach the destination may take some time and that is called fading. The receiver is bound to receive higher interference when placed near an EM wave source as it is completely blocked. This causes a large-scale fading. Some potential solutions are introducing a sparse matrix as an algorithm (A sparse matrix that has more entries as zero than no of non-zero entities) or converging with AI. It can "adapt" to the situation by reducing the time for service to disconnect. This is also a very complex environment. Applying this concept in our daily transportation improves traffic management and also increases traffic safety. We can also apply this concept in public parking facilities, but with all these advantages there are also problems to overcome to make sure that with proper communication the desired components achieve their function by coordinating efficiently [18]. There are many suggestions to improve this set-up like setting up a collision avoidance system along with self-driving vehicles for transportation in short distances and also the humongous amount of data captured altogether helps in researching in many other fields of similar applications. In IoV, we use a cloud unit called RSU (Roadside Unit) as the main part of this vehicle network. Now there are several ways to ensure security in this connection setup. One simple thing that we can do is use a verification process to identify and allow the network to connect among themselves. RSU issues a verification ID which is used for further communication along the network, but one issue we face is constant maintenance of this set-up. Just like in cryptography keys, for small-scale maintenance of this vehicle network we can use private key pairs, but the issue is we can use them only on a small scale. As the list of the networks increases, the entire security of the network becomes tedious.

15.6 Convergence of IoT with Cyber-Physical Systems

Cyber-physical systems use IoT as their foundational or supporting technology. Cyber-physical systems are the IoT's progress in terms of full conception and perception and they have a significant capacity for physical world control. Traditional embedded and control systems are also a part of cyber-physical systems, which have evolved into cutting-edge techniques [19]. While IoT demonstrates the interconnectedness of many end devices that communicate with one other through the internet, CPS is the integration of physical and cyber systems in which embedded computing devices operate as physical entities, while communication, calculation, and control are cyber entities. IoT aims to provide decentralized control among the linked endpoints. Some of the functions shared by CPS and IoT include sensing, processing, storing, and networking. To comprehend their distinct roles in the system where they are utilized, it is necessary to identify, become familiar with, and compare some of the similarities between CPS and IoT that occurred when CPS was introduced alongside IoT. The system is now safe, secure, and sustainable, which are the core characteristics of the CPS thanks to new communication and control mechanisms. For dependable transmission and information processing, IoT links information collecting devices including sensors, RFID (Radio Frequency Identification) wireless sensor networks, and cloud computing technologies [20]. In contrast, CPS is a control technique that combines computation, communication, and IoT control. It is scalable and dependable. Wireless Sensor Networks (WSN) are an interesting concept. They are a sensor network that used to measure conditions of the environment by physically measuring the necessary parameters and outputting the conditions. This is a perfect example of cyber-physical systems and basically from this, we can infer that it is a set of computing networks that coordinate perfectly to give the desired outputs. They are physically engineered systems whose operations are coordinated properly by the process of computation and components (sensors and actuators) are interconnected like transducers. One more thing is that we are ensured that these computations for a particular event are coordinated to run along in real-time with the physical world so that they can monitor physical entities. The working of IoT is that it interconnects all physical devices and operates from a networking point of view. But, in the case of Cyber-Physical Systems (CPS), it monitors physical entities and captures the observations and focuses on data exchange. It is like a set of rules to be followed almost like a close-end setup. The Internet of Things and CPS both have the same architecture, which makes them comparable, however, CPS places a larger priority on separating physical from computational parts. CPSs are explicitly portrayed as a framework of interacting elements with physical input and output rather than independent devices, in contrast to standard embedded systems [21]. Healthcare and industrial automation, control technologies, distributed energy systems, aviation control, and other uses for CPS are among them. Engineering's physical systems will alter thanks to CPS, which will also improve economic wellbeing. When CPS is coordinated with the intricacies of IoT, the result is a very efficient CPS that has high adaptability, scalability, and increased security that has a range of applications in economic sectors. It also enables the feature of modularity that employs some creative uses of this in the industries. Modern-day vehicles have driver-assist functionalities that are based on AI systems. AI could be used to manage traffic and in turn, reduce it, thus also economically and environmentally helping by reducing energy emissions [22]. The combination of Environmental IoT and cyber-physical clouds produces a complex distributed system that aids in managing and understanding interdependencies in the natural world. Abstraction representation, Network-centric representation, and Node-centric representation are implemented in a three-stage approach. By enhancing one another, these technologies offer greater opportunities in the application domain. Numerous tools for use in medical applications are made possible by the integration of wireless devices, wearables that are quick and have a functional real-time algorithm, and cyber-physical systems. While using the idea of blockchain is insightful, cloud technologies are immensely advantageous when it comes to scalability and usability.

The terms cyber-extortion [23, 24], man-in-the-middle, brute-force [25], spoofing, insider, or malware threats are often used to characterise next-generation intrusions. Each one brings a distinct variety of challenges and calls for equally potent next-generation cyber safeguards [26]. Since they can penetrate and spread fast and covertly, 5th generation threats pose a greater threat than attacks [27] from previous generations.

15.7 Conclusion

IoT, blockchain, and AI are three technologies that may be coupled in many ways. We contend that as business models, goods, and services stand to gain from the fusion of these technologies, there will be a convergence of these advances. Any autonomous agents, including sensors, vehicles, machinery, trucks, cameras, and other Internet of Things (IoT) devices, might be used with such business models. These agents would be able to send and receive money on their own, as well as use AI and data analytics to act as independent economic actors.

Such business model development and the digital transformation of industrial enterprises will be fueled by convergence. To achieve significant efficiency improvements, executives need to engage with these technologies. A new era of digitalization will be ushered in by blockchain technology, IoT, and AI. The interoperability, security, and standardization of the aforementioned technical sectors must be made feasible by considering this new convergence because there is a connection between data and the electronic systems that gather, classify, store, examine, and safeguard it.

These convergent technologies must be viewed as a whole since they reinforce one another. A technology's full potential for use in commercial applications cannot be explored by concentrating just on its technological developments. As they develop and grow more complicated and as firms embrace additional disruptive technologies, their interactions will multiply exponentially. That's how all major technology areas come together in the digital era, inside the structure of a "data-centric" strategy, making it feasible to establish. But, as we've seen, the intellectual, operational, and practical connections between Data, AI, Blockchain, Cloud, IoT, and cybersecurity.

References

1. W. Yang, E. Aghasian, S. Garg, D. Herbert, L. Disiuta, and B. Kang, "A survey on blockchain-based internet service architecture: Requirements, challenges, trends, and future," IEEE Access, vol. 7, pp. 75 845–75 872, 2019.
2. D. Tapscott and A. Tapscott, "Realizing the potential of blockchain: A multi stakeholder approach to the stewardship of blockchain and cryptocurrencies," http://www3.weforum.org/docs/WEF Realizing Potential Blockchain.pdf, (Accessed on 20-May-2020).
3. A. Deshmukh, N. Sreenath, A. K. Tyagi and U. V. Eswara Abhichandan, "Blockchain Enabled Cyber Security: A Comprehensive Survey," 2022 International Conference on Computer Communication and Informatics (ICCCI), 2022, pp. 1-6, doi: 10.1109/ICCCI54379.2022.9740843.
4. N. Lasla, M. Younis, W. Znaidi, and D. B. Arbia, "Efficient distributed admission and revocation using blockchain for cooperative ITS," in 2018 9th IFIP International Conference on New Technologies, Mobility and Security (NTMS), Paris, France, February 2018.
5. Shi, S., He, D., Li, L., Kumar, N., Khan, M. K., & Choo, K.-K. R. (2020). Applications of Blockchain in Ensuring the Security and Privacy of Electronic Health Record Systems: A Survey. Computers & Security, 101966. doi:10.1016/j.cose.2020.101966
6. Esposito, C., De Santis, A., Tortora, G., Chang, H., and Choo, K. K. R., Blockchain: a panacea for healthcare cloud-based data security and privacy?. IEEE Cloud Computing 5(1):31–37, 2018.
7. Mishra S., Tyagi A.K. (2022) The Role of Machine Learning Techniques in Internet of Things-Based Cloud Applications. In: Pal S., De D., Buyya R. (eds.) Artificial Intelligence-Based Internet of Things Systems. Internet of Things (Technology, Communications and Computing). Springer, Cham. https://doi.org/10.1007/978-3-030-87059-1_4

8. Tyagi, Amit Kumar; Nair, Meghna Manoj "Internet of Everything (IoE) and Internet of Things (IoTs): Threat Analyses, Possible Opportunities for Future, Journal of Information Assurance & Security (JIAS), Vol. 15 Issue 4, 2020.
9. A. Deshmukh, N. Sreenath, A. K. Tyagi and S. Jathar, "Internet of Things Based Smart Environment: Threat Analysis, Open Issues, and a Way Forward to Future," 2022 International Conference on Computer Communication and Informatics (ICCCI), 2022, pp. 1-6, doi: 10.1109/ICCCI54379.2022.9740741.
10. M. H. Miraz, M. Ali, P. S. Excell, and R. Picking, "A Review on Internet of Things (IoT), Internet of Everything (IoE) and Internet of Nano Things (IoNT)", in 2015 Internet Technologies and Applications (ITA), pp. 219–224, Sep. 2015, DOI: 10.1109/ITechA.2015.7317398.
11. M. S. Ali, M. Vecchio, M. Pincheira, K. Dolui, F. Antonelli, and M. H. Rehmani, "Applications of blockchains in the internet of things: A comprehensive survey," IEEE Communications Surveys & Tutorials, 2018.
12. K. K. Patel, S. M. Patel, *et al.*, "Internet of things IOT: definition, characteristics, architecture, enabling technologies, application future challenges," International journal of engineering science and computing, vol. 6, no. 5, pp. 6122–6131, 2016.
13. Mano, Y., Faical B. S., Nakamura L., Gomes, P. G. Libralon, R. Meneguete, G. Filho, G. Giancristofaro, G. Pessin, B. Krishnamachari, and Jo Ueyama. 2015. Exploiting IoT technologies for enhancing Health Smart Homes through patient identification and emotion recognition. Computer Communications, 89.90, (178-190). DOI: 10.1016/j.comcom.2016.03.010.
14. Tyagi A.K., Fernandez T.F., Mishra S., Kumari S. (2021) Intelligent Automation Systems at the Core of Industry 4.0. In: Abraham A., Piuri V., Gandhi N., Siarry P., Kaklauskas A., Madureira A. (eds) Intelligent Systems Design and Applications. ISDA 2020. Advances in Intelligent Systems and Computing, vol 1351. Springer, Cham. https://doi.org/10.1007/978-3-030-71187-0_1
15. Y. Rahulamathavan, R. C. Phan, M. Rajarajan, S. Misra and A. Kondoz, "Privacy-preserving blockchain based IoT ecosystem using attribute-based encryption," *2017 IEEE International Conference on Advanced Networks and Telecommunications Systems (ANTS)*, 2017, pp. 1-6, doi: 10.1109/ANTS.2017.8384164.
16. Ahmed, Ejaz & Yaqoob, Ibrar & Gani, Abdullah & Imran, Muhammad & Guizani, Mohsen. (2016). Internet of Things based Smart Environments: State-of-the-art, Taxonomy, and Open Research Challenges. IEEE Wireless Communications. 23. 10.1109/MWC.2016.7721736.
17. Ahram, T., Sargolzaei, A., Sargolzaei, S., Daniels, J., Amaba, B., 2017. Blockchain technology innovations. In: 2017 IEEE Technology & Engineering Management Conference (TEMSCON). IEEE, pp. 137–141.
18. H. J. D. Lopez, M. Siller, and I. Huerta, "Internet of vehicles: cloud and fog computing approaches," in 2017 IEEE International Conference on Service Operations and Logistics, and Informatics, SOLI, Bari, Italy, 2017.

19. M. Priyan and G. U. Devi, "A survey on Internet of vehicles: applications, technologies, challenges and opportunities," International Journal of Advanced Intelligence Paradigms, vol. 12, no. 1/2, pp. 98–119, 2019.
20. M. Kalverkamp and C. Gorldt, "IoT service development via adaptive interfaces: Improving utilization of cyber-physical systems by competence-based user interfaces," in 2014 International Conference on Engineering, Technology and Innovation (ICE), 2014, pp. 1-8.
21. A. Ahmadi, C. Cherifi, V. Cheutet, and Y. Ouzrout, "A review of CPS 5 components architecture for manufacturing based on standards," in 2017 11th International Conference on Software, Knowledge, Information Management and Applications (SKIMA), 2017, pp. 1-6.
22. A. Ahmadi, A. H. Sodhro, C. Cherifi, V. Cheutet, and Y. Ouzrout, "Evolution of 3C cyber-physical systems architecture for industry 4.0," in International Workshop on Service Orientation in Holonic and MultiAgent Manufacturing, 2018, pp. 448-459.
23. Rawat, R., Mahor, V., Álvarez, J. D., & Ch, F. (2023). Cognitive Systems for Dark Web Cyber Delinquent Association Malignant Data Crawling: A Review. *Handbook of Research on War Policies, Strategies, and Cyber Wars*, 45-63.
24. Rawat, R., Chakrawarti, R. K., Vyas, P., Gonzáles, J. L. A., Sikarwar, R., & Bhardwaj, R. (2023). Intelligent Fog Computing Surveillance System for Crime and Vulnerability Identification and Tracing. *International Journal of Information Security and Privacy (IJISP)*, 17(1), 1-25.
25. Rawat, R., Sowjanya, A. M., Patel, S. I., Jaiswal, V., Khan, I., & Balaram, A. (Eds.) (2022). Using Machine Intelligence: *Autonomous Vehicles Volume 1*. John Wiley & Sons.
26. Rawat, R., Bhardwaj, P., Kaur, U., Telang, S., Chouhan, M., & Sankaran, K. S. (2023). *Smart Vehicles for Communication, Volume 2*. John Wiley & Sons.
27. Mahor, V., Bijrothiya, S., Rawat, R., Kumar, A., Garg, B., & Pachlasiya, K. (2023). IoT and Artificial Intelligence Techniques for Public Safety and Security. *Smart Urban Computing Applications*, 111.

16

Robots, Cyborgs, and Modern Society: Future of Society 5.0

Shashank Sharma[1], Sanjana Chelat Menon[2] and Amit Kumar Tyagi[1,3]*

[1]School of Computer Science and Engineering, Vellore Institute of Technology, Chennai, Tamil Nadu, India
[2]Department of Electrical and Electronics, PSG Institute of Technology and Applied Research, Coimbatore, Tamil Nadu, India
[3]Department of Fashion Technology, National Institute of Fashion Technology, New Delhi, Delhi, India

Abstract
Well-known scientists and technologists have expressed concerns that robots (a part of Robotics) may rule the world because of significant growth in the field of Artificial Intelligence. Concerns like robots having the potential to take over human jobs and leave billions of people unemployed don't stop human beings from gaining in-depth knowledge of bodily technologies and becoming cyborgs that have superior capabilities to Robots. Types of cyborgs include human beings with mass Imagineered body hacks, human beings with mass-produced biomedical implants, and human beings with mass-customized Insideables. The findings of this chapter are related to debates about the future of society. In particular, opportunity versus exploitation, utopia versus dystopia, and emancipation versus extinction. Furthermore, the chapter throws light on the futuristic aspects of robots and the possible debates between robots and cyborgs. Indeed, the future must also focus on the extensive potential and efficiency of cyborgs and robots which are likely to replace humans.

***Keywords*:** Body hacking, cyborgs, mass imagineering, robots, sensors, technology domestication

**Corresponding author*: amitkrtyagi025@gmail.com

Romil Rawat, Rajesh Kumar Chakrawarti, Sanjaya Kumar Sarangi, Rahul Choudhary, Anand Singh Gadwal and Vivek Bhardwaj (eds.) Robotic Process Automation, (243–258) © 2023 Scrivener Publishing LLC

16.1 Introduction

For tens of thousands of years, humans have ruled over our world as our planet's only intelligent and self-aware species, but the rise of artificial intelligence like cyborgs means that could change soon and homo sapiens may be eradicated from this world. In this chapter, we will shed some differences on human intelligence vs artificial intelligence and their future together. We will then explore the cyborgs or cybernetic humans and then what it feels like to be living with species that succeed in our intelligence.

It's highly likely that the future could be a witness to humans being persistently in contact with computers and processors being stored within their nervous systems as put forth in the Borg society of Star Trek: New Generation. This could be considered an alienation form of human society. At last, we will draw implications discussed concerning debates about the future of society concerning opportunity versus exploitation, utopia versus dystopia, and emancipation versus extermination [1]. Rapid revolutionary processes have led to the key solution being narrowed down to experimental penalties caused by the reality of cyborgs and robots [2].

16.2 Comparing Humans, Cyborgs, and Robots

When discussing cyborgs or cybernetic humans, we find them fascinating as well as strange, as we know they are different from us yet so similar. Some researchers are trying to understand these feelings and this research is important as cyborgs and robots are going to be as involved in our futures as computers are in our present.

We know that humans are the product of the evolutionary process. The human being is a biological creature that has reached a degree of complexity through natural selection. As we know that humans evolved from animals through biological evolution, we can say the same thing for cyborgs and robots, as they evolved from us (Humans) through technological evolution [3]. Today we can easily imagine the human mind uploading and downloading into a machine where we can live forever and can become immortals, as it was mentioned by Sheldon Cooper, a TV show character (who was a genius by the way) in an episode of "Big Bang Theory". Cyborgs are descended from the idea of cybernetics, which is defined as the fusion of technology and living things. Cyborg refers to an organism that is partially mechanical and partially human. Cyborgation is the term used to describe the entire cyborgization process. For an organism to be referred to as a cyborg, it often needs to possess abilities that go beyond those displayed by either its biological or technical aspects alone. This may be the result of literary works that depict cyborgs as having significantly superior mental and/or physical capacities to humans. A cyborg is a literal robotic firm that assists people in doing daily duties. They can be utilised for military, civilian, or medicinal objectives. Service robots help people by taking care of chores like housekeeping, laundering, and even cooking. Although service robots are built to follow human directions, such instructions must be specific for the robot to complete the task. Robots come in several different forms [4]. A robot, though, is a computer. A robot is a vacuum. Robots either perform tasks for humans or make them easier for us to complete. Cyborgs and robots appear to be science fiction staples and, in some ways, they are. But, the majority of people are unaware that robots and cyborgs exist, albeit not in the same form as they are portrayed in popular culture. The existence of life distinguishes cyborgs from robots as their main distinction. A robot is essentially a highly developed machine. It frequently involves automation and hardly any connection with people. Cyborgs, in contrast, are a hybrid of a machine and living beings. It doesn't always have to be a person; it may be a dog, a bird, or anything else that lives. A cyborg is different from a robot since it has a life component. This suggests that in contrast to robots, cyborgs are alive. Even though certain robots can mimic some elements of biological things, they can never fully be alive. A robot can only carry out pre-programmed tasks, whereas cyborgs, particularly human cyborgs, have free will over their actions.

There are some excellent instances of robots. Robots that do monotonous duties in manufacturing are one of them. Because they are extremely swift and never grow weary, these robots are superior to humans at certain tasks. Even though we might not conceive of cyborgs as people, they do indeed exist. In films like Gattaca or Brave New World, part of a person's social organization and conceivable outcomes needs to do straightforwardly with their genetic material, which is subjected to an engineering technique designed to serve specific purposes, make the individual more skilled, and have complex mental capacity to exercise or, more appropriately, to obey tasks. The globalization of technology and the internet makes us believe that cyborgs are going to be there in a foreseeable future and humans are making robots as their helpers or we can say, servants. Humans or human-cyborgs that cohabit the same space as that of smart robots and machines often put out a large scope of research which needs to be explored. Robots, like the industry robots discussed above, can be quite complicated. But, there are also pretty straightforward robots. Young children can access basic robotic kits to pique their curiosity and introduce them to the fundamental concepts of robotics. In contrast, a cyborg's machine component is frequently quite sophisticated since it interacts with the organic component to function. Some prosthetic arms can perform almost all of the functions of a natural limb and some users can even reach out and grab objects [5]. The cyborg is now no longer a sci-fi word as many humans are now cyborgs because of some techno-scientific criteria. The era where human cyborgs and intelligent robots exist is not that far off.

16.3 Some Philosophical Aspects

The mind is an element that enables humans to be aware of the world and their experiences. The mind emerges from the higher level of brain functioning that enables us to interact with our body and the neighbouring environment. Neurons cannot handle these things alone. Neurons cannot fully represent the interaction of human beings with the environment or their surroundings, which includes desires and emotions and all things that make us humans as they are deficit in terms of the purpose and intention towards world affairs [6]. We want to point out that when we philosophically or scientifically analyse the relationship between the mind and its support (its vat) in cyborgs and robots, what we are doing is a task of "naturalizing" a phenomenological account of the mind besides and beyond the reductive concept of thinking as an exclusive human process. We can say that the human brain is a non-biological or physical vat that we can see in cyborgs and robots in the form of computer chips, but the mind is a biological vat that only we humans or animals have possessed until now. If mind and body are one, then we can consider that it is the part of the biological structure that we gain from this never-ending loop of evolution. We often forget the fact that mind and body are the same and think of the mind as a part of our body, as highlighted by Damasio who states that it is the fundamental human-like functions that lead to complex cognitive functions. We can consider that mind can be uploaded or downloaded in cyborgs and robots in the future. When we say humans are different from cyborgs and robots, we state that based on precluding philosophical and scientific differences scientist have researched. However, without this assumption, it is difficult to pose the question incurring in an avoidable circular argumentation [7].

16.4 Reproduction or Replication

Another interesting point is that researchers do not consider is philosophical and scientific problems associated with the reproductive aspects.

Haraway (1991) presented the cyborg metaphor as a way to overcome sex limitations. Human beings reproduce through biological processes, which means natural selection can act on human beings which is good for adaptation aspects. On the other hand, cyborgs were made by installing things after their birth and not all cyborgs will react biologically in the same way. Some of them may use their biological functions more efficiently than their artificial transplant functions or vice versa, some may reject the transplant because of some immunological discrepancies or other incompatibilities. Any part of the human body, but particularly the area where the majority of neurons are coupled and transmit and receive electronic impulses, can have a silicon chip implanted. This silicon chip's construction enables it to pick up nerve signals, amplify them, and transform the signal into an electronic medium that enables appropriate computer accessibility. A wireless communication channel is favoured since it eliminates the need for cables to connect the cyborg and computer [8]. A glass tube houses this implant. One houses the power source, which is a copper coil that is energised in response to signals from the cyborg to create an electric current using radio waves. Three miniature PCBs will transmit and receive signals at the other end. The prosthesis is joined to the body via a band that encircles the nerve fibres, and it is attached to the glass capsule by a very thin wire. The implant's chips will instantly convey signals that are picked up by the nerve fibres and sent to a computer. For instance, an electronic signal from the brain activates the muscles and tendons that control the hand when a finger is moved. The finger will continue to receive these nerve impulses. Since the implant's signal will be analogue, it must be converted to digital before being stored in the computer. However, several studies on human-machine dyads, networks, computerised information systems (CIS), organisational culture, and work show a desire to investigate [9]. Like this, many other methods could be employed to alter the cyborgs. This is a particular point of substrate wherein natural selection may affect reproductivity in terms of capacity and can even be altered. On the contrary, replicating a new robot will need a large amount of money with expensive complex raw materials and manpower.

16.5 Future of our Society

The first reflection underscores the importance of introducing a common code that humans can use when interacting with different types of robots, while the second emphasizes the urgency with which scientists must position themselves to be aware of the constraints that their viewpoint entails. It explains why the debate about the future of society is flawed when it focuses on robots and cyborgs. The latter debate merely looks at the potential of cyborgs (robots) to replace humans. James Lovelock, the famous British environmentalist and futurist, described cyborgs as the self-sufficient, confident descendants of today's robots and artificial intelligence. As we explain below, the general observation that man is becoming a cyborg is not new. Commentators have used the term cyborg to capture the descriptive material of what they view as an unprecedented fusion of man and machine and express their concern about how the body and the brain become places of control and commodification [10]. We are already witnessing the mainstream use of household robots that can handle complex and demanding domestic tasks and the introduction of cyborg technology that integrates machines with humans, enabling us to expand our senses and intelligence. Humans replace robots in some professions and supplement them in many others. It will take centuries before we expect robots to be better at some tasks than humans, but the money credited for these tasks will pay well for robots and their owners. For example, if one prefers a future robotic society with decentralized security mechanisms, one could try to foster such a future by encouraging the development and introduction of these mechanisms today. The larger changes in the organisation of the robotic society seem more plausible because machines differ in relative costs and productivity from humans, making different arrangements more efficient [11]. In future, the world dominated by robots could in principle become a world dominated by humans. Robots are displacing millions of people in various industries. Some experts say that the more robots outperform humans, the more people are forced to keep up. The machine will outperform humans in a certain task, but humans will no longer beat computers, for example, in chess. In factory and warehouse environments, some say it's beneficial to have robots that look like humans. Most famous science fiction robots look like humans, but some of them, even the most dazzling models, are rudimentary in their abilities. Cyborgs are often portrayed as more robotic than organic. Many robot development projects try to develop robots that behave like humans and look like humans. Some articles have raised concerns that sex robots isolate people from each other and give unrealistic expectations of what a human partner should be, but they quickly throw out concerns about loneliness when it comes to people with disabilities.

David Levy, 45, for example, believes humans will marry robots in the near future. Sending humans into space is a dangerous task, but the use of various cyborg technology can reduce risk in the future.

Such technologies already exist with the development of Brain-Machine Interfaces (BMIS) that allow humans to control robotic arms through their thoughts and specially designed exoskeleton suits that enable paralyzed people to walk again (Elon Musk's Neuralink company plans to implant neural electrodes next year into paralyzed people). Researchers in a novel niche called soft robotics are also working on mimicking human movements [12]. In their series of experiments, they demonstrate that humans tend to behave in the same way as computers with other humans. This includes leaving gender stereotypes and prejudices intact, even if the robot receives a female or male voice. The art of reproduction raises new questions about the boundaries between the human and cyborg soul: if we assume the human soul cannot be reproduced, this explains why cyborgs can download new forms and create new individuals.

It seems that our society will emerge as a world where human cyborgs and humanoid robots will coexist and maybe they will end up engaging in war as we see in many Roman films or cyborgs might end up getting their freedom and humans and cyborgs will then live in sperate zones. The concept states that a cyborg's hybrid state, a blend of biological and artificial components, defines a cyborg. In the history of film, cyborgs have been portrayed in a variety of ways, especially in science fiction and horror, and because of their intrinsic ambiguity, they are commonly combined with aliens, mutants, androids, copies, robots, and monsters. Body modification, practice, and culture has spread across the globe and made significant progress in fusing the human body with technology. It is also possible that robots may implant human skin on them to look more human than us like we see in the series of movies *Terminator*. In order to understand and analyse the concept of AI and transforming humans into cyborgs, it is essential to consider the metaphors of the network paradigm and the Dedalus paradigm developed by Garrido (2007) and recovered by Coca and Valero (2010).

As per the previous statements, new interactions between robotic systems and humans will be a common practice in the future, leading to a new society.

16.5.1 Eugenics or Genetic Engineered Humans

Due to the advancement of innovation, today it is feasible to update one's own body totally through various clinical and corrective mediations. One can change the morphology of their face, different real aspects, orientation, or skin tone. It isn't just about beauty care products. Sometimes vital signs are at stake; whether we have the privilege of replacing the malfunctioning organ with a new prosthesis can decide the length, quality, or end of our life [13]. Today, the possibility of cyborgization of the body is a matter of class privilege. There are some ways to evolve humans and make them more perfect than today. The concept of making cyborgs by implanting bio-chips or by biotechnological engineering is one way, but there is an organic way too. Genetics and genomics will definitely pitch in to improve and enhance capabilities just like a complementing equation. The very first proposal of improving humans through genetics was rightly put forward by Galton through Eugenics methodology. Galton and Pearson developed a laboratory, a community, and even a newspaper that handles this topic essentially. Eugenics is a process of conceiving a human by the mating of two different types of genetic people which can lead to a human superior to them. He/she may be immune from diseases and might have intelligence of a whole new level. The main idea of eugenics was spread in the USA in the early 19th century where people of some racial or ethnic background are treated as superior. The worse impact of Eugenic doctrines was in Germany during the Nazi regime [14]. All these historic facts converge the eugenic theory as a racial or perverse theory. However, it is likely that researchers are able to scale up further and propose the GEH (Genetic Engineered Human) and its definition. The improvement in humans caused by the broad-sense genetics engineer is what will be termed as GEH. He/she would be 100% organic and of the human constitution. Thus, evolution by natural selection will be possible in these humans compared to cyborgs who were implanted by technology after their birth. Evolution with the help of cyborgs is way too advanced and expensive for us; we should consider eugenics as our priority option for betterment, but it needs some in-depth debates on GEH and some ethical consideration.

16.5.2 Why Do We Need Cyborgs?

In a heart-warming speech delivered by the most popular billionaire technology expert, Elon Musk said that human burgeoning citizens are at risk of being dwarfed or fooled by artificial intelligence and must develop the ability to communicate with machines. The CEO of Tesla and SpaceX said that humans must merge with machines and become a kind of cyborg. Musk and other technology leaders have long warned that the growth of artificial intelligence is a necessary part of the development of sophisticated cyborg technology. One fear is that AI-driven cyborg technology will eliminate human empathy and reduce morality to a set of algorithms based on logic and statistics. Part human, part robot, cyborgs are all around us, and if humans can complement their bodies with microchips, prostheses, and skull implants, we can unlock the potential of cyborg technologies [15]. Humans have seen more cyborgs than we have in the past, but we still have a long way to go before most of us can use mechanical implants to expand our capabilities.

If we are to remain relevant in a future full of artificial intelligence, humans need to improve their brains, Tesla founder Elon Musk said. Enthusiasts and transhumanists envision the next stage of human evolution as a technological future in which each generation is expected to strengthen bones and improve brain performance thanks to cybernetic improvements. This type of technology raises all sorts of ethical questions about whether we should improve our brains in a way that involves brain chips (which does not sound pleasant) and whether it creates an insurmountable gap between those who can afford to become cyborgs and those who can't. Musk says that as artificial intelligence advances, humans will have to expand their brains with the power of digital technology to prevent them from becoming irrelevant. Nicolelis argues that, contrary to what Musk and singularity advocate Ray Kurzweil said, the brain is unpredictable and that human consciousness is the result of the unpredictable and nonlinear interactions of billions of cells. Humans will not become irrelevant when machines will be able to replicate the human brain, which is not thought possible.

James Lovelock, the famous British environmentalist and futurist, describes cyborgs as self-sufficient, confident descendants of today's robots and artificial intelligence systems. Nowadays, computers process data as if we were independent of artificial intelligence he says, but tomorrow the cyborg will be a million times smarter than us. Nicolelis agrees with Musk that if we interact with machines, we will make quantum leaps in digital infrastructure, but that is far from what we predicted today and humans will retain ultimate control. To answer the question posed in the title of this article, cyborg technology is the next step in human evolution [16]. To understand what the next human cyborg will mean, we need to define what a cyborg is. A cyborg is a human being, a machine, a system of control mechanisms, a human part, a genetically modified drug regulation device, or a being living in a different environment from a normal one. New cyborgs are constantly being born all around us as people push the boundaries of what their bodies can do with the help of technological advancements and expand the definition of what constitutes a human body. Tim Hawkinson promoted the idea that the body is a machine that combines a human trait with technology to create a cyborg [17]. Researchers are working on implanted brain-computer interfaces (BCIs) of all kinds that would enable humans to control electronic devices via the Internet of thought, but one of the biggest hurdles to future cyborgs is the fact that our bodies are not designed to house electronics for long periods. The manipulation of electrical signals in the body is a fascinating step in the direction of communication between man and machine, but implanting the brain itself could change our perception of reality.

Professor Panagiotis Artemiadis of Arizona State University is trying to do more with this bandwidth by using 128 electrodes in an EEG cap to allow people to control flying robots without a human brain swarm. Musk's use of the bandwidth is intriguing because it can be applied to future human brain-machine user experiences. If humans fail to find a way to mitigate the effects of global warming, cyborgs will have to do it. Elon Musk believes that if humans want to avoid becoming house cats for robots, they need to expand to compete with robots. Musk believes the human mind needs to evolve to access information that can be tapped by artificial intelligence. Elon Musk's Neuralink company has expressed ambitions to improve humanity by merging artificial intelligence (AI) with our brains [18]. Musk even took the time to speak about his latest bet, Neuralink, with his plans for brain-computer interfaces that would let humans compete with artificial intelligence.

16.6 Implications

16.6.1 Opportunity vs Exploitation

Human enhancement of bodily technologies introduces debates like: is it subjected to opportunity or exploitation. Body hacking suppliers could have a dangerous scope in the future because there may be exploitations in these companies as workers are getting less paid in comparison to what they generate. Specifically, body hackers often enact the roles of pioneering scientists from early centuries by taking up individual experiments without any financial assistance.

Human enhancement of bodily technologies also introduces a new topic of digital exploitation. Becoming a cyborg exploits privacy as we have body hacks installed in us which can be tracked and they might install some cameras in us that we are not aware of, which could pose serious danger in terms of privacy and security.

In conclusion, human enhancement in bodily technologies introduces both individual opportunities and individual exploitation. So, we need a great knowledge in the depth of these things to make a fortune in this field or we can make a fortune by exploiting them, but those techniques are unethical.

16.6.2 Utopia vs Dystopia

When discussing the social effects of new technology, the issue can be framed in terms of utopia vs. dystopia. The rejection of the present society and the claim that another society is feasible and desirable are all part of utopian rhetoric. Then, utopian activities attempt to reproduce at least some of the characteristics of utopian discourse in the hopes of spreading them across society. Arguments that technological advancements should bring about utopias where human beings transcend biological restrictions and become transhuman or even posthuman are examples of techno-utopia claims.

There is an idea that when cyborgs or robots with artificial intelligence will become so developed that they see humans as inferior beings to them and treat us like animals. As we have seen in the movie series the "Matrix," humans are only harvested for their blood and they were put in constant sleep in the dream of worlds where they are living their false and happy life are examples of techno-dystopia claims [19].

These are realities we will have to face: in future, a cyborg's evolution is going to make our planet into a utopia of opportunities or a dystopia of exploitation.

16.6.3 Emancipation vs Extermination

When concerns relating to the social effects of cyborgs and robots spread across humanity, it can often be reframed to question if technology will bring back emancipation or extermination of humans. For example, arguments for human emancipation are that robots will take over all the human work and pay for their entertainment. Meanwhile, arguments for human extermination are that if robots are instructed to make a football, they might convert all the matter into footballs, including human beings.

In no time, people can enhance the performance of the brain or any other body parts by implanting them. An appreciable case of the brain implant case is that of the brain gate which consists of a sensor being embedded in the brain which is linked to an external decoder that can connect it to other prosthetics or external objects. The sensor takes the form of a multi-electrode array which contains hundreds of extremely thin electrodes that can sense the electromagnetic signatures emitted by the neurons in certain areas of the brain. The sensor comes in handy when it needs to translate the activity into electrically charged signals [20] which are transmitted to an external gadget like a robotic arm, cursor, etc. Thus, the brain gate allows the person to manipulate his surroundings using his brain only. These brain implants can enhance the capabilities of humans themselves, which reduces the need for robots in the future [21].

Hence, those who consider human emancipation or extermination due to robots forget humans enhancing into cyborgs can make the need for robots in the future disappear. Further, few other interesting remarks like need of Society 5.0 and issues in post covid era, etc., can be found be found in [22–27].

"Society 5.0" is a community where online [28] and physical space are integrated and where economic expansion and social issue [29] solving are balanced. But in this "society 5.0", there may be new security [30] concerns, attacks [31], and vulnerabilities. Developments in cyber security [32] for vital infrastructure and how it will change in this so-called "Society 5.0" era.

16.7 Conclusion

Mass production, mass adaptation, and mass imagination involve bodies and technologies that improve humans through better usage of resources, greater material and energy use, and efforts to build robots that go beyond the natural level of human intelligence. Such a scenario is an advanced robot programmed to produce paperclip-optimized results, with the goal of turning everything that matters (including humans) into paperclip machines. In other words, people will project their anthropomorphic interpretations of social robots and view them as human beings with human qualities. Looking at the future of humanity in terms of robot emancipation or eradication does not mean recognizing that human improvement of the cyborg body through technology and widespread practice will diminish the comparative advantage of robots over humans in terms of the need for robots. It is difficult to predict which scenario society will follow, but the humanization of social robots seems inevitable and is on the agenda of developers and marketers. What follows is an overview of possible ethical challenges we will face with intelligent systems and robots in our society, followed by countermeasures related to technology and risks to be taken, including machine ethics and precautions for designers. Tomorrow's robots will be dwarfed by advances in artificial intelligence, machines, and deep learning. On the one hand, artificial intelligence has the potential to facilitate the future development of robots as emotional and intellectual entities, transforming them into human agents from independent creatures. The way in which man and machine connect will change and there will be many new kinds of man-machine symbiosis. Humans will replace robots in some jobs and supplement them in many others.

Currently, artificial intelligence (AI) and human brain interfaces are experimenting with brain hacking devices that are being researched to demonstrate the potential of real cyborgs and products and systems that are available for testing and dissemination. The implementation of the technology is in line with my neologisms "Neural Plug Compatibility" and "Softwiring" for the realization of AI systems, robots, android avatars, and cyborgs. Whether it is the ability to surpass the figure of the "robot cyborg" (Palese, 2011) or our modern world of advanced scientific progress, which combines artificial realities with human and natural ones, the technology of embodying us becomes an extension of the body.

Daniel Berninger, an Internet pioneer who conducted the first VoIP deployments by Verizon, HP, and NASA and the founder of the Voice Communication Exchange Committee (VCXC) says that luminaries who claim that artificial intelligence will surpass human intelligence encourage robots out of awe and imagine how improved computation will push machines towards self-realization of science fiction into reality. This means that when we look to the near future, we will see thinking robots with brains not unlike humans. Such brain implants could restore or improve human function and reduce the human need for robots in the future.

As cyborgization progresses, we will be faced with a constant choice between investing in machines that can integrate and measure human rights or taking human rights away from people and expecting them to become machine parts. While science fiction films and dystopian depictions of future technologies can have an impact on human extinction, there can also be positive effects on our awareness of potential vulnerabilities that can be proactively addressed. A society that sees itself as a cyborg is engaged in a very different cultural spectrum from an idea of electronic surveillance or a society that sees itself as composed of people with tools. Hubert Dreyfus, an American philosopher and professor of philosophy at Berkeley University, makes a persuasive critique of artificial intelligence in *The Limitations of Robot Intelligence*. He argues that computers will never be able to replace humans and live on an equal footing with them. We refer to this in human-robot culture as a recursive influence on the cultural values of human society, where the development of robots influences the cultural values of the robot vis-à-vis humans. Today's computers can process data faster than we can with independent artificial intelligence, he says, but tomorrow's cyborgs will be a million times smarter than us. And of course, Arnold Schwarzenegger proved to be a virtuous cyborg in the Terminator movie.

Based on the various reasons and statements exposed in this chapter, we consider that amplified humans, cyborgs, and robots will definitely play a significant role in the coming years.

References

1. Brynjolfsson, E.; McAfee, A. The Second Machine Age: Work, Progress, and Prosperity in Time of Brilliant Technologies; W.W. Norton & Company, Inc.: New York, NY, USA, 2014.
2. Francesc Mestres and Josep Vives-Rego, Behind and Beyond Of The Cyborgs And Robots Ideas And Realities: Some Techno-Scientific And Philosophical Hints, L.K. Who Goes First? The Story of Self-Experimentation in Medicine; University of California Press: Berkeley, CA, USA, 1998.
3. Liao, Y.; Leeson, M.S.; Cai, Q.; Ai, Q.; Liu, Q. Mutual-Information-Based Incremental Relaying Communications for Wireless Biomedical Implant Systems. Sensors 2018, 18, 515.
4. Clarke, R.A. Information technology and dataveillance. Commun. ACM 1988, 31, 498–512.
5. Bossy, S. The utopias of political consumerism: The search of alternatives to mass consumption. J. Consum. Cult. 2014, 14, 179–198.
6. Kurzweil, R. The Singularity Is Near: When Humans Transcend Biology; Viking: New York, NY, USA, 2005.
7. Abbott, R.; Bogenschneider, B. Should Robots Pay Taxes? Tax Policy in the Age of Automation (13 March 2017). Harvard Law & Policy Review. 2018, Volume 12. Available online: https://ssrn.com/abstract=2932483 (accessed on 9 May 2018).
8. Bostrom, N. Ethical issues in advanced artificial intelligence. In Cognitive, Emotive and Ethical Aspects of Decision Making in Humans and in Artificial Intelligence; Smit, I., Lasker, G.E., Eds.; International Institute for Advanced Studies in Systems Research and Cybernetics: Windsor, ON, Canada, 2003; Volume 2, pp. 12–17.
9. Willett, F.R.; Pandarinath, C.; Jarosiewicz, B.; Murphy, B.A.; Memberg,W.D.; Blabe, C.H.; Saab, J.;Walter, B.L.; Sweet, J.A.; Miller, J.P. Feedback control policies employed by people using intracortical brain–computer interfaces. J. Neural Eng. 2016, 14, 016001.
10. Meghna Manoj Nair, Amit Kumar Tyagi, Richa Goyal, Medical Cyber Physical Systems and Its Issues, Procedia Computer Science, Volume 165, 2019, Pages 647-655, ISSN 1877-0509, https://doi.org/10.1016/j.procs.2020.01.059.
11. Moravec, H. Mind Children; Harvard University Press: Cambridge, MA, USA, 1988.
12. Pfeifer, R.; Bongard, J. How the Body Shapes the Way We Think: A New View of Intelligence; MIT Press: Cambridge, MA, USA,2006
13. Arai, Kohei; Bhatia, Rahul; Kapoor, Supriya (2019). Advances in Intelligent Systems and Computing. Proceedings of the Future Technologies Conference (FTC) 2018 Volume 881 (Volume 2) || Wetware and the Cyborg Era: The Future of Modifications on the Human Body According to Science Fiction., 10.1007/978-3-030-02683-7 (Chapter 4), 34–52. doi:10.1007/978-3-030-02683-7_4

14. Stojnić, A. (2013). Cyborgs from Fiction to Reality: Marginalized Other or Privileged First? Identities: Journal for Politics, Gender and Culture, 10(1–2), 49–53. https://doi.org/10.51151/identities.v10i1-2.278
15. Warwick, K, Gasson, M, Hutt, B, Goodhew, I, Kyberd, P, Schulzrinne, H and Wu, X: "Thought Communication and Control: A First Step using Radiotelegraphy", IEE Proceedings on Communications, 151(3), pp.185–189, 2004.
16. Haraway, D. A Cyborg Manifesto: Science, Technology, and Socialist-Feminism in the Late Twentieth Century, in Simians, Cyborgs and Women: The Reinvention of Nature; Routledge: New York, NY, USA, 1994; pp. 150-182.
17. Reinares-Lara, E., Olarte-Pascual, C., & Pelegrín-Borondo, J. (2018). Do you Want to be a Cyborg? The Moderating Effect of Ethics on Neural Implant Acceptance. Computers in Human Behavior, 85, 43–53. https://doi.org/10.1016/j.chb.2018.03.032
18. Park, Enno. (2014). Ethical Issues in Cyborg Technology: Diversity and Inclusion. NanoEthics, 8(3), 303–306. https://doi.org/10.1007/s11569-014-0206-x
19. Graaf, M. M. A. de, Allouch, S. Ben, & Klamer, T. (2015). Sharing a life with Harvey: Exploring the acceptance of and relationship-building with a social robot. Computers in Human Behavior, 43, 1– 14. https://doi.org/10.1016/j.chb.2014.10.030
20. DiSalvo, C. F., Gemperle, F., Forlizzi, J., & Kiesler, S. (2002). All Robots are not Created Equal: The Design and Perception of Humanoid Robot Heads. In DIS 2002 Designing Interactive Systems 2002 (pp. 321–326). London: ACM. https://doi.org/10.1145/778712.778756
21. M. M. Nair, A. K. Tyagi and N. Sreenath, "The Future with Industry 4.0 at the Core of Society 5.0: Open Issues, Future Opportunities and Challenges," 2021 International Conference on Computer Communication and Informatics (ICCCI), 2021, pp. 1-7, doi: 10.1109/ICCCI50826.2021.9402498.
22. Tyagi A.K., Fernandez T.F., Mishra S., Kumari S. (2021) Intelligent Automation Systems at the Core of Industry 4.0. In: Abraham A., Piuri
23. V., Gandhi N., Siarry P., Kaklauskas A., Madureira A. (eds) Intelligent Systems Design and Applications. ISDA 2020. Advances in Intelligent Systems and Computing, vol 1351. Springer, Cham. https://doi.org/10.1007/978-3-030-71187-0_1
23. Goyal, Deepti & Tyagi, Amit. (2020). A Look at Top 35 Problems in the Computer Science Field for the Next Decade. 10.1201/9781003052098-40
24. Madhav A.V.S., Tyagi A.K. (2022) The World with Future Technologies (Post-COVID-19): Open Issues, Challenges, and the Road Ahead. In: Tyagi A.K., Abraham A., Kaklauskas A. (eds) Intelligent Interactive Multimedia Systems for e-Healthcare Applications. Springer, Singapore. https://doi.org/10.1007/978-981-16-6542-4_22
25. Mishra S., Tyagi A.K. (2022) The Role of Machine Learning Techniques in Internet of Things-Based Cloud Applications. In: Pal S., De D., Buyya R. (eds) Artificial Intelligence-based Internet of Things Systems. Internet of

Things (Technology, Communications and Computing). Springer, Cham. https://doi.org/10.1007/978-3-030-87059-1_4

26. Nair, Meghna Manoj; Tyagi, Amit Kumar "Privacy: History, Statistics, Policy, Laws, Preservation and Threat Analysis", Journal of Information Assurance & Security. 2021, Vol. 16 Issue 1, p24-34. 11p.
27. Tyagi, Amit Kumar; Nair, Meghna Manoj; Niladhuri, Sreenath; Abraham, Ajith, "Security, Privacy Research issues in Various Computing Platforms: A Survey and the Road Ahead", Journal of Information Assurance & Security. 2020, Vol. 15 Issue 1, p1-16. 16p.
28. Rawat, R., Sowjanya, A. M., Patel, S. I., Jaiswal, V., Khan, I., & Balaram, A. (Eds.). (2022). *Using Machine Intelligence: Autonomous Vehicles Volume 1.* John Wiley & Sons.
29. Rawat, R., Bhardwaj, P., Kaur, U., Telang, S., Chouhan, M., & Sankaran, K. S. (2023). *Smart Vehicles for Communication, Volume 2.* John Wiley & Sons.
30. Mahor, V., Bijrothiya, S., Rawat, R., Kumar, A., Garg, B., & Pachlasiya, K. (2023). IoT and Artificial Intelligence Techniques for Public Safety and Security. *Smart Urban Computing Applications*, 111.
31. Mahor, V., Bijrothiya, S., Mishra, R., & Rawat, R. (2022). ML Techniques for Attack and Anomaly Detection in Internet of Things Networks. *Autonomous Vehicles Volume 1: Using Machine Intelligence*, 235-252.
32. Mahor, V., Bijrothiya, S., Mishra, R., Rawat, R., & Soni, A. (2022). The Smart City Based on AI and Infrastructure: A New Mobility Concepts and Realities. *Autonomous Vehicles Volume 1: Using Machine Intelligence*, 277-295.

17

Security and Privacy of Blockchain-Based Robotics System

Pratham Jaiswal[1] **and Amit Kumar Tyagi**[1,2*]

[1]*School of Computer Science and Engineering, Vellore Institute of Technology, Chennai, Tamil Nadu, India*
[2]*Department of Fashion Technology, National Institute of Fashion Technology, New Delhi, Delhi, India*

Abstract

The area of cryptocurrencies has been fundamentally transformed by blockchain technology. Blockchain is a novel and effective way to store information, perform transactions, and so on, but when it comes to adopting blockchain-based cryptocurrencies, the security and privacy of blockchains are constantly at the forefront of the argument. There is a misbelief that cryptocurrencies are an anonymous means of payment. The blockchain is a public, fully transparent ledger, so anybody can browse the data of a blockchain using a block explorer and see the amount, at what time, and what address transferred money. But even on a fully transparent ledger, we can transact on blockchain privately using an address, as an address doesn't carry any information about its owner. We can say that addresses act like pseudonyms. This chapter will provide a comprehensive review of blockchain-based cryptocurrency security and privacy. This chapter begins with an introduction to blockchain and its applications in cryptocurrency, followed by a description of the security and privacy properties required to build a blockchain-based robotics system, and finally a review of the security and privacy techniques required to achieve these properties in blockchain-based cryptocurrencies, such as consensus algorithms, zero-knowledge proof, etc. The readers of this chapter will obtain a thorough grasp of blockchain in cryptocurrency security and privacy.

Keywords**:** Blockchain, cryptocurrency, robotics, security, privacy

**Corresponding author*: amitkrtyagi025@gmail.com

Romil Rawat, Rajesh Kumar Chakrawarti, Sanjaya Kumar Sarangi, Rahul Choudhary, Anand Singh Gadwal and Vivek Bhardwaj (eds.) Robotic Process Automation, (259–274) © 2023 Scrivener Publishing LLC

17.1 Introduction

Blockchain is like a linked list as it doesn't keep data in a huge continuous ledger, but splits the data into nodes known as blocks. Each block contains several elements which contain the block header and its transactions. The transactions in a block account for almost all of the data, while the block header contains a timestamp, hash of the preceding block, Merkle root hash, and other such essential metadata. What makes a blockchain different from a linked list is that in a blockchain the hash of the previous block, also known as a reference, is cryptographically enciphered, hence tamper-evident, and new data can only be added in the form of new blocks which will be linked with previous blocks of data (refer to Figure 17.1).

The main technology behind Cryptocurrency is Blockchain. The key elements of blockchain include hashing for security and immutability, peer-to-peer (P2P) networks for transaction verification, data structures for storing and managing the transactions, smart contracts for corporate bond transfers, consensus protocols for decentralization and avoiding double-spending issues, and incentive mechanisms for secure transactions [1].

Cryptocurrencies based on blockchains (refer Figure 17.2) have become a new form of money in the last few years. Instead of depending on centralized authorities like the bank to manage money, cryptocurrencies depend on mathematical design and complex cryptographic protocols. Since most cryptocurrencies are fully decentralized, no individual or organization can keep track of or prevent the transfer of funds. Cryptocurrencies grew from just being an idea and model to being a worldwide prodigy with millions of individuals and organizations investing in them.

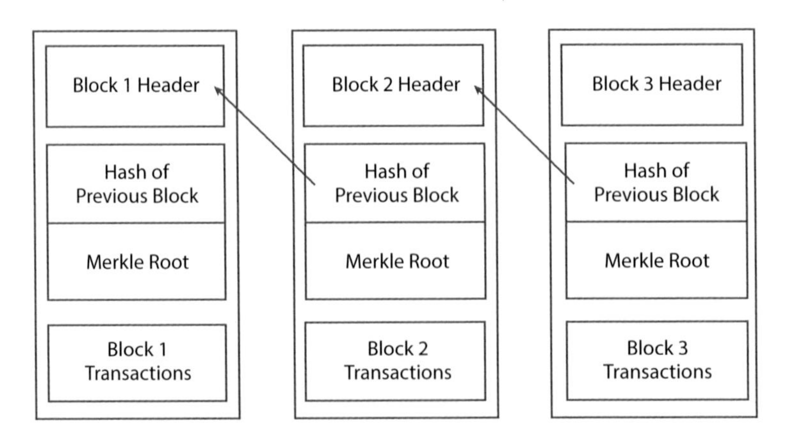

Figure 17.1 Architecture of blockchain.

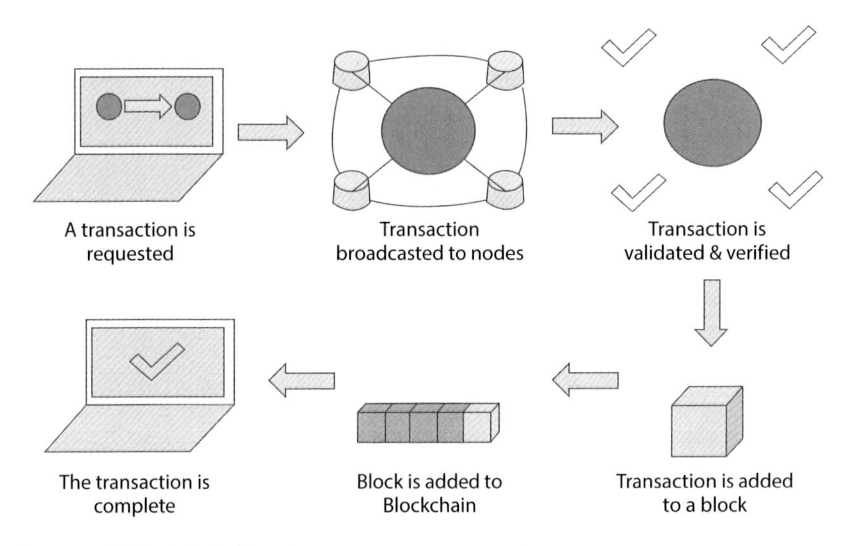

Figure 17.2 Blockchain-based cryptocurrency transaction.

HashCash (1997) was created as a measure against spam and the idea of b-money (1998) was proposed to conceive the ideas of Proof-of-Work and Proof-of-Stake for achieving consensus across the blockchain network. Satoshi Nakamoto, in 2008, released the Bitcoin white paper, based on HashCash and b-money. Bitcoin, the first cryptocurrency, became popular because it was being controlled by a decentralized network and could avoid double-spending. Currently, Bitcoin is the leading cryptocurrency, having 51% (as of June 2021) of the total market share among 5000 altcoins.

As of 2017, the total market cap of the cryptocurrency market was $760 billion [2] and rose above $2 trillion in 2021 [3] because many big businesses companies and entities have started investing in cryptocurrencies. But cryptocurrencies still being in their early stages creates a lack of trust among many stakeholders.

The most common misbelief about cryptocurrencies is that they are anonymous. Most cryptocurrencies are pseudonymous, which means that the real identities of people are represented by addresses, so the two can be connected through data analysis. But some cryptocurrencies follow certain approaches for making it difficult to trace the transactions and link addresses on the blockchain to real-world identities or follow more advanced concepts, which allow transactions to be completely private even on public blockchains [4].

The current studies on blockchain technology lack a proper analysis of the issues and challenges that affect trust in blockchain-based cryptocurrencies because the research on blockchain technologies is majorly done by industry practitioners and not researchers. So, in this research chapter we will describe the security and privacy concerns users might face when using blockchain-based cryptocurrencies, the security and privacy requirements, and privacy and security techniques used in blockchain-based cryptocurrencies.

17.2 Security and Privacy Concerns

Cryptocurrencies face some serious security concerns and risks, which can cause frequent dramatic price drops. These concerns can be destructive to any cryptocurrency. Some of such concerns are briefly described below.

17.2.1 Double Spending

Double-spending is a process in which the same single unit of a cryptocurrency can be spent more than once. It occurs when a cryptocurrency is stolen by altering or damaging the blockchain network. The transaction could be erased or a copy of the transaction would be sent by the hacker to make it look authorized. Most commonly, the hacker will send numerous packets to the network for reversing a transaction, which will make it looks like it never happened [5].

17.2.2 Vulnerable Wallets

A wallet should protect our money and privacy, but cryptocurrency wallets are very vulnerable to hacking attacks and theft. Using malware, the wallet can be prevented from communicating with the PC, hence breaching the security. This affects the privacy of its users and their transactions can now easily be redirected to different accounts.

17.2.3 Cyber-Attacks

A disastrous cyber attack on not only blockchain but also cryptocurrency exchanges is one of the major concerns of its users. There have been major attacks on exchanges before, which resulted in the loss of people's money as well as the downfall of cryptocurrency value. In 2014, the Mt. Gox heist took place, in which the hackers went away with 850,000 [6] bitcoins which is equivalent to USD 47.03 billion as of October 2021. Distributed Denial of Service (DDoS) attacks are also a threat to cryptocurrency exchanges. Bitfinex, which is among the top currency exchanges, reported that it faced continuous DDoS attacks in 2017 which impacted trading operations and resulted in a drop in Bitcoin value [7]. There have been many such heists, which is why people don't trust the security of blockchain-based cryptocurrency.

17.2.4 Sybil Attack

In a Sybil Attack, numerous fake identities are created and controlled by a single entity to manipulate a peer-to-peer network. Various fake nodes gather around a node so that it can't connect to the other nodes on the network, which then prevents the user from sending or receiving information to the blockchain.

17.2.5 Selfish Mining

Due to the proof-of-work consensus mechanism, certain cryptocurrencies can be threatened by the selfish mining of the major mining pools. Cryptomining is the process in which transactions of cryptocurrency are verified and confirmed, with miners earning cryptocurrencies in return for their computational effort [8]. For selfish mining, greedy miners hide their generated blocks from the main blockchain [8] and later reveal them to earn more revenue. With this combined with the Sybil attack, miners can invalidate transactions on the network with their power [9].

17.2.6 51 Percent Attacks

A 51 percent attack is an attack by a group of miners controlling the majority of the blockchain's computing power. The attackers can reverse transactions that were completed resulting in double-spending or can prevent new transactions from getting confirmed. In 2018, hackers pulled off more than $18 million worth of Bitcoin Gold through a 51 percent Attack [10].

17.3 Security and Privacy Requirements

The following is a quick description of the security and privacy attributes necessary in blockchain-based cryptocurrencies.

17.3.1 Integrity of Data

Blockchains must be designed for data integrity, otherwise the data will be vulnerable or completely non-functional. The blockchain should be used to collect and manage precise, authentic, and timely data, so it is useful to the users.

17.3.2 Tamper-Resistant Data

The data in a blockchain should be able to resist any type of damage. The data which is stored on the blocks of the blockchain cannot be modified anyhow.

17.3.3 Preventing Double-Spending

Double spending occurs when a particular unit of currency is spent more than once. For transactions performed with a decentralized blockchain-based cryptocurrency, the blockchain should have strong security measures to prevent the double-spending of a coin.

17.3.4 Anonymous User Identities

If the user data is shared with various financial institutions, then the user's identity may be disclosed. Also, in a transaction, the two users might be unwilling to disclose their real identities to each other [11]. So, the blockchain must have strong security and privacy methods to make user identity anonymous or at the very least pseudonymous.

17.3.5 Transaction Unlinkability

If all the transactions of a user could be linked, then the user's identity and other information can be deduced. So, the blockchain should have security measures to provide unlinkability of transactions.

17.3.6 Transaction Confidentiality

The blockchain must have security measures so that the user's data cannot be accessed or disclosed without his or her permission, even under unexpected failures [11].

17.3.7 DDoS Attack Resistant

A distributed denial-of-service (DDoS) attack is a malicious attempt to flood the internet traffic of a network or server and to take advantage of its security vulnerabilities. A heavy DDoS attack could be used to knock off a blockchain network, so the blockchain should have security measures to prevent or tackle it.

17.3.8 51% Attack Resistant

Malicious miners can conspire and launch various security and privacy attacks like illegal transfer of cryptocurrency or reversing transactions. So, the blockchain should have security measures to prevent or tackle it.

17.4 Consensus Algorithms

Blockchain, being a distributed decentralized network with no central authority present to validate and verify the transactions, is considered to be secure only because of the consensus protocols, a core part of blockchain networks.

A consensus mechanism refers to methods used to achieve agreement, trust, and security across a decentralized blockchain network. These consensus protocols help all the nodes in the network to verify the transactions.

These protocols will now be reviewed.

17.4.1 Proof of Work (PoW)

In 1993, the idea of Proof of Work was first proposed by Moni Naor and Cynthia Dwork, while the algorithm was termed "proof of work" by Ari Juels and Markus Jakobsson in a publication in 1999 [12]. In 2008, Satoshi Nakamoto applied the Proof of Work algorithm to the Bitcoin whitepaper. The Proof of Work algorithm is a technique used by many cryptocurrencies. Once all the nodes are brought into an agreement, the transactions will get validated and the new block will be forged on the blockchain.

In the Proof of Work consensus algorithm, miners solve a complex computational problem to create and add new blocks to the blockchain. The verification and organization of transactions in a block, and the introduction of the newly mined block to the blockchain network needs much less time and energy than solving the 'complex computational problems' required to add the new block to the blockchain. When a miner successfully solves the 'complex computational problem', the nodes broadcast the block to the blockchain and the miner receives some of the cryptocurrency as a reward.

Although proof of work is an efficient mechanism, it has some disadvantages, like it increases the chance of 51% attacks, finding the correct solution to 'complex computational problems' is very time-consuming, and money and electricity consumption is too high [12].

17.4.2 Practical Byzantine Fault Tolerance (pBFT)

The idea of practical Byzantine Fault Tolerance was first proposed by Barbara Liskov and Miguel Castro in 1999. The objective of the practical Byzantine Fault Tolerance mechanism is to optimize aspects of Byzantine Fault Tolerance in a blockchain.

Byzantine Fault Tolerance (BFT) is an algorithm that is used to reach an agreement even in the presence of some malicious or faulty nodes in the network. The objective of a BFT mechanism is to employ collective decision making for reducing the effect of the faulty nodes.

The failures of faulty nodes are of two types: fail-stop and arbitrary-node failure. In the fail-stop failure, the node fails and stops operating, while in arbitrary node failure, the node fails to return a result or returns an incorrect result [13].

One node in a practical Byzantine Fault Tolerant enabled blockchain is set as the primary node and others as secondary nodes. A practical Byzantine Fault Tolerant system functions if the number of malicious or faulty nodes is lesser than or equal to one-third (or 34%) of all the nodes in the system.

Although the practical Byzantine Fault Tolerant mechanisms are energy and time-efficient, there are a few limitations to it as it is open to Sybil attack and it does not scale well.

17.4.3 Proof of Stake (PoS)

In 2012, the Proof of Stake was first used for a cryptocurrency named Peercoin. The Proof of Stake mechanism states that the mining power is directly dependent on the number of coins staked. Proof of Stake is one of the most common alternatives to Proof of Work.

Proof of Stake mechanism requires users to stake their coins to become a validator in the network rather than buying expensive hardware or wasting resources to mine. Validators are randomly chosen to propose new blocks and validate proposed blocks when they are not chosen. Validators get rewarded for doing so but validating malicious blocks will result in losing stakes.

Although energy and money efficient, the Proof of Stake has some disadvantages such as a staked coin can't be sold until the staking period is over, the staking reward much is lesser than the mining reward, the users holding a large number of coins can have a huge influence on the mechanism, and this mechanism is still in its early stage and the privacy and security of Proof of Stake are not proven to be even as good as that of Proof of Work [14].

17.4.4 Proof of Burn (PoB)

Proof of Burn, first proposed by Iain Stewart in 2012, is an algorithm in which a miner uses a virtual rig to burn (permanently erase) their coins. The more coins they burn, the better virtual mining rig they get. So, Proof of Burn is Proof of Work without the high energy consumption.

When burning the coins, they are sent to a verifiably unspendable address. Miners are allowed to burn the coins as specified and get rewarded with tokens of that particular cryptocurrency [15].

The Proof of Burn mechanism uses the burning of cryptocurrency coins periodically to increase the mining power. The value of burnt coins reduces with every newly mined block to keep miners regularly engaged.

The Proof of Burn mechanism is more sustainable as the power consumptions are very low and expensive mining hardware isn't needed, which can make the miners stay dedicated to it for long periods. But the verification process is slow and its efficiency and security are yet to be confirmed on a larger scale.

17.4.5 Proof of Capacity (PoC)

The Proof of Capacity algorithm enables mining by using the hard drive space. In Proof of Capacity, a list of possible solutions is stored on the hard drive of the mining device. The number of possible solution values increases if the available space in the hard drive is more, which increases the chances of getting the correct solution, hence increasing the chances to win the mining reward.

The Proof of Capacity mechanism has two steps: plotting, and mining. In plotting, through repeated hashing of data, all the possible nonce values are listed and stored in the hard drive. In mining, a miner calculates a scoop number. For each nonce in the hard drive, the process is repeated to calculate its deadline. Then, the miner chooses the one with the minimum deadline. A deadline is the amount of time that has to be passed after the creation of a block for a miner to create a new block. A miner can create a block and gets the block reward if no one else has created a block within that time [16].

Proof of Capacity uses just a hard drive, it is more efficient than Proof of Work and Proof of Stake mechanisms, and expensive hardware is not needed. But Proof of Capacity is still new and not adopted by many and hackers can take advantage of it using malware.

Table 17.1 provides a detailed comparison of various consensus algorithms in detail. Further, several interesting works of blockchain technology in many other applications can be found in [20–30].

Table 17.1 Comparison of the consensus algorithms.

Consensus algorithms	Working	Pros	Cons
Proof of Work	Solving a complex computational problem to verify transactions and to create and add new blocks to the blockchain	• Attacking blockchain networks become increasingly complex and expensive • Allows miners to earn cryptocurrency as a reward	• Requires a lot of computing resources, which is wasted a every time an equation is computed • Excessive energy use is demanded, resulting in environmental damage • Mining blocks necessitate the purchase of expensive equipment • High transaction fees
Practical Byzantine Fault Tolerance	All honest nodes contribute to establishing a consensus on the state of the system	• Energy consumption is decreased since hashing energy is not required to enter the next block • Low transaction fees	• Prone to Sybil Attacks • Poor scalability
Proof of Stake	Allows cryptocurrency owners to stake coins and set up their own validator nodes for transaction verification	• Doesn't require a lot of computational power, thus it's energy-efficient • Transactions are processed quickly and at a comparatively low cost	• Less secure than Proof of Work • Staked coins cannot be sold for a certain amount of time • Validators holding a large number of coins have more influence on transaction verification

(Continued)

Table 17.1 Comparison of the consensus algorithms. (*Continued*)

Consensus algorithms	**Working**	**Pros**	**Cons**
Proof of Burn	Miners burn their coins using the virtual mining rig	• Low power consumption • Low hardware requirements	• Security hasn't been confirmed • Slow verification process
Proof of Capacity	Allows mining devices in the network to validate transactions using their available hard drive capacity	• More energy-efficient than Proof of Work • No need for dedicated hardware • Mined data can easily be deleted	• Mining is prone to malware • Mining competition leads to buying higher-capacity hard drives

17.5 Privacy and Security Techniques

There are various techniques a blockchain system uses or can use to achieve or enhance its privacy and security. Such techniques will be discussed below.

17.5.1 Change Addresses

A cryptocurrency transaction needs at least one input and output. The existing funds used for sending a transaction are called input and when the funds are received it's called output. When a user gives an input fund more than the transaction cost, the remaining funds (or the change) are sent back to the user as a change, but to a newly generated Change Address. But, if the input fund is exactly equal to the transaction cost, then a change address is not needed. Most of the wallets generate change addresses, automatically creating a transaction. Change address makes it difficult to track the user's transaction history, hence improving privacy.

17.5.2 Coin Mixing

Coin mixing protocols like CoinJoin, SharedCoin, or Mixcoin were an evolutionary step to improve privacy. In Coin Mixing, several input coins, in a mixing pool, are combined and then sent to their receivers' addresses. This makes tracking the transactions more difficult.

Coin mixers (refer to Figure 17.3) are software companies serving as a middleman between parties looking to send and receive cryptocurrencies. If a user wants a transaction to be untraceable, he will send a particular

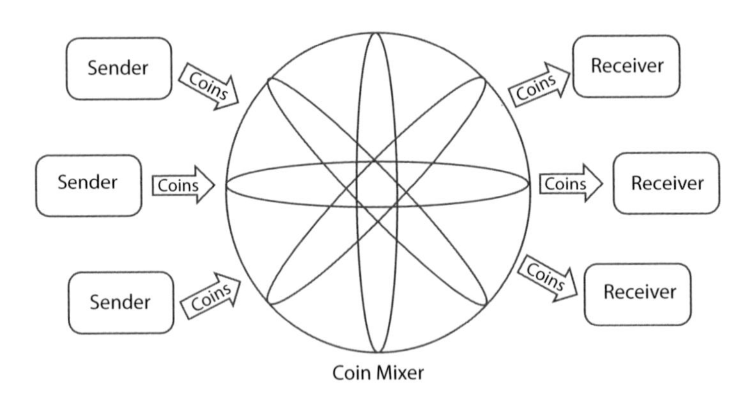

Figure 17.3 Coin mixer.

number of coins to the coin mixer, who will then combine it with many other transactions of the same currency, and then redistribute it to the receiving addresses [17]. A fee is charged for the mixing services by coin mixers.

Coin mixing increases the level of privacy more than regular transactions, but addresses can be linked by observing the amounts of coins in a mixing pool and many mixing services are centralized.

17.5.3 Ring Signatures

Ron Rivest, Adi Shamir, and Yael Tauman Kalai first introduced the Ring Signature in 2001. A ring signature is a digital signature that can be created by any member of a group and each member has their own keys. So, it is not possible to determine which group member created the signature. In a ring signature, we have a group of people, each of them having public and secret keys. To sign a message, a member of the group has to use his secret key as well as the public keys of others in the group. The public keys of the group are used to validate that person signing the message is a member of the group, but it is not possible to figure out who signed the message from the group [18]. Ring Signatures are great for private transactions, but if the secret key of a person is compromised, any of his signed messages can be modified.

17.5.4 Homomorphic Encryption

Homomorphic encryption allows users to perform computations on encrypted data, so decrypting the data is not needed. The decrypted form of the computed results will be the same as the result of the same operations performed on the unencrypted data.

Homomorphic encryption is used to ensure that the data stored on the blockchain is encrypted. The homomorphic encryption technique is used to preserve privacy and to allow access to the encrypted data over the public blockchain for validation. But homomorphic encryption requires a special client-server application to work functionally [19].

17.5.5 Zero-Knowledge Proofs (zk-Proofs)

Charles Rekkofom, Shafi Goldwasser, and Silvio Micali first proposed the idea of Zero-Knowledge Proof in 1985. Zero-knowledge proof lets someone prove the truth regarding something to the verifier without revealing any additional information. In blockchain-based cryptocurrency, the

Zero-Knowledge Proofs are used for private transactions. The sender will have to create a proof, without revealing any of the actual transaction data, that the transaction will be considered valid by a verifying node. This allows the identities of both the parties and the amount to be private [4]. Zero-Knowledge Proofs do not require any complicated encryption methods, increase the privacy of users, and strengthen the security of information, but a tremendous amount of computing power is needed and if the user forgets their information, all the data associated with it will be lost.

A 53 percent threat [31, 32], which takes full advantage of how blockchain [33] platforms work, aims to take over more than half of the channel's mining strength. This will let the culprit manipulate [33, 34] and alter the transactional ledger. Blockchain is frequently misinterpreted as a safe design method since it is based on well-established cryptographic techniques. These intrinsic cryptographic features, however, are not strong enough to fend off every cybersecurity attack [35].

17.6 Conclusion

With many big organizations and institutions investing in blockchain-based cryptocurrency, the security and privacy of blockchains have attracted huge interest. Proper knowledge of the security and privacy properties and techniques of blockchain plays a major role in building up or increasing the trust regarding the security and privacy of blockchain-based cryptocurrencies among the users. Following and improving the above-mentioned techniques like consensus algorithms, mixing, etc. and coming up with better techniques is believed to be a key to the future development of blockchain-based cryptocurrencies. Above that, the users can and should use strong passwords, two-factor authentication, VPN (Virtual Private Network), encrypted emails, etc. to protect their privacy.

References

1. Muhammad Habib ur Rehman, Khaled Salah, Ernesto Damiani, and Davor Svetinovic, "Trust in Blockchain Cryptocurrency Ecosystem," IEEE Transactions on Engineering Management, vol. 67, no. 4, November 2020.
2. Zack Voell "Total Cryptocurrency Market Value Hits Record $1 Trillion," Jan 7, 2021.
3. Hassan Maishera, "Total Cryptocurrency Market Cap Stays Above $2 Trillion Despite Recent Losses," August 26, 2021.

4. Horizen Academy, "Privacy on the Blockchain".
5. Jake Frankenfield, "Double Spending," June 30, 2020.
6. Jake Frankenfield, "Mt. Gox" March 25, 2021.
7. Kate Rooney, "Cryptocurrency exchange Bitfinex briefly halts trading after cyberattack," June 5, 2018.
8. Jake Frankenfield, "Selfish Mining" March 31, 2021.
9. Mark Schwaz, "Security Concerns and Risks Related To Bitcoin," March 8, 2018.
10. Jake Frankenfield, "51% Attack" August 25, 2021.
11. Rui Zhang, Rui Xue, and Ling Liu, "Security and Privacy on Blockchain" ACM Computing Surveys, Vol. 52, No. 3, Article 51, July 2019.
12. Parikshit Hooda, "Proof of Work (PoW) Consensus" January 9, 2019.
13. Parikshit Hooda, "practical Byzantine Fault Tolerance(pBFT)" December 12, 2019.
14. Harriet Phelps, "Ethereum and Proof of Stake (The Pros and Cons of PoS)" July 23, 2021.
15. Jake Frankenfield, "Proof of Burn (Cryptocurrency)" September 22, 2021.
16. Adam Hayes, "Proof of Capacity (Cryptocurrency)" June 29, 2021.
17. Alexandaria, "Coin Mixer".
18. Prof Bill Buchanan OBE, "Ring Signatures And Anonymisation" August 13, 2018.
19. Ameesh Divatia, "The Advantages and Disadvantages of Homomorphic Encryption" February 27, 2019.
20. Shabnam Kumari, Amit Kumar Tyagi, Aswathy S U, "The Future of Edge Computing with Blockchain Technology: Possibility of Threats, Opportunities and Challenges", in the Book "Recent Trends in Blockchain for Information Systems Security and Privacy", CRC Press, 2021.
21. Tibrewal I., Srivastava M., Tyagi A.K. (2022) Blockchain Technology for Securing Cyber-Infrastructure and Internet of Things Networks. In: Tyagi A.K., Abraham A., Kaklauskas A. (eds) Intelligent Interactive Multimedia Systems for e-Healthcare Applications. Springer, Singapore. https://doi.org/10.1007/978-981-16-6542-4_1
22. Agrawal, D., Bansal, R., Fernandez, T.F., Tyagi, A.K. (2022). Blockchain Integrated Machine Learning for Training Autonomous Cars. In: *et al.* Hybrid Intelligent Systems. HIS 2021. Lecture Notes in Networks and Systems, vol 420. Springer, Cham. doi.org/10.1007/978-3-030-96305-7_4
23. A K. Tyagi, D. Agarwal and N. Sreenath, "SecVT: Securing the Vehicles of Tomorrow using Blockchain Technology," 2022 International Conference on Computer Communication and Informatics (ICCCI), 2022, pp. 1-6, doi: 10.1109/ICCCI54379.2022.9740965.
24. K. V, A. K. Tyagi and S. P. Kumar, "Blockchain Technology for Securing Internet of Vehicle: Issues and Challenges," 2022 International Conference on Computer Communication and Informatics (ICCCI), 2022, pp. 1-6, doi: 10.1109/ICCCI54379.2022.9740856.

25. Amit Kumar Tyagi, "Analysis of Security and Privacy Aspects of Blockchain Technologies from Smart Era' Perspective: The Challenges and a Way Forward", in the Book "Recent Trends in Blockchain for Information Systems Security and Privacy", CRC Press, 2021.
26. Amit Kumar Tyagi, G Rekha, Shabnam Kumari "Applications of Blockchain Technologies in Digital Forensic and Threat Hunting", in the Book "Recent Trends in Blockchain for Information Systems Security and Privacy", CRC Press, 2021.
27. S. Mishra and A. K. Tyagi, "Intrusion Detection in Internet of Things (IoTs) Based Applications using Blockchain Technology," 2019 Third International conference on I-SMAC (IoT in Social, Mobile, Analytics and Cloud) (I-SMAC), 2019, pp. 123-128, doi: 10.1109/I-SMAC47947.2019.9032557.
28. R, Varsha *et al.* 'Deep Learning Based Blockchain Solution for Preserving Privacy in Future Vehicles'. International Journal of Hybrid Intelligent System, Vol 16, Issue 4: 223 – 236, 1 Jan. 2020.
29. A M. Krishna and A. K. Tyagi, "Intrusion Detection in Intelligent Transportation System and its Applications using Blockchain Technology," 2020 International Conference on Emerging Trends in Information Technology and Engineering (ic-ETITE), 2020, pp. 1-8, doi: 10.1109/ic-ETITE47903.2020.332.
30. Deshmukh, N. Sreenath, A. K. Tyagi and U. V. Eswara Abhichandan, "Blockchain Enabled Cyber Security: A Comprehensive Survey," 2022 International Conference on Computer Communication and Informatics (ICCCI), 2022, pp. 1-6, doi: 10.1109/ICCCI54379.2022.9740843.
31. Mahor, V., Bijrothiya, S., Mishra, R., & Rawat, R. (2022). ML Techniques for Attack and Anomaly Detection in Internet of Things Networks. *Autonomous Vehicles Volume 1: Using Machine Intelligence*, 235-252.
32. Mahor, V., Bijrothiya, S., Mishra, R., Rawat, R., & Soni, A. (2022). The Smart City Based on AI and Infrastructure: A New Mobility Concepts and Realities. *Autonomous Vehicles Volume 1: Using Machine Intelligence*, 277-295.
33. Mahor, V., Pachlasiya, K., Garg, B., Chouhan, M., Telang, S., & Rawat, R. (2022, June). Mobile Operating System (Android) Vulnerability Analysis Using Machine Learning. In *Proceedings of International Conference on Network Security and Blockchain Technology: ICNSBT 2021* (pp. 159-169). Singapore: Springer Nature Singapore.
34. Mahor, V., Garg, B., Telang, S., Pachlasiya, K., Chouhan, M., & Rawat, R. (2022, June). Cyber Threat Phylogeny Assessment and Vulnerabilities Representation at Thermal Power Station. In *Proceedings of International Conference on Network Security and Blockchain Technology: ICNSBT 2021* (pp. 28-39). Singapore: Springer Nature Singapore.
35. Rawat, R., Gupta, S., Sivaranjani, S., CU, O. K., Kuliha, M., & Sankaran, K. S. (2022). Malevolent Information Crawling Mechanism for Forming Structured Illegal Organisations in Hidden Networks. *International Journal of Cyber Warfare and Terrorism (IJCWT)*, 12(1), 1-14.

18

Digital Footprints: Opportunities and Challenges for Online Robotic Technologies

Sudhir Kumar Rathi[1*], Pritam Prasad Lata[2†], Nitin Soni[2‡], Sanat Jain[3] and Shrikant Telang[4]

[1]*Faculty of CSE, Poornima University, Jaipur, India*
[2]*Department of Computer Applications, Sobhasaria College, Sikar, India*
[3]*Department of Computer Science and Engineering, Manipal University Jaipur, Rajasthan, India*
[4]*Department of Information Technology, Shri Vaishnav Vidyapeeth Vishwavidyalaya, Indore, India*

Abstract

The conception of digital vestiges is constantly applied and stands for a miracle of the ultramodern digital period. The users of digital services produce, designedly or intentionally, a kind of digital imprint which contains particular sensitive information. Particular data can be fairly tracked by digital service providers and latterly reused for marketable purposes, generally for targeted advertising, or misused for illegal purposes. Thus, particular data shall be regarded as an implicit trouble to existent's sequestration. It should be kept in mind that mindfulness about digital safety within society is still low and social websites encourage users to partake in particular sensitive data with an undisclosed range of connections. Some settings of internet cyber surfers allow them to track eyefuls or bare visiting websites enables technical programs to produce a comprehensive behavioral profile of one's private life, customs, social status, or consuming preferences. Current trends in digital security legislation seek for a balanced result between the right to sequestration and marketable interests of particular data processors. European legislation has begun to constitute consumer protection laws on particular data

**Corresponding author*: rathisudhir@gmail.com
†*Corresponding author*: pritamlata@gmail.com
‡*Corresponding author*: nsoni6789@gmail.com

Romil Rawat, Rajesh Kumar Chakrawarti, Sanjaya Kumar Sarangi, Rahul Choudhary, Anand Singh Gadwal and Vivek Bhardwaj (eds.) Robotic Process Automation, (275–284) © 2023 Scrivener Publishing LLC

protection and separate case law. This encompasses the operation of Google, Google Scholar, and different specific disciplines of certain countries.

***Keywords*:** Digital footprints, information mining, data mining, content mining, hacking

18.1 Introduction

A digital footmark is a commodity that's left behind every time a user pierces the internet or any other electronic system. It's produced when its services are penetrated. A digital footmark is also known as a shadow which is digital and refers to one's unique set of traceable digital [1, 2] conditioning, conduct, or dispatches that are manifested on the internet. Unlike the vestiges left in the sand at the beach, online visits and data frequently stick around long after the tide has gone out. Most of the users on the internet are ignorant to what particular information is available online about them and roughly one third of internet users point out that some pieces of information are available online like their dispatch address, home address, hearthstone phone number, or their employer's details. One quarter to one third of internet users claim they have no idea if their data points are available online. One quarter of internet users assert a print, names of groups they belong to, or effects they've mentioned that have their name on it. Many internet users expose that their particular information and cell phone number or videos and data appear online [3].

Social networking sites like Facebook, Twitter, Google, and LinkedIn are dominating the entire web. People can post nearly everything they can on the internet, like how they're feeling, what they're doing, their pursuits, what they like, what they don't like, etc. Everything is posted online. Since these conditions are posted online, they must be kept neatly and by notoriety. This can be achieved by using private servers, grid computing, or pall computing. Notoriety can be the companies like Facebook, Google or some other companies on which we're doing online exertion. Wherever the information is stored from, our online exertion is nothing but our online exertion log or digital footprint. A digital footmark can be anything that we partake in directly or laterally online, whether it's simple text, image, audio, or video; anything which is traceable back to us. In Figure 18.1, we can see different ways through which digital footprints come into life and how they are reused and employed. It can be evolved by content we post online like commentary, likes, image shares, etc. It can also evolve through

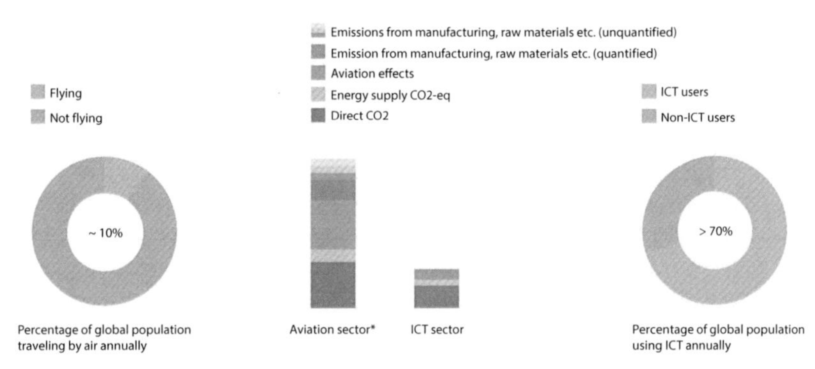

Figure 18.1 Carbon footprint comparison.

the content that our friends have posted about us like image markers, hash markers of our name, etc.

Users can cover their digital footmark by paying due attention to what they post online, abstaining from transferring dispatches which are suspicious, and by not posting questionable content.

18.1.1 Types of Digital Footprint

In the digital world, digital footprints have been classified into two types:

- Active Digital Footprints [4, 5]
- Passive Digital Footprints

18.1.1.1 Active Digital Footprints

Active digital vestiges are created when particular data is released by the user himself for the purpose of participating in sharing information about oneself by means of social point or social runner. In an online digital world, it can be stored by the user logging into a website when making a post or change, with the registered name being connected to the edit. In an offline digital world, the footmark may be stored as lines when the proprietor of the computer uses a Keylogger [6], so logs can reflect the conduct performed on the machine and who performed them. One of the features of crucial jack is to cover the action being performed on a device.

18.1.1.2 Passive Digital Footprints

A passive digital footprint is created when data is collected without bringing it into the notice of its proprietor. There are numerous ways of restoring digital vestiges. In an online digital world, a digital footmark may be stored as a "megahit" [7] in an online database. This footmark can indeed track a user's IP address, when it was created, and where it was created with the footmark latterly being anatomized. In an offline digital world, a footmark may be stored as lines which can be penetrated by the director to have a look on the conduct performed on the machine, without being suitable to see who performed them. While most Americans don't laboriously manage their online presence, there is a number of internet users who have jobs that encourage them to use their name in digital world or make information about themselves available online. As a matter of fact, most users seek for E-libraries in order to negotiate and enhance their exploration related tasks, hence they leave traces which might be dangerous. As one might anticipate, those motivated by work-related prospects are much more likely to use a search engine to track their digital footprints.

18.1.2 Evolution of Digital Footprint

In the early 20th century, Dr. Edmond Cargo remarked that "every contact leaves a trace" and that principle is no less true today in computer forensics. Any exertion which is performed leaves an IP address, a list of services taken used, and log lines created and whenever there's a conscious or unconscious online attack, the adversary leaves a digital footmark. On the base of overmentioned reflections and arguments, it can be refocused that one could use these footprints to improve the structure being used and help prevent attacks also. This is exactly the type of allowing in RISKIQ. We talked to a lot of the public about the problems around Advanced Persistent Pitfalls (APT) which, as users know, are a fast growing area in cyber security, according to Elias Manousos, author and CEO of RiskIQ. What people aren't really talking about is how the bad guys get down with this and how they can hideout on the Internet. RiskIQ [8] was erected with the motive to give associations with the same sense of controlling security, but outside of the firewall on the end points where guests and workers might be targeted with bait tactics (sites that might look real) or mischief apps (apps that look licit but are not). RiskIQ is continuously working for tracing an association's approved digital footmark: what belongs, what disciplines, what apps, what social media, and also the sites and apps that do not belong so they can be shut down. Rather than coming in the front door,

Manousos said, hackers will frequently find a copycat mobile operation or produce a bait sphere name with common types of brand name and use that as part of their social fashion of engineering, either in social media through a mobile or through a website. "We are all in a hurry," he justified, adding that associations aren't well defended from these attacks because out on the open Internet we're prey. We traditionally use commercial security controls. Digital vestiges (aka cyber or digital shadows) are digital shadows and traces on the internet that a user leaves behind as a result of web conditioning. Websites collect information as users use them, with or without the knowledge of the user. Whenever a user visits and enters data into a web point, they should be apprehensive that the data they've entered could be stored or used by the web point. The data can be anything from probing web runners, phone calls, online shopping, updates and uploads on Facebook and Twitter, and emails, to word quests on search engines such as Google, Bing, and Yahoo. In the cyber world, anything one performs, places which are visited, and online contents that are read and written are in some way stored and can be traced back to find the conditioning of the user. The principal handicap is getting worse, Manousos [9] asserts, because the incremental costs for bad guys to register a malicious website or develop a malicious mobile app is veritably low. Digital vestiges aren't a digital identity, but the content and useful data collected impacts upon Internet sequestration, trust, security, and digital character. According to the digital world, aspects of life, power, and rights of data have become important and necessary.

18.1.3 Managing Digital Footprints

Here are some free tools which can be utilized right away to track both active and passive digital footprints:

18.1.3.1 Google Alerts

The below mentioned free tools allow covering the web for content that's specified. The company's advice is to set up alerts for their username, as well as that of their academy and/or quarter. In this case, when a user was a professor he'd have an alert set up for New Milford High School so that any time content was posted specifically to his academy on the web, he could reply if necessary. Now, users have alerts set up for different duplications of the user's name including Eric Sheninger, Mr. Sheninger, and Professor Sheninger. Each day, he receives a dispatch with news of what people write about him on the web from Google [10].

18.1.3.2 Mention

Mention is considered one of the biggest competitors of Google alerts. It lets the users cover any keywords related to the user, user's [6] professional brand, school, sections, or anything additional information the user wants to cover. The alert settings are much more dependable and secure than that of Google alerts. Not only can user set it up to cover the web (news, blogs, vids, forums, images), but the user can also have it cover mentions on Facebook, Twitter, or any other social media services. What's better about Mention is the variety of ways users can see and be notified of new alerts.

18.1.3.3 Tweet Deck and Hootsuite

These applications not only enhance users' Twitter experience, but they also allow the users to create different columns or categories in each respective dashboard. Each column or category, in a sense, becomes a search based on the keywords which are identified. The above tools are very helpful in tracking digital footprint, especially in terms of what other people post about us. The user has to be very cautious while posting contents online and he should take care of the following points:

1. Keep it professional and focus on work.
2. Don't ignore the indispensable role of the user within the school and/or professional community.
3. Think before posting.
4. Be consistent.
5. Don't be afraid to engage.

18.1.4 Impact of Digital Footprint on Personal Privacy

User's digital footmark is anything about our personality that appears online, from websites and blogs to commentary posted on review sites. Indeed, if the user has substantially favorable online results, it's still a good idea to get a review on a web cybersurfer [7] on a regular basis. Effects can change snappily online. All it takes is one bad review to change our online perception. While it's unrealistic to avoid doing anything online if the user isn't expert, there are pros and cons to leaving a digital footmark.

18.1.4.1 Pros

Although a digital footmark may sound like a negative aspect of online exertion, as a log of nearly every activity is maintained actively or passively, there are numerous advantages of digital footmarks which help in making online experience more friendly and direct. Without a digital footmark online, the world may not be easy like it is today. Google Ads uses users' digital footmarks to serve them with applicable advertisements. However, this makes no sense to users and it will be seen as unsavory for the company, which will cause users to avoid these websites. If Google has not used the user's digital footmark data, it may also be that he's searching content related to a job and he's seeing an announcement of transnational vacation packages. Hence, with the help of a digital footmark Google can target the applicable user for showing the commercial, which is beneficial for both parties. YouTube is also one of the products handled by Google that uses our digital footmark [6] to ameliorate our affiliated videos list. For example, if the stoner searches for a movie, a suggested video is shown to him. Facebook also uses a digital footmark to give us a better social networking experience. It displays the announcement based on the user's preferences. It displays a suggested friend list based on the user's friend list and mutual friend list. Nowadays, users have GPS enabled mobile phones and tablets which add geographical information to our digital footmark. It also helps the user to better serve web services like Google Charts to find locations more directly with the help of searches and current GPS positioning.

18.1.4.2 Cons

A digital footmark helps the user have a better work experience, but one of the disadvantages of the digital footmark is that most of the time data is collected about a customer's online activity without the customer knowing, especially in an unresisting digital footmark. Today, users can browse numerous webpages in a nanosecond. We visit different types of websites with different terms and conditions which are long to read and most of time, users just agree with terms and conditions without reading them and start using the website and services, which results in giving legal rights for a website to collect their online use information. According to Microsoft's 2013 sequestration Survey Results, a majority of people don't bother about the terms and conditions of websites. Users [5] can control their digital footmark if they do proper sequestration setting and take precautions while probing, but today our digital footmark has become important to observe and maintain. A digital footmark can help others to prognosticate a users' personality which they don't want to partake with online. Terms of sequestration might mean users are surprised by the results shown by the Google search and they will be more surprised if they search their name in the image search tab. It might show a picture that we posted a long time ago on some social networking point and forgotten about. Social sites give users the privilege to control online participating effects, but occasionally users don't understand what's happening. Social networking websites like Facebook and Google give users control over what they partake in online through their sequestration setting features and in terms and conditions, they also mention that directly or laterally the overall rights of the content we partake in will belong to these social networking sites only. The users may kill their account, but there's no guarantee whatever they participated in online before killing the account will be destroyed permanently after killing the account. As most of the social networking sites do not destroy their stoner information, it may not appear on the web runner of the social networking spots, but they keep the data of every single user even if the user has left the site. They use data of those users to analyze what kind of users have left the website, how long they used the website, what their activity was before leaving, etc.

According to Microsoft's 2013 sequestration check, results guarding consumer's online sequestration is a participated responsibility which consists of individualities, companies, and governments.

18.1.5 Robots Reduce Carbon Footprint

Investments in modern robotics technology will also be driven by the requirement for a smaller carbon footprint. Modern robots are energy-efficient, thus directly reducing the energy consumption of manufacturing. Through higher precision, they also produce fewer rejections and substandard goods, which has a positive impact on the ratio of resource input over output. In addition, robots help in the cost-efficient production of renewable [4, 5] energy equipment such as photo voltaic or hydrogen fuel cells.

18.2 Proposed Methodology

The research deals with challenges of digital footprints and its applications. So, we need to revisit the old system used for the research on how digital footprints work and how we can track them with specified techniques and allow users to alert to better use their digital footprints.

Data [11, 12] left behind by users after they access the web [13] is termed a digital footprint. It comes in two classes [13, 14]: passive and active, and is used by businesses to offer you customised advertisements. But it may also be used by cybercriminals [15, 16] to gather intelligence about companies and launch customised cyberattacks [17] like phishing, identity theft, and tailgating.

18.3 Conclusion

It's no longer sufficient to simply dissect original computers and associated media when trying to roster a person's life works. Ever decreasing communication, particular documents, and published workshops are migrating to the web space. Social Networking spots contain prints, videos, and particular communication. Blog spots contain particular ramblings and narratives, both named and anonymous. E-mail and conversations, as well as particular videos, are also migrating to the web. There may be many disadvantages of digital footmarks, but in future they are going to play a pivotal part in serving the right content to web user, as well as help online web service providers to understand their guests and serve them. Every alternate active or unresisting digital footmark is generated online, and it'll keep generating as long as someone is using some online content, as it helps in furnishing affiliated web content to its applicable user. According to Microsoft's 2013 sequestration check, results guarding consumer's online sequestration is a participated responsibility which consists of individualities, companies, and governments.

Acknowledgement

I would also like to thank Dr. Sudheer Rathi for her advice and assistance in keeping my progress on schedule. My grateful thanks are also extended to my department faculties for help in doing the meteorological data analysis and helping me calculate the wind pressure coefficient.

I would also like to extend my thanks to the technicians of the laboratory of the Computer Science department for their help in offering me the resources in running the program.

Finally, I wish to thank my parents for their support and encouragement throughout my study.

References

1. Taylor, C. R. (2015). Creating win–win situations via advertising: new developments in digital out-of-home advertising. *International Journal of Advertising, 34*(2), 177-180.
2. Záleský, V. (2021). The Emergence of Digital Footprints and Digital Heritage in the Age of Big Data. *Megatrendy a médiá, 8*(1), 618-629.
3. Arya, V., Sethi, D., & Paul, J. (2019). Does digital footprint act as a digital asset?–Enhancing brand experience through remarketing. *International Journal of Information Management, 49*, 142-156.
4. Hicks, B., Culley, S., Gopsill, J., & Snider, C. (2020). Managing complex engineering projects: What can we learn from the evolving digital footprint?. *International Journal of Information Management, 51*, 102016.

5. Hinds, J., & Joinson, A. (2019). Human and computer personality prediction from digital footprints. *Current Directions in Psychological Science, 28*(2), 204-211.
6. McCaddon, David. Following Digital Footprints. Austin Macauley Publishers Limited, 2016, 2016.
7. Boudlaie, H., Nargesian, A., & Nik, B. K. (2019). Digital footprint in web 3.0: social media usage in recruitment. *AD-minister*, (34), 131-148.
8. Ray, B., Saha, K. K., Biswas, M., & Rahman, M. M. (2020, December). User perspective on usages and privacy of ehealth systems in bangladesh: A dhaka based survey. In *2020 IEEE Asia-Pacific Conference on Computer Science and Data Engineering (CSDE)* (pp. 1-5). IEEE.
9. Pasquini, L. A., & Eaton, P. W. (2021). Being/becoming professional online: Wayfinding through networked practices and digital experiences. *New Media & Society, 23*(5), 939-959.
10. Khalilov, M. C. K., & Levi, A. (2018). A survey on anonymity and privacy in bitcoin-like digital cash systems. *IEEE Communications Surveys & Tutorials, 20*(3), 2543-2585.
11. Mahor, V., Pachlasiya, K., Garg, B., Chouhan, M., Telang, S., & Rawat, R. (2022, June). Mobile Operating System (Android) Vulnerability Analysis Using Machine Learning. In *Proceedings of International Conference on Network Security and Blockchain Technology: ICNSBT 2021* (pp. 159-169). Singapore: Springer Nature Singapore.
12. Mahor, V., Garg, B., Telang, S., Pachlasiya, K., Chouhan, M., & Rawat, R. (2022, June). Cyber Threat Phylogeny Assessment and Vulnerabilities Representation at Thermal Power Station. In *Proceedings of International Conference on Network Security and Blockchain Technology: ICNSBT 2021* (pp. 28-39). Singapore: Springer Nature Singapore.
13. Rawat, R., Gupta, S., Sivaranjani, S., CU, O. K., Kuliha, M., & Sankaran, K. S. (2022). Malevolent Information Crawling Mechanism for Forming Structured Illegal Organisations in Hidden Networks. *International Journal of Cyber Warfare and Terrorism (IJCWT)*, 12(1), 1-14.
14. Rawat, R., Rimal, Y. N., William, P., Dahima, S., Gupta, S., & Sankaran, K. S. (2022). Malware Threat Affecting Financial Organization Analysis Using Machine Learning Approach. *International Journal of Information Technology and Web Engineering (IJITWE)*, 17(1), 1-20.
15. Rawat, R., Mahor, V., Chouhan, M., Pachlasiya, K., Telang, S., & Garg, B. (2022). Systematic literature Review (SLR) on social media and the Digital Transformation of Drug Trafficking on Darkweb. In *International Conference on Network Security and Blockchain Technology* (pp. 181-205). Springer, Singapore.
16. Rawat, R., Bhardwaj, P., Kaur, U., Telang, S., Chouhan, M., & Sankaran, K. S. (2023). *Smart Vehicles for Communication, Volume 2.* John Wiley & Sons.
17. Mahor, V., Bijrothiya, S., Rawat, R., Kumar, A., Garg, B., & Pachlasiya, K. (2023). IoT and Artificial Intelligence Techniques for Public Safety and Security. *Smart Urban Computing Applications*, 111.

19

SOCIAL MEDIA: The 21st Century's Latest Addiction Detracted Using Robotic Technology

Sudhir Kumar Rathi[1]*, Pritam Prasad Lata[2†] and Nitin Soni[2‡]

[1]*Faculty of CSE, Poornima University, Jaipur, India*
[2]*Department of Computer Applications, Sobhasaria College, Sikar, India*

Abstract
Today, internet technology shows quick progress and social networks increase their number of users each day. Rapid and deep penetration of the internet in India has impacted every aspect of life across all periods. One of the main indexes of the technology period is social networking, which attracts people of all periods, while the virtual world goes beyond real life via the operations it offers. Social Networking spots (SNSs) are virtual communities where users can produce individual public biographies, interact with real-life musketeers, and meet other people with similar interests. Young people especially show a violent interest in social media, which is an expansive example of internet technology. Social media dependence is growing both in India and around the world. This study aims to determine the position of social media dependence in young people and how many of them are addicted and make suggestions on the forestallment of the dependence. The current landscape of social media dependence and social media is examined in depth to determine causes of the dependence.

Keywords: Social media, hacking, digital world, digital social people, robotics software design, service robotics, medical robotics

**Corresponding author*: rathisudhir@gmail.com
†*Corresponding author*: pritamlata@gmail.com
‡*Corresponding author*: nsoni6789@gmail.com

Romil Rawat, Rajesh Kumar Chakrawarti, Sanjaya Kumar Sarangi, Rahul Choudhary, Anand Singh Gadwal and Vivek Bhardwaj (eds.) Robotic Process Automation, (285–296) © 2023 Scrivener Publishing LLC

19.1 Introduction

Blockchain is the world's most stormy distributed tally technology. In this exploration, the work concentrates on the process of DCS and SCADA [1] on the dark web platform for applying the new technology.

The internet is a worldwide system of computer networks, a network of networks in which users at any one computer can, if they have authorization, get information from any other computer.

A decade ago, the world knew little to nothing about the internet. It was a private cooperative of computer scientists and experimenters who used it to interact with associates in their separate disciplines. Today, the internet's magnitude is thousands of times what it was only a decade ago. It is estimated that about 60 [2] million host computers on the internet at present serve about 200 million users in over 200 countries and homes.

A popular name for the internet is the information superhighway. Whether people want to find the most recent fiscal news, exchange information with associates, browse through library canons, or join lively political debate, the internet is the tool that will take you beyond faxes, telephones, and insulated computers to a burgeoning network information frontier. The internet shrinks the world and brings information and knowledge on every subject imaginable straight to the computer. The internet has combined the technology [2] of dispatches and calculating to give instant connectivity and global information services to all its users at a veritably low cost.

People with a dependence aren't in control of the effects that are occurring. The dependence therefore reaches a point that's supposedly dangerous and has a dangerous hold over a person physically, emotionally, and socially and in turn affects the good of a person. Dependences not only include physical attributes that people consume, similar to alcohol or medicines, but also in that overconsumption of anything that can have similar abstract effects, from gambling to chocolate. In simpler terms, dependence may relate to a substance dependence (e.g., medicine dependence) or behavioral dependence (e.g., gambling dependence) [5].

19.2 Proposed Methodology

Young people especially show an intense interest in social media, which is an extension of internet technology. Social media addiction is increasing both in India and all around the world. This study aims to determine the level of social media addiction in young people and how many of them are addicted and also to make suggestions on the prevention of the addiction. The current environment of social media addiction is examined in depth to determine causes of the addiction.

19.3 Importance and Value of Internet

The internet is revolutionizing society, frugality, and technologic systems. No one knows for certain how far, or in which direction, the internet will evolve, but no one should underrate its significance. The internet has become a huge part of people's lives. Numerous people today count on the internet to do different tasks fluently with a few clicks. In fact, these days people are holding apps [6] and using the internet to play games or search for things that they want. But of course, the internet isn't just about entertainment. It's also useful in numerous other ways as well. The Figure 19.1 shows about the social network overview and the Figure 19.2 shows about the user activity design.

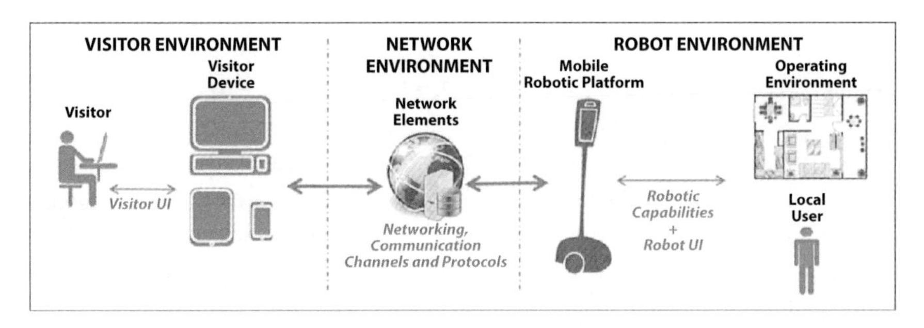

Figure 19.1 Social network overview.

Social media is veritably important when it comes to communication. Before, when people wanted to speak with someone who lived in a distant place, they would have to reach a phone and make a phone call or they would write a letter, which generally takes a many days to arrive. Now, through emails and social media, people can quickly communicate with their loved ones. People can also make video calls and see the person even if he or she is on the other side of the world! This benefits other diligence as well, particularly entrepreneurs and business owners. Now, business owners who would have to travel overseas [5, 6] to speak to a customer can make accommodations from the comfort of their own office.

Recently, social media, more than anything else, has significantly impacted most people's daily lives. Social media [7] connects people with the world and informs them of numerous things like news, history, entertainment, etc. Social media apps, like Facebook, Twitter, LinkedIn, Vine, and so on, make it simple to stay connected in people's lives. People can fluently catch up with friends with their status, photos, and videos they post.

Social media has become an essential tool for marketing and advertising. A business can present itself to customers with the use of a website or online announcements. Numerous businesses now use the internet as a means of making customers excited about their current promotions.

This can be good for businesses that are targeting young followership. At the present time, every business needs to work with the proper social media channels in the most stylish possible way not because it's the in thing or it sounds simple, but because their target followership is hanging around the popular social networks. By giving the social media touch to business, people can only induce more business and connect with their guests more to serve them at a higher level. It makes their online marketing easier. 71% of consumers are more likely to recommend a brand to another if they've had a positive experience with it on social media [8].

19.4 Effects of Online Addiction on Society

Social media has had an atrocious impact on culture in business and on the world-at-large. Today, the most popular stomping ground on the internet is social media websites [9]. Social media websites have streamlined the way people communicate and fraternize on the web. With most teens and adolescents, there's a thin line between casual internet use and dependence. For social commerce, information, and entertainment, the internet has become an important tool. Social media dependence is considered inadequately controlled internet use and can lead to impulse-control diseases. Social media dependence, especially among adolescents, has been honored as an important social issue in many countries because of the high frequencies of depression, aggressiveness, psychiatric symptoms, and interpersonal problems associated with this dependence. Adolescents are more vulnerable to social media dependence than grown-ups and because of this dependence, the social performance, psychology, and life habits of social media addicts can be affected. A multitude of cross-sectional studies have shown that social media dependence has an adverse effect on several life-related factors in adolescents; it can affect social habits, extend time spent on social networking sites, decrease physical inactivity, shorten duration of sleep, and increase use of alcohol and tobacco. The change in lifestyle-related factors caused by heavy social networking use could have an adverse impact on the growth and development of addicts.

Dependence on social media is a global problem affecting all generations and exploration has shown that substantial internet operation can have an extensively negative impact on people's internal and emotional health. Networks such as Twitter, Facebook, and Instagram continue to rise and the popularity of social media sites continuously grow. Three positive goods of social media include advertising, networking, and tone-expression. From the perspective of advertising, social media offers a great outlet for businesses, charities, and individuals to promote themselves. Social media also allows for tone-expression and can serve as a creative outlet for individuals to express themselves and share their opinion on specific motifs and their artwork. Being able to express yourself in a healthy way is a veritably important part of the mortal experience and social media can be a great outlet for young adults. Can following inspirational social media accounts, similar to fitness or health-inspired Instagram accounts, be motivational to followers?

Social media has positive goods, including advertising businesses [10], promoting mindfulness of specific causes, and helping foster fellowship between individuals who may have no way to meet without social networking. The over-usage of social media networks can also lead to negatives. For people who want to be productive, the internet offers numerous openings to pursue anything, but there are also vicious individuals who want to exploit the internet for selfish motives. Social media has both positive and negative effects on society. Positively, they can boost communication, make connections, improve education, help us keep in touch, be a source of literacy and tutoring, save time and plutocrat, allow communication with family, friends, and cousins, allow for empathy and to encourage community participation and ameliorate tone-confidence. On the other hand, dependence increases violence, affects tone- regard, can cause cyber-bullying and distraction, increase alcohol use, create insulation, spend further plutocrat, and increase pitfalls.

19.4.1 Positive Effects of Social Media

COMMUNICATION WITH FAMILY, FRIENDS, AND RELATIVES IS FASTER

It has never been easier to make friends than it is right now thanks to social media. Several decades ago, it was rather tough for people to connect with others if they didn't go out and make connections with others. Today, connecting people is easy with the changes in technology. It is completely possible to have thousands of friends on social networking spots like Facebook or Twitter. Services like Skype and Gmail [5] enable people to communicate with one another in nearly any part of the globe where internet services are available. Now, it is common for family members to keep up with cousins after they move to another state and to address them fluently with Skype, emails, and face to face communication, to make moving from home much easier via keeping in contact.

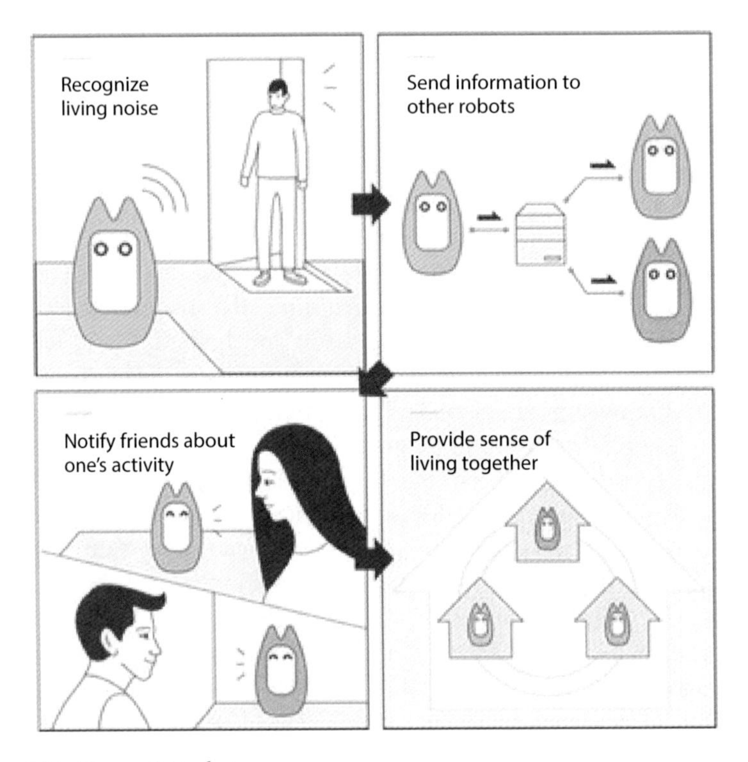

Figure 19.2 User activity design.

SOURCE OF LEARNING AND TEACHING

The goods of social media are huge for both scholars and preceptors because now we can go to the internet and learn or change knowledge. We can also get lots of motifs and sources there. By using those motifs, we can have great subjects and motifs to ameliorate knowledge at an academy. Participation on social networking sites gives people a chance to share instructional papers and videos with each other.

SAVE TIME AND MONEY

Obviously, we can stay at home and do workshops and search for useful information without going to the store or school. Older people can continue learning via online courses. With a network at home, teenagers or youth ameliorate knowledge. We can also buy products through sites.

BOOST COMMUNICATION

In today's busy society, time is being stretched between family and work commitments, yet social media will offer people a chance to communicate in an effective and quick way. We need just 20 seconds to write an update and that update will reach our friends.

ENCOURAGE COMMUNITY PARTICIPATION

Through forums, people, in general, and teenagers in particular, can contribute their own opinions as well as studies on certain subjects and themes that intrigue them. By taking part in conversations, they will engage in positive tone-expression.

19.4.2 Negative Effects of Social Media

ADDICTION

It is a reality today that social media is addictive. We can spend hours on the internet playing games, reading news, or working. Numerous people cannot go far from the computer. Rather than learning on social sites, youth can use networks to do other work and they can stay on the internet all day long.

INCREASED VIOLENCE [10]

It is inarguable that the internet gives a new approach to technology, as well as ultramodern life. Still, the pitfalls of violence and coitus abuse can increase due to some social media spots. Teenagers or youth can watch videos or read bad news and relate to negative action.

MAKE US UNHAPPY

Social media can reduce our moods and it can make people unhappy. There are numerous reasons that can make people unhappy when using social networks. Simply, online relations won't last like when you have a true fellowship with friends in real life.

AFFECT SELF-ESTEEM

Commentary and likes on Facebook can boost our tone-regard, but in some cases it also negatively impacts our tone-regard. Specifically, they singled out the blow to tone-regard that comes from comparing themselves to peers on Facebook and Twitter as the topmost downfall. People regularly look at others' social networking sites in respect to their clothes, life, or connections. Occasionally, they wish their own lives were as great as others. This problem isn't taken into consideration and people will end up feeling miserable, which could at times lead to low tone-regard and depression.

CYBER-BULLYING [9]
Today, cyber-bullying has become a crucial issue for teenagers. The proximity created by social media is available to both friends and foes. The desolation of the online attacks could leave deep-emotional scars. In some well-publicized cases, victims are driven to suicide. The insults hurled online could bring out dark impulses which may otherwise be suppressed. Cyber-bullying is popular among youth in recent years.

DISTRACTION [10]
One of the main negative effects of social networking is the distraction in diurnal life. People cannot go an hour without checking for updates and this could beget people to get detracted from what's actually important in life. Thus, rather than concentrating on work, family, and school, people spend the majority of their time on looking at status updates, photos, or captions. Social networking sites have shifted from a place where people can learn to a place where groups can become distracted because it's quickly changing from a platform where people communicate with other people into a platform where people announce themselves to others.

19.5 Challenges to Reduce Social Media Addiction

Using social media too much can be a problem. Social media is delightful and everyone is using Facebook, Instagram, YouTube, etc., but it has started to become a trend that can consume people's daily lives. It's hard to regulate when using social media; one click turns into another and another and all of the unforeseen hours pass by. Internet vacuity at low cost is one of the reasons for the social media dependence. Today, everyone can use social networking sites at a veritably low cost. Another reason why people can't quit social networking is that people follow brands and people to see what is trending or simply to influence others with their own prints and status. People who are feeling sad are prone to get pulled into the social media platform. Another reason why users can't stop using social media is because it is a one stop shop to consume news, catch-up with friends, and partake in particular information.

19.6 Challenges Future Impact of Social Media Addiction on Youth

The world is quickly moving towards ubiquitous connectivity that will further change where and how people associate, partake and gather information, and consume media.

According to statistic analysis, it is predicted that the number of social network users in India will be 258.27 million. It's a drastic rise from roughly 168 million users in 2016 (Nayar 2017). Now, with the internet period of digital and internet marketing, every company is trying to reach a good online presence. Social media marketing has a fantabulous future and it's clear that social media is here to stay in a strong way. In the future, the use of social media will be ubiquitous and integrated into people's daily lives in multiple of ways. It is anticipated that social media will be included into wearables that track people's habits. When news breaks in the future, it'll be covered by a multitude of observers streaming live video. These aqueducts will knit together into a single immersive videotape, enabling the bystander to nearly witness the event in real time. For better or worse, the world will feel like it's right around the corner.

- Cory Bergman, Breaking News Co-founder (Bergman 2004) [3, 4]

Regarding the future of social media, the primary generalities are the connection between social media sites and internal health. The dependence of media and technology has a negative effect on the health of children and teens by making them more prone to anxiety and depression and numerous other health problems. Feeling the need to see posts and constantly check announcements can cause gratuitous stress for adolescents.

Online social network (OSN) [11–13] credentials are frequently used by attackers [14–16] during the surveillance stage of a social engineering [17] or phishing operation. OSN can provide attackers with the knowledge they need to conduct subsequent exploits, such as social engineering and spamming [18], or a platform to pose as reliable individuals or companies. The OSN platform that is being attacked [19] determines the offensive strategy. An attacker would frequently friend a specified user's friends or submit a friend request straight to a chosen user in order to view their comments because Facebook [20] participants can keep their photographs and opinions secret. According to the number of linked friends, the likelihood that the targeted user would accept a friend request increases if an attacker [21] can link to many of the intended user's friends.

19.7 Conclusion

The overall result proves that social media addiction will rise in the future and users of these networking sites will increase because the internet becomes much cheaper and accessible. Specific and targeted efforts need to counter online risk in order for youth to benefit from the opportunities offered by the internet. Thus, social media is very important for people to get information, connecting to family and friends, encourage community participation, and boost communication for business and it is impossible to separate social media from the online world. Social network is a revolutionary idea with a bright future with further scope for advancement and the future of social networking seems promising and has a positive effect on society only if this technology will be used in the proper and right way.

Acknowledgement

We would also like to thank Dr. Sudheer Rathi for her advice and assistance in keeping my progress on schedule. My grateful thanks are also extended to my department faculties for help in doing the meteorological data analysis and helped me calculate the wind pressure coefficient.

I would also like to extend my thanks to the technicians of the laboratory of the Computer Science department for their help in offering me the resources in running the program.

Finally, I wish to thank my parents for their support and encouragement throughout my study.

References

1. Crammond, R., Omeihe, K. O., Murray, A., & Ledger, K. (2018). Managing knowledge through social media: Modelling an entrepreneurial approach for Scottish SMEs and beyond. *Baltic Journal of Management.*
2. Appel, G., Grewal, L., Hadi, R., & Stephen, A. T. (2020). The future of social media in marketing. *Journal of the Academy of Marketing Science, 48*(1), 79-95.

3. Sheth, J. N. (2018). How social media will impact marketing media. In *Social media marketing* (pp. 3-18). Palgrave Macmillan, Singapore.
4. Jain, V. (2021). An overview of facebook. *ACADEMICIA: An International Multidisciplinary Research Journal, 11*(12), 782-788.
5. Eldridge, R. (2019). Analytic Philosophy of Film:(Contrasted with Continental Film Theory). In *The Palgrave Handbook of the Philosophy of Film and Motion Pictures* (pp. 237-258). Palgrave Macmillan, Cham.
6. Kalia, P., Kaur, N., & Singh, T. (2018). E-Commerce in India: evolution and revolution of online retail. In *Mobile commerce: Concepts, methodologies, tools, and applications* (pp. 736-758). IGI Global.
7. Duffy, B. E., & Chan, N. K. (2019). "You never really know who's looking": Imagined surveillance across social media platforms. *New Media & Society, 21*(1), 119-138.
8. Hamutoglu, N. B., Topal, M., & Gezgin, D. M. (2020). Investigating Direct and Indirect Effects of Social Media Addiction, Social Media Usage and Personality Traits on FOMO. *International Journal of Progressive Education, 16*(2), 248-261.
9. Sahin, C. (2018). Social Media Addiction Scale-Student Form: The Reliability and Validity Study. *Turkish Online Journal of Educational Technology-TOJET, 17*(1), 169-182.
10. Lee, S., Yoon, C., Kang, H., Kim, Y., Kim, Y., Han, D., ... & Shin, S. (2019, February). Cybercriminal minds: an investigative study of cryptocurrency abuses in the dark web. In *26th Annual Network and Distributed System Security Symposium (NDSS 2019)* (pp. 1-15). Internet Society.
11. Rawat, R. (2023). Logical concept mapping and social media analytics relating to cyber criminal activities for ontology creation. *International Journal of Information Technology*, 15(2), 893-903.
12. Rawat, R., Mahor, V., Álvarez, J. D., & Ch, F. (2023). Cognitive Systems for Dark Web Cyber Delinquent Association Malignant Data Crawling: A Review. *Handbook of Research on War Policies, Strategies, and Cyber Wars*, 45-63.
13. Rawat, R., Chakrawarti, R. K., Vyas, P., Gonzáles, J. L. A., Sikarwar, R., & Bhardwaj, R. (2023). Intelligent Fog Computing Surveillance System for Crime and Vulnerability Identification and Tracing. *International Journal of Information Security and Privacy (IJISP)*, 17(1), 1-25.
14. Rawat, R., Sowjanya, A. M., Patel, S. I., Jaiswal, V., Khan, I., & Balaram, A. (Eds.). (2022). *Using Machine Intelligence: Autonomous Vehicles Volume 1.* John Wiley & Sons.
15. Rawat, R., Bhardwaj, P., Kaur, U., Telang, S., Chouhan, M., & Sankaran, K. S. (2023). *Smart Vehicles for Communication, Volume 2.* John Wiley & Sons.
16. Mahor, V., Bijrothiya, S., Rawat, R., Kumar, A., Garg, B., & Pachlasiya, K. (2023). IoT and Artificial Intelligence Techniques for Public Safety and Security. *Smart Urban Computing Applications*, 111.

17. Mahor, V., Bijrothiya, S., Mishra, R., & Rawat, R. (2022). ML Techniques for Attack and Anomaly Detection in Internet of Things Networks. *Autonomous Vehicles Volume 1: Using Machine Intelligence*, 235-252.
18. Mahor, V., Bijrothiya, S., Mishra, R., Rawat, R., & Soni, A. (2022). The Smart City Based on AI and Infrastructure: A New Mobility Concepts and Realities. *Autonomous Vehicles Volume 1: Using Machine Intelligence*, 277-295.
19. Mahor, V., Pachlasiya, K., Garg, B., Chouhan, M., Telang, S., & Rawat, R. (2022, June). Mobile Operating System (Android) Vulnerability Analysis Using Machine Learning. In *Proceedings of International Conference on Network Security and Blockchain Technology: ICNSBT 2021* (pp. 159-169). Singapore: Springer Nature Singapore.
20. Mahor, V., Garg, B., Telang, S., Pachlasiya, K., Chouhan, M., & Rawat, R. (2022, June). Cyber Threat Phylogeny Assessment and Vulnerabilities Representation at Thermal Power Station. In *Proceedings of International Conference on Network Security and Blockchain Technology: ICNSBT 2021* (pp. 28-39). Singapore: Springer Nature Singapore.
21. Rawat, R., Gupta, S., Sivaranjani, S., CU, O. K., Kuliha, M., & Sankaran, K. S. (2022). Malevolent Information Crawling Mechanism for Forming Structured Illegal Organisations in Hidden Networks. *International Journal of Cyber Warfare and Terrorism (IJCWT)*, 12(1), 1-14.

20

Future of Digital Work Force in Robotic Process Automation

P. William[1]*, Vishal M. Tidake[2], Sandip R. Thorat[3] and Apurv Verma[4]

[1]*Department of Information Technology, Sanjivani College of Engineering, SPPU, Pune, India*
[2]*Department of MBA, Sanjivani College of Engineering, SPPU, Pune, India*
[3]*Department of Mechanical Engineering, Sanjivani College of Engineering, SPPU, Pune, India*
[4]*Department of Computer Science and Engineering, MATS University, Raipur, India*

Abstract

Robotic Process Automation (RPA) represents an advance in the state of the art in terms of technology. Robotic Process Automation (RPA) is one of the most cutting-edge technologies in the fields of information technology, computer science, electrical engineering, and mechanical engineering. This system combines both hardware and software, as well as networking and automation, in order to make doing mundane tasks as easy and stress-free as is humanly feasible. Secondary data was analysed in light of this, which may be accessible on the internet and in academic databases. From January 1, 2018 until June 30, 2018, the research was conducted. RPA found and utilised a small number of empirical publications, white papers, and blogs in the creation of this study. Because of the contemporary phenomenon, this inquiry has an exploratory nature. RPA, Robots, AI, and Blue Prism were some of the search phrases used to access the database of information. The result of the study was that robots and RPA technologies are becoming a need for businesses across the globe to run their operations. With Robotic Process Automation (RPA), fundamental company activities including payroll, new hire onboarding, employee status changes, accounts receivable/payable, invoice processing and inventory management, report production, and software installation may be handled promptly. Healthcare and pharmaceuticals are just some of the many industries where RPA may be put to use. It can also be employed in banking and financial services, as well as retail and energy enterprises and fast-moving consumer products (FMCG). Robotic process automation (RPA) is gaining ground

**Corresponding author*: william160891@gmail.com

Romil Rawat, Rajesh Kumar Chakrawarti, Sanjaya Kumar Sarangi, Rahul Choudhary, Anand Singh Gadwal and Vivek Bhardwaj (eds.) Robotic Process Automation, (297–314) © 2023 Scrivener Publishing LLC

in the business world thanks to the convergence of artificial intelligence, machine learning, deep learning, HR analytics, virtual reality (second life), home automation, and blockchain technology. Numerous start-ups and established companies using RPA solutions are included in this resource. This text will provide academics, researchers, students, and practitioners with a general understanding.

***Keywords*:** Automation, robotic automatic process, RPA, blue prism, business process outsourcing

20.1 Introduction

"Watson [1] is building a new kind of human-computer cooperation that will expand, scale, and speed up human knowledge acquisition".

J. Presper Eckert and John Mauchly of the University of Pennsylvania [2] conceived the ENIAC, the world's first electronic digital computer, in 1943. However, construction on the ENIAC [1] did not begin until 1946. Some people believe the abacus was the first computer ever created. Several decades ago, this is how the age of computers got its start. When it comes to desktop computers, servers, laptops, and smartphones, the world of information technology is undergoing a metamorphosis like that of a chameleon. The introduction of rolltops might be considered a recent development. At the forefront of high-bandwidth networking, significant advancements in software have been achieved in the areas of operating systems, applications, utilities, and computational power. Organizational applications such as punch cards, spreadsheets, office software, management information systems, and enterprise resource planning are some examples of the types of applications that have helped enable the growth of business applications. Robotic Process Automation (RPA) [2] is now gaining momentum in the business world. As a result, "Robotic Process Automation" was proclaimed in the year of 2018. "Robotic Process Automation," or "RPA," is a new area of technology that's being introduced to the business world. The day-to-day operations of a business can no longer function without RPA and its implementation must be hastened. A company that does not utilise this technology in its operations may not be able to compete in the near future. Using software robots or an artificial intelligence (AI) [3] workforce, RPA is a developing business process automation approach. In today's commercial world, RPA is the new standard. This is the most powerful technology of the 21st century. New technology, software, and smart devices will revolutionise business, government, and everyday life. They will make it simpler for everyone to carry out their responsibilities and maintain connections with one another. The lives of people all over the world are being profoundly altered as a result of global collaboration, the existence of international organisations, and recent developments in IT/ITes that make use of RPA technology [4]. Along with the Internet of Things, Big Data Analytics, Deep Learning, Artificial Intelligence, and Machine Learning, Robotic Process Automation (RPA) is quickly becoming one of the most significant disruptive technologies. Research reveals that both human-robot interaction (HRI) and robot companionship have favourable psychological effects on individuals, which might be used as a solution for the scarcity of human resources [5]. The results of a recently conducted study on robotic process automation are valuable to businesses of all sizes, from Fortune 500 corporations to fledgling start-ups. A definition of robots, automation, and the significance of robotic automation is presented at the beginning of the RPA-specific information.

20.1.1 Robots

It is possible to train robots, which are mechanical machines controlled by computers, to carry out a series of more complex tasks on their own. The accomplishment of a robot's task depends on its ability to move about in the physical environment in which it operates. These sophisticated robots are outfitted with state-of-the-art perception-to-action connecting systems. If the connection between the two is going to be intelligent, then Artificial Intelligence (AI) will need to play a significant part in robotics [6]. The creation of robotics requires contributions from a broad variety of technical and scientific fields, such as mechanical engineering, computer science, and electrical engineering, amongst others. One such example is Hanson Robotics' social humanoid robot named Sophia, which was developed in Hong Kong by Hanson Robotics. After being activated on April 19, 2015, Sophia made her first appearance in public during the Southwest Festival (SXSW) in Austin, Texas, United States [7], in the middle of March 2016. SXSW is held annually in Austin, Texas, United States (Harriet Taylor, 2016). According to Jump Up (2018), Sophia is capable of exhibiting more than fifty unique emotions on her face. According to Gershgorn, reviewers of open-source code agree that the most accurate description of Sophia is that of a Chabot with a face (2016). Sophia [8], the world's first humanoid robot, has been granted citizenship in Saudi Arabia. She is the first robot of her kind in the kingdom. By the year 2020, ten percent of the world's major firms that are reliant on supply chains will have a chief robotics officer on staff (Douglas, and Ankush, 2017) [9]. In recent years, there has been a meteoric rise in the number of start-ups in the robotics business. Manufacturing facilities often make use of robots and other equipment that include artificial intelligence. Using wireless networking, big data, cloud computing, statistical machine learning, and open-source software are just some of the ways that robots may increase their performance in a broad variety of applications. There are several other ways that they can do this as well (Kehoe, B. *et al.*, 2015). The phrase "swarm robotics" comes to mind when thinking about the process of coordinating the actions of a huge number of very basic robots. Palgrave and colleagues suggest that it is possible to build autonomous robots using evolutionary robotics in a way that is completely automated (2000). With this approach, robots are envisioned as self-sufficient artificial creatures who may develop their own abilities via intimate contact with their surroundings without the intervention of humans. The purpose of ambient robotics is to combine physical and informational modifications in order to close the gap that currently exists between service robots and situations that are beneficial [10]. Facility planners are able to view the system before it is built and come up with alternative designs by using 3D graphics. Additionally, facility planners can use 3D graphics to programme robot paths, gather system layouts, collect data for discrete event simulations, and generate a cell control programme (Gunasekaran, 1999). In the not-too-distant future, robots will turn into an indispensable component of human civilization.

20.1.2 Process

All sectors and individuals use the term "Process" on a daily basis and the term's meaning is universally recognised. It is a vital component of any system or organisation, as well as the method of carrying out a mission or undertaking. It's possible that people, objects, or some mix of the two may finish the process. Regardless of whether the system is closed or open, the process receives inputs from a variety of devices or persons and then carries out its instructions in accordance with the predetermined rules in order to generate the desired output. Only the transformation of input to output is involved in the process. When it comes to the length and costs of a system's or a process's implementation, there is no one-size-fits-all. All kinds of processes are familiar to people, from biological to chemical to admissions. As an example, consider a network of computer processors. A keyboard, joystick, mouse, and voice recognition may all be used to control it. It uses a central processing unit (CPU) or a graphics processing unit (GPU) [11] to carry out activities as instructed and outputs the results to a screen, printer, or an invoice format. Processors with a single or several cores may do a single or multiple tasks at the same time. Concurrent procedures may be performed with remarkable speed and accuracy in today's contemporary computer environment. Thanks to cutting-edge technology that organisations have access to, the latest versions of hardware, software, and CPUs are found in all smart devices, including personal digital assistants. Intel's 8th-generation Core CPUs (Coffee Lake) and AMD's Ryzen 3 and 5 chipsets with RX Vega 11 graphics are the latest processors available. Even in the remarkable technology known as "Robotic Process Automation," processes/processors are the most important duties.

20.1.3 Automation

When a product, process, or system is made to function automatically, it is called "automation" or "automated." Despite this, the advantages of automated applications are already being felt by the general public. A system's ability to process information is a key component of its automation. People and systems must be integrated to achieve automation. It is quite typical for human considerations in system design to be misunderstood or ignored entirely throughout the design process [12]. The development of an application program to control an industrial automation mechanism by using logic, motion, and/or process control components, also includes running, monitoring, and troubleshooting the application [13]. The purpose of the energy monitoring system that is put in the typical home is to cut down on the amount of electricity that is used. Some of the most recent advancements in home automation technology include the Lutron Dimmer Light Switches, the Sonos Wireless Speaker System, Philips Hue Smart Light Bulbs, the Belkin WeMo Switch and Motion Sensor, the Amazon Echo device Alexa, and the Amazon Echo Dot. Because of the software and technology that is built in these intelligent devices, they are able to complete tasks without human intervention. As a result of their ability to enhance human well-being and promote operational efficiency, these intelligent devices have become an essential part of our daily lives [5]. Automated processes simplify, speed up, and improve the efficiency of routine tasks, freeing people from the burden of repetitive labour and boredom. Driverless/autonomous cars use cutting-edge automation technology such as machine vision and artificial intelligence, which are detailed in detail. Internet of Things-connected autos and collaborative robots use cognitive computing. Automation has already been used in the industrial sector, most notably via the use of "Industry 4.0" technology [6]. Robotic production lines are becoming more commonplace.

20.2 Robotic Process Automation

The word "Process" is well-known to everyone and is related with people's daily lives and widespread across all industries. It is an important component of every system or organisation and the process of completing a job. People, objects, or a mix of both may complete the procedure. Regardless of whether the system is closed or open, the process gets inputs from different devices or individuals and executes according to established rules in order to produce the expected output. Only the transformation of input to output is involved in the process. When it comes to the length and costs of a system's or a process's implementation, there is no one-size-fits-all. All kinds of processes are familiar to people, from biological to chemical to admissions. As an example, consider a network of computer processors. Controlling it might be done with anything from a keyboard and mouse to a joystick and even voice recognition. It does work by using a central processing unit (CPU) or a graphics processing unit (GPU) according to the instructions that are given and then it shows the results on a screen, printer, or invoice form. Processors having a single or several cores are being used to do single or numerous tasks concurrently. In today's modern computing environment, businesses have access to cutting-edge technology that allows them to perform concurrent processes with great speed and precision within the defined restrictions. Personal digital assistants, as well as all other smart gadgets, are equipped with the latest versions of hardware, software, and CPUs. Currently, the most modern CPUs are Intel Core I3 - I7 8th Gen (Coffee Lake) and AMD Ryzen 3 and 5 with RX Vega 11 Graphics. Even in the spectacular "Robotic Process Automation" technology, processes/processors are the most significant responsibilities in this astonishing new technology [7].

When a product, process, or system is made to function automatically, it is called "automation" or "automated." Despite this, the advantages of automated applications are already being felt by the general public. A system's ability to process information is a key component of its automation. People and systems must be integrated to achieve automation. Misunderstanding and overlooking of human factors in system design is a common problem (Sheridan, 2002). The process of constructing an application program to operate an industrial automation mechanism using logic, motion, and/or process control components, as well as executing, monitoring, and debugging it [8]. Reducing power consumption is the goal of the energy monitoring system installed in the average house. Sonos Wireless Speaker System, Philips Hue Smart Light Bulbs, Belkin WeMo Switch and Motion Sensor, Amazon Echo device Alexa, and Lutron Dimmer Light Switches are some of the most recent home automation smart goods. These intelligent devices are able to carry out tasks on their own thanks to embedded software and hardware. As a result of their ability to enhance human well-being and promote operational efficiency, these intelligent devices have become an essential part of our daily lives [9]. Automated processes simplify, speed up, and improve the efficiency of routine tasks, freeing people from the burden of repetitive labour and boredom. Driverless/autonomous cars use cutting-edge automation technology such as machine vision and artificial intelligence, which are detailed in detail. Internet of Things-connected autos and collaborative robots use cognitive computing. Automation has already been used in the industrial sector, most notably via the use of "Industry 4.0" technology. Robotic production lines are becoming more commonplace.

20.3 Robotic Process Automation Operations

There are currently no standard operating models available for robotic process automation (RPA) operations. Mr. Kristina Romero and his technical team present the Robotic Process Automation (RPA) operations indicated in Figure 20.1, as reported by InfoCap Networks LLC (InfoCap) in San Diego. Automating content-enabled manual labour is the primary goal of this technological paradigm, which will replace the whole business system's operations (Kristina Romero, 2017). In this perspective, "Digital Labor's" primary benefit is to cut costs, limit mistakes, and remove hazards. The RPA offers several operational benefits to multifunctional and diverse businesses.

Credible Business Transformation: Processes in businesses will be completely redesigned as a result of the introduction of RPA. Now, with the help of Robotic Process Automation (RPA), businesses can boost their workforce's productivity significantly while also providing more reliable, cost-effective digital labour. As a result, companies are able to save costs, reduce errors, and eliminate risk. The Figure 20.2 shows about the system compliance.

Figure 20.1 RPA modelling.

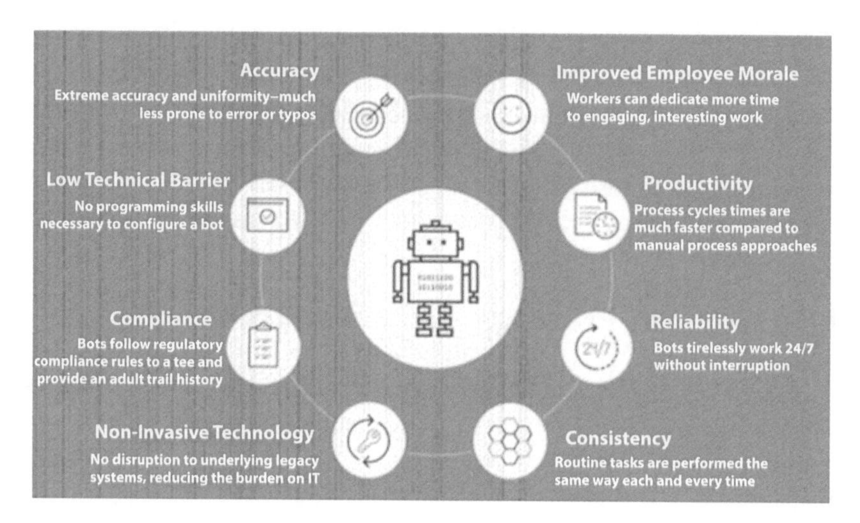

Figure 20.2 System compliance.

Content Migrations: Every company is churning out an enormous amount of content. The number of people required to collect, review, and create a report may rise as the complexity of daily tasks rises. Businesses and organisations may only benefit from Robotic Process Automation by speeding up application consolidation and legacy application integration by transferring information or connecting to older systems more quickly and with less effort.

Web Crawling/OSINT: Robotic Process Automation (RPA) [7] uses a wide range of devices to gather data in any format from any source. Text, photos, audio, and video are all examples of media. The data may be organised, semi-structured, or unstructured, depending on your preferences. This Robotic Process Automation system can collect deep web data using deep learning approaches. In addition, artificial intelligence, big data analytics, and other web analytics will be used to complete the mining process.

IT Department Enabler: A blog post by the authors of the book describes "Robots" as "software programs that mimic human-computer interactions, execute a repetitive process, rules-based tasks such as collecting, comparing data from multiple systems, reading/writing to databases [8], or extracting/formatting data for reports and dashboards." They keep an eye on the hardware, software, and networking in order to identify problems and ensure smooth operations.

20.4 RPA-Operating Model Design

Business process automation based on artificially intelligent software robots or employees is called Robotic Process Automation (RPA). When Rodger Howell and Tom Torlone [10] wrote a noteworthy essay in 2017, they focused on the development of Robotic Process Automation technologies. According to them, pilot programs are helping to bring Robotic Process Automation to life. RPA models should be created by organisations themselves in order to maximise operational efficiency and save costs. RPA operational models might differ across industries and enterprises, according to the authors of this paper. Operating frameworks for Robotic Process Automation aren't, in their opinion, a "one size fits all" proposition. A successful RPA operational model, on the other hand, prioritises three key functions. Contributors to the design of future-state processes are made possible through robotic process automation. The technique consists only of work completion. It is linked to several software components, such as Interactive Voice Response, hardware components, such as robots, and little human participation. In addition to long-term and short-term processes/round robin, there are other kinds of processes, such as Priority (FCFS)/First In First Out (FIFO)/Last In First Out (LIFO) (LIFO) [12]. All of these procedures in both centralised and decentralised process systems must be defined by the process architects. They must first grasp the present system flow and identify the system's missing components, tasks, time requirements, cost savings, and overall system efficiency. In certain cases, business analysts are responsible for the automation of business processes. Standardized technique, protocol, and standards for robotic procedure automation are established by the process architects.

20.4.1 Technologists

Who is responsible for writing the code that translates business logic into a robot's actions? The term "technologist" refers to a scientist or engineer who focuses on a particular technology or who uses technology in a particular industry or profession. Firms now have additional options available to them. Those who write the code and program it follow the Software Requirement Specifications (SRS) [13] and work with functionalists, designers, and executors to gather feedback (SRS). In certain cases, these computer programs can do routine activities without human intervention. To further enhance a Robotic Process Automation program, the authors observe that a Robotic Process Automation Center of Excellence will often deploy an enthusiastic team of developers in a low-cost location.

Compared to conventional application development, the technical abilities needed for RPA tools are comparatively less advanced.

20.4.2 Ongoing Support and Maintenance Staff

How do you go about automating your processes and implementing critical software upgrades? An annual maintenance contract (AMC) with the software vendor or provider is commonly used to do this. Debugging system software and application errors and failures is simple. If the AMC is active, they provide technical support at any given time of the day or night throughout the whole year. This kind of technical help significantly saves the cost and time required to acquire technical personnel for internal usage. In addition, there would be no training expenses. The firms' continuing support packages are targeted to the requirements of the customer's company, not the demands of the masses. Whether a consumer needs help with a single issue or weekly evaluations and guidance, the company offers a package to meet their demands, giving them complete peace of mind.

20.5 Who is Who in RPA Business?

Robotic Process Automation (RPA) has the potential to completely revamp the way business operations are conducted as a result of developments in both hardware and software, particularly Artificial Intelligence (AI) [11]. According to Willcocks and Lacity (2016), Robotic Process Automation (RPA) and cognitive automation have the potential to be very disruptive to the more conventional people-centric outsourcing paradigm. This is the paradigm on which offshore outsourcing organisations and captive centres were formed. RPA and cognitive automation have been observed as having this potential. Customers will make new demands on cloud computing, robotic process automation, and as-a-service with each contract renewal cycle from 2017 to 2020 and service providers will adjust their products to reflect these changes. Customers will make these new requests during the contract renewal cycles [12]. There has been a steady growth in the usage of Robotic Process Automation (RPA) by BPO providers over the last several years and the technology is now being adopted by a growing number of end-user organisations. RP Automation refers to the automation of formerly human-performed tasks via the use of technology. User actions, like browsing through an application or filling out a form, are mimicked by the technology according on a set of rules [10]. RPA combines a number of new techniques to artificial intelligence, including virtual agents, machine learning, computer vision, and natural language classification, among others. Two potential applications of artificial intelligence technology in the insurance industry are image categorization for the purpose of claims processing and text analytics for the purpose of responding to consumer inquiries. The automation and improvement of insurance processes will be hastened by the introduction of these new technologies.

Blue Prism has developed a digital workforce operating system, Blue Prism Version 6. In the field of Robotic Process Automation, this company is a trailblazer. As a consequence of this, they think that effective digital strategies need a connected organisation in which technologies such as artificial intelligence (AI), machine learning (ML), and sentiment analysis can be readily linked with company activities in order to generate value. A "digital workforce" may be created by using the Blue Prism v 6 software platform, which automates back-office administrative tasks, saving money and enhancing accuracy while creating a flexible and cost-effective business operation.

For the following problems, Blue Prism's Robotic Process Automation software is an excellent solution:

Inputs that are digitally organised, such as credit card activation or fraud detection, are processed according to predetermined rules:

- Repeated transactions, such as switching SIM cards or completing invoices
- In order to conduct complex and mission-critical tasks, such as financial reconciliation and pension redemptions
- A large amount of transactions, such as invoices or new phone orders
- Process adherence/quality issues, such as the renewal and relocation of policies
- Demand fluctuations or backlogs, such as when a new product is introduced
- Employee on-boarding or the release of a new online product without integration are examples of "swivel chair" methods

Due to the company's ability to deploy software robots quickly, M/s UiPath is at the forefront of the international digital business transformation, which it is helping to drive through improving corporate efficiency, compliance, and front- and back-office customer service. RPA systems can, in the words of UiPath, log into programmes, transfer files and directories, copy and paste data, fill out forms, extract structured and semi-structured data from documents, scrape web browsers, and do a great deal of other tasks as well. UiPath was valued at $1.1 billion as of March 6, 2018, after the receipt of an investment of $153 million from Accel [3], CapitalG, and Kleiner Perkins Caufield & Byers. RPA, or robotic process automation, is one of the technologies that is used the most in today's workplaces and Automation Anywhere is one of the most effective instances of this technology. With the help of Automation Anywhere's Bot Store, companies can swiftly expand their digital workforces thanks to the start-up's ground-breaking technology. Human workers benefit from this since it frees them up to work on projects that are more human-centred. Several of the world's largest healthcare, financial services, technology, manufacturing, and insurance firms have relied on M/s Automation Anywhere for over a decade to supply the most sophisticated robotic process automation and cognitive technologies.

One of the leading automation and cloud consulting firms in the United Kingdom is Endpoint Automated Services (EAS). M/s Endpoint Automated Services (EAS) extends its extensive automation expertise to the Robotic Process Automation (RPA) field. For robots, automation entails automating a whole business process from data collection through data entry into an accounting system. Automated processes have been included in even the most basic functions of Microsoft Office. RPA is a well-known code-free automation technique. Screen anchors may be used as an alternative screen scraping method for robots instead of pixels. Robotic Process Automation, according to M/s Endpoint Automated Services, creates instant profits while boosting accuracy in all industries and organisations. As with human beings, software robots may do a broad range of monotonous tasks by reading text, initiating responses, and connecting with other systems. Unlike a human employee, a robot doesn't need sleep, is error-free, and costs less than one.

Founded in 1985 by David Silver and Dean Hough, M/s Kofax Inc., located in Irvine, California, is a global provider of process automation software. Kofax provides automated techniques for extracting and categorising data from both paper and digital media. As a consequence of this effective data processing, human error and contact are minimised to the greatest extent possible. Document Workflow, Imaging, Intelligent Capture, and Business Process Management are all products that Kofax created in response to the increasing need for mobile apps.

"O2" [6] is a telecommunication company based in London, England (UK). It is one of the most prominent digital communications organisations in the United Kingdom due to its more than 25 million users, more than 450 retail locations, and many advertisers. Tesco Mobile and O2

Wireless Fieldality are two of the mobile networks that O2 operates in the United Kingdom. Both of these networks are owned by O2, the corporation that operates the networks. O2 has eliminated the need for 150 full-time employees over the course of the three years between 2012 and 2015 by using a single piece of automated software. An employee of O2 operations named Mr. Wayne Butterfield said that the majority of the company's customer assistance is now handled by automated systems. In addition to that, he discussed "SIM swaps," which include changing cell numbers, moving from prepaid to contract service, and unlocking an O2 phone. This is a working solution for RPA that operates in real time. The in-house robots at Cap Gemini concluded their work in 2015 after having completed 1.5 million transactions, which is the equivalent of having 200 people. In the research, process design had a bigger influence on ROI than the technology utilised. RPA technology has been used by a bank to reform its claims process and handle 1.5 million requests yearly, according to Deloitte LP's managing director David Schatsky, who spoke at a conference on the subject. More than 200 full-time employees may be added to the bank's capacity for a fraction of the expense of hiring more staff [5]. These companies are: AT&T, Walgreens, American Express, Deutsche, Ernst & Young and Vanguard, and Anthem. An increasing number of companies are using RPA technology, including Global Business Travel.

For example, RPA may be used in a variety of industries, including the following: healthcare and pharmaceuticals, financial services, outsourced services, retail (including telecommunications and utilities), fast-moving consumer goods (FMCG), real estate, and others. Automated marketing campaigns and data cleansing are only two of the many tasks that may be performed by RPA technology in the areas of finance and compliance, as well as in the treasury and marketing departments. The rise of robotic process automation has resulted in the creation of a number of new professional categories, including those in the fields of transcribing, financial analysis, taxi driving, home automation, and elder care. Early adopters of Robotic Products solutions may provide customers with a wide range of choices, including non-disruptive alternatives, on-premise or cloud-based solutions, managed service capabilities, trained professionals to help with automation, and experience in process consultation. The solution improves speed and consistency while reducing costs significantly and none of these solutions necessitate the replacement of existing systems. Infosys BPM, an Indian company's RPA solution, reimagines business process management and delivers an innovative business process service stack by integrating automation and artificial intelligence. That has been honed over many thousands of hours in many challenging customer situations. Products and services linked to RPA are offered by a number of multinational companies. Even start-ups are developing RPA technology and using it in their businesses. However, other related technologies such as artificial intelligence, machine learning, deep learning, data analytics, HR analytics, virtual reality (second life), home automation, blockchain technologies, and 4D printing are working in the background to put robotic process automation (RPA) in its appropriate place in business operations.

Future of Robotic Process Automation: According to M/s Everest Group, for instance, Robotic Process Automation (RPA) can be utilised in a wide variety of industries, including banking and financial services, manufacturing, high technology, and telecommunications, to generate bills of materials (BoMs), fraud claims discovery, insurance claims processing and new business preparation, healthcare sector reports automation and system reconciliation, and so on. The processing of accounts payable, accounts receivable, and general ledger transactions is one of the most prevalent applications of robotic process automation (RPA), which is used most often in manufacturing and banking. On the other hand, the card activation, fraud detection, claim processing, new business development, report automation, and system reconciliation processes in the banking, finance, and healthcare sectors all have a lot riding on them. In Figure 20.3, this can be observed. As a result, Everest Group believes that the worldwide market for RPA applications will develop tremendously in the near future.

Potential for RPA Low High

illustrative process with High-Potential

Function	F & A	Procurment	Human Resource	Contract Centre	Industry Specific Purposes	
Industry	Accounts Payable, Accounts Receivables, General Leadger	Invoice processing, Requesting to Purchasing Orders	Payroll, Hiring, Candidate Management	Customer Service		
Banking & Financial Services						✔ Cards activation ✔ Frauds Claims Discovery
Insurance						✔ Claim Processing ✔ New Business Preparation
Healthcare						✔ Reports Automation ✔ System Reconciliation
Manufacturing						✔ Bills of Material (BoM) Generation
High Tech & Telecom						✔ Service order Managment ✔ Quality Reporting
Energy Utilities						✔ Account setup ✔ Meter Reading Validation

Figure 20.3 RPA charting.

RPA technology products and services will be worth more than $400 million by 2017 [3]. It is expected to reach $1.2 billion in revenue by 2021, growing at a compound annual growth rate of 36%. Installation and consulting services to improve an organization's RPA capabilities make up the direct services market. The future of business and banking operations will be shaped by the use of bots, algorithms, and other cutting-edge technologies. In the future, automation and artificial intelligence (AI) will undoubtedly transform the financial sector. It means that financial institutions and banks will install a huge number of automated equipment to do routine activities, which would virtually remove operational expenses, 24-hour online services, and security. These bots are very user-friendly, even for native speakers who are illiterate in their own language. Bots have been deployed by the State Bank of India in the past to provide financial services in India. As of 2020, Gartner expects that 85 percent of consumers' interactions with firms will take place without them ever speaking to a human being at all. As a result, almost every firm will be computerised to some degree.

With online [14, 15] working becoming the new standard in today's society and an unparalleled dependence on technology, cybersecurity [16, 17], threat, and vulnerability parameters, fingerprint identification is more sensitive and crucial than ever. More issues exist as a result of the persistence of digital [18] divisions and the degree of digital change for which mankind is inadequate. If the acceleration of digitalization [19] had already started before the epidemic, despite ambiguous levels of public mistrust and commercial difficulties, the coronavirus-induced revolution bears discussion since it has an impact on all spheres of life, especially the corporate sector.

20.6 Conclusion

Robots and Robotic Process Automation technologies are increasingly obligatory for organisations to conduct commercial operations. In addition to standard repetitious work, they will be responsible for the recruiting, screening, and selection of personnel, the induction and training of newly-hired employees, and the search for the organization's employee welfare programs. In certain businesses, robots will take the place of human workers totally. The best aspect is that they'll perform tedious tasks and even risk their own lives to help others, so that companies' human resources may focus on strategic decision-making and intelligent processes, rather than wasting time on mundane tasks. Like Sophia, future humanoids will aid humans in their daily life. For the elderly and the handicapped, this Robotic Process Automation technology is especially beneficial since it is easier to use at any time of the day or night. The technique of robotic process automation that is being covered in this article is now being used in a wide number of sectors. These industries include manufacturing, chemical plants, healthcare, aviation, and many more. Accept the use of robots with artificial intelligence as a means to make life better for people, for example to enhance quality of life and care for the elderly, acquire 24-hour services, avoid hazardous jobs, minimise workflow inefficiencies, raise productivity, and be freed from completing repetitive tasks using this technology.

The field of computer science and information technology is home to some of the most advanced technologies now available, including robotic process automation. Automation of Robotic Processes is a relatively new phenomenon. No operational definitions, connotations, tested models, or derived hypotheses exist and hence, no one is aware of its precise meaning. It gives the impression that one is prepared to take chances. Therefore, the research study relies on a wide range of secondary (internet) resources for its data and findings, including academic publications, company white papers, and expert blogs. It was gathered during April and June of this year, that in order to find research publications, researchers use the keywords "robots," "robotic process automation," "Artificial Intelligence," and "AI." For database articles, Google and Google scholar are utilised as search engines. To better comprehend the phenomena of Robotic Process Automation, supplementary data was acquired, compiled, evaluated, and thematically narrated. This is an exploratory study that is similar to descriptive research.

References

1. Ssu Chieh Lin, Lian Hua Shih, Damon Yang, James Lin, Ji Fu Kung, "Apply RPA (Robotic Process Automation) in Semiconductor Smart Manufacturing",2018 e-Manufacturing Design Collaboration Symposium (eMDC), 7th September 2018 in IEEE.
2. Moffitt, Kevin C., *et al.* "Robotic Process Automation for Auditing." *Journal of Emerging Technologies in Accounting*, vol. 15, no. 1, July 2018, pp. 1–10. *Crossref*, doi:10.2308/jeta-10589.
3. Ruchi Isaac, Riya Muni, Kenali Desai, "Delineated Analysis of Robotic Process Automation Tools." *2018 Second International Conference on Advances in Electronics, Computers and Communications (ICAECC)*, IEEE, 2018, pp. 1–5. *Crossref*, doi:10.1109/ICAECC.2018.8479511.
4. Audrey Bourgouin, Abderrahmane Leshob, Laurent Renard. "Towards a Process Analysis Approach to Adopt Robotic Process Automation." *2018*

IEEE 15th International Conference on E-Business Engineering (ICEBE), IEEE, 2018, pp. 46–53. *Crossref*, doi:10.1109/ICEBE.2018.00018.

5. Audrey Bourgouin, Abderrahmane Leshob, Laurent Renard "Robotic Process Automation: Dynamic Roadmap for Successful Implementation", Reykjavik University, June, 2018.
6. Capgemini "Robotic Process Automation: Gearing up for greater integration" , 2017 Storyful, 11th August 2017.
7. Everest Research Group. "Robotic Process Automation: Technology vendor landscape with fit ma-trix assessment- Technologies for building virtual workforce", December 2016.
8. Rebecca Dilla, Heidi Jaynes, and Lauren Livingston ,"Introduction to Robotics Process Automation a Primer", 2015 by the Institute for Robotic Process Automation, December 2015.
9. Anagnoste, Sorin. "Robotic Automation Process - The next Major Revolution in Terms of Back Office Operations Improvement." *Proceedings of the International Conference on Business Excellence*, vol. 11, no. 1, July 2017, pp. 676–86. *Crossref*, doi:10.1515/picbe-2017-0072.
10. Mary C. Lacity and Leslie P. Willcocks. "A new approach to automating services" MIT Sloan Management Review, October 2016.
11. International Auditing and Assurance Standards Board (IAASB). 2016. "Exploring the Growing Use of Technology in the Audit, with a Focus on Data Analytics." New York, NY: IFAC.
12. C. Mendis, C. Silva, and N. Perera, "Moving ahead with Intelligent Automation", 2016.
13. UiPath Community Forum - Robotic Process Automation, "UIPath vs Automation Anywhere-Blue Prism", 2017.
14. Mahor, V., Bijrothiya, S., Mishra, R., Rawat, R., & Soni, A. (2022). The Smart City Based on AI and Infrastructure: A New Mobility Concepts and Realities. *Autonomous Vehicles Volume 1: Using Machine Intelligence*, 277-295.
15. Mahor, V., Pachlasiya, K., Garg, B., Chouhan, M., Telang, S., & Rawat, R. (2022, June). Mobile Operating System (Android) Vulnerability Analysis Using Machine Learning. In *Proceedings of International Conference on Network Security and Blockchain Technology: ICNSBT 2021* (pp. 159-169). Singapore: Springer Nature Singapore.
16. Mahor, V., Garg, B., Telang, S., Pachlasiya, K., Chouhan, M., & Rawat, R. (2022, June). Cyber Threat Phylogeny Assessment and Vulnerabilities Representation at Thermal Power Station. In *Proceedings of International Conference on Network Security and Blockchain Technology: ICNSBT 2021* (pp. 28-39). Singapore: Springer Nature Singapore.
17. Rawat, R., Gupta, S., Sivaranjani, S., CU, O. K., Kuliha, M., & Sankaran, K. S. (2022). Malevolent Information Crawling Mechanism for Forming Structured Illegal Organisations in Hidden Networks. *International Journal of Cyber Warfare and Terrorism (IJCWT)*, 12(1), 1-14.

18. Rawat, R., Rimal, Y. N., William, P., Dahima, S., Gupta, S., & Sankaran, K. S. (2022). Malware Threat Affecting Financial Organization Analysis Using Machine Learning Approach. *International Journal of Information Technology and Web Engineering (IJITWE)*, 17(1), 1-20.
19. Rawat, R., Mahor, V., Chouhan, M., Pachlasiya, K., Telang, S., & Garg, B. (2022). Systematic literature Review (SLR) on social media and the Digital Transformation of Drug Trafficking on Darkweb. In *International Conference on Network Security and Blockchain Technology* (pp. 181-205). Springer, Singapore.

21

Evolutionary Survey on Robotic Process Automation and Artificial Intelligence: Industry 4.0

P. William[1*], Siddhartha Choubey[2], Abha Choubey[2] and Gurpreet Singh Chhabra[3]

[1]*Department of Information Technology, Sanjivani College of Engineering, SPPU, Pune, India*
[2]*Department of CSE, Shri Shankaracharya Technical Campus, CSVTU, Bhilai, India*
[3]*Department of CSE, GITAM School of Technology, GITAM University, Visakhapatnam, India*

Abstract
Taking into consideration the evolution of technical things during the last decade and the proliferation of information systems in the society, we see that the vast majority of services provided by companies and institutions as digital services. Industry 4.0 is the fourth industrial revolution, where technologies and automation are asserting themselves as major changes. Robotic Process Automation (RPA) has numerous advantages in terms of automating organizational and business processes. Allied to these advantages, the complementary use of Artificial Intelligence (AI) algorithms and techniques allows for improvement in the accuracy and execution of RPA processes in the extraction of information and in the recognition, classification, forecasting, and optimization of processes. In this context, this paper aims to present a study of the RPA tools associated with AI that can contribute to the improvement of the organizational processes associated with Industry 4.0. It appears that the RPA tools enhance their functionality with the objectives of AI being extended with the use of Artificial Neural Network algorithms, Text Mining techniques, and Natural Language Processing techniques for the extraction of information and consequent process of optimization and of forecasting scenarios in improving the operational and business processes of the organization.

**Corresponding author*: william160891@gmail.com

Romil Rawat, Rajesh Kumar Chakrawarti, Sanjaya Kumar Sarangi, Rahul Choudhary, Anand Singh Gadwal and Vivek Bhardwaj (eds.) Robotic Process Automation, (315–328) © 2023 Scrivener Publishing LLC

21.1 Introduction

Digital services are becoming more and more available at the enterprise level. This is because people are using information systems more and more in society and technology is getting better and better at many different levels. People, businesses, and organizations have started to share digital information as their main way to talk to each other. Since there is so much information and digital documents exchanged, it's hard to reply quickly and maintain track of internal activities in a timely manner.

Robotic Process Automation (RPA) is a "technology that enables administrative, scientific, or industrial activities to be done on their own" and we emphasize its significance in this manner. To put it another way, robotics is the study and use of automata (robots) in lieu of humans to perform various activities [1]. To summarize, RPA tools are a collection of tactics aimed at increasing productivity by eliminating or automating routine operations [2]. Algorithms and approaches in Artificial Intelligence (AI) enable more precise execution of automated procedures in conjunction with RPA. To further advance AI automation in business processes and activities, the concept of "Industry 4.0" focuses on a collection of technologies and sensors. This boosts productivity and provides new avenues for growth.

This paper's primary contribution is a study of how AI and RPA fit into Industry 4.0. The research also examines and compares the capabilities of a variety of commercial and open-source systems. This paper is organized as follows. In Section 21.2, the general idea of Robotic Process Automation is explained. In Section 21.3, the general ideas of Artificial Intelligence and Industry 4.0 are given. Tools and their characteristics are presented in Section 21.4 so that Section 21.5 may be utilized to detail the tools more thoroughly. After the findings are presented in Section 21.6, the sources supporting this research are provided.

21.2 Robotic Process Automation

Service operations that appear like they would be performed by a human may be automated using Robotic Process Automation (RPA). Employees who are proficient at repetitive activities, such as software robots or artificial intelligence (AI), are used to automate processes. Using a snapshot of the screen and variables, the developer works out how to accomplish the task. Data entry, copying and pasting, opening emails, and filling out forms are all examples of these tasks [4]. According to Van Aalst *et al.*, RPA is a catch-all name for tools that operate on the user interface of other computer systems. Though RPA uses the computer's user interface in the same manner that previous methods of automation (screen recording, scraping, and macros) do, RPA's primary purpose is to identify elements, not screen coordinates [4] or XPath selections. This usually results in a more intuitive and intelligent user interface. RPA technology suppliers have seen an upsurge in demand since 2016 [3, 4]. Digital forensics, audits, and industry all show evidence of these technologies being utilized to automate their processes [4, 5]. In the fourth industrial revolution (Industry 4.0), RPA technology and smart device data are paving the way for new approaches to automate tedious rules-based business operations. Robotic process automation (RPA) is the process of using robots to do routine human job activities (where those tasks are done quickly and profitably). An outsider's perspective is used in this endeavor to try to replace humans with technology. Because RPA is not a component of the information infrastructure, it differs from more conventional approaches. It is less obtrusive and potentially cost saving because it just rests on top of it. According to certain research, utilizing RPA technology may reduce transactional service costs by up to 50% [8].

21.3 Artificial Intelligence and Industry 4.0

A few years ago, AI was a notion that had been broken down into a number of distinct subdomains. Natural language processing, automated programming, robotics, computer vision, automatic theorem proving, intelligent data retrieval, and so on were some of the examples of these technologies. There are now so many different uses for these technologies that each might be considered a separate discipline. Currently, AI is best described as a set of basic concepts that underlie many of these applications [9]. Smart factories and industries are based on the premise that AI can be used by robots to do difficult jobs, save costs, and improve the quality of goods and services. 4.0 [10] Artificial intelligence (AI) and cyber-physical systems (CPS) are finding their way into manufacturing. When AI is used in manufacturing, it makes the industry smarter and able to deal with new problems, such as customizable specifications, less time to get to the market, and more sensors being used in equipment. Using flexible robots and AI together makes it easier to make different things. AI methods (like data mining) can handle a huge amount of real-time data from many sensors [12].

21.4 RPA Tools with IA Support

There has been increasing success in the real world with the use of artificial intelligence (AI) algorithms and machine learning (ML) approaches. Machine learning (ML) [14] is used to train computers on how to better cope with data. For this, it uses AI algorithms (or methodologies) that mimic the ways in which rational humans learn, such as connectionist, genetics, statistical probability, case-based learning, etc. You may categorize, connect, optimize, group, forecast, detect patterns, etc. using AI algorithms and ML methodology. RPA has been gradually integrating AI-based techniques for classification, identification, and categorization into its automation tools since AI is valuable in so many ways (e.g., Enterprise Resource Planning, Accounting, Human Resources). Several academic studies on the problems and prospects of RPA and AI have been published in the last few years, as well as several case studies. Case studies in the areas of automated data discovery and transformation [15], business process management [17], and productivity process optimization [18] are all examples of this. Delloite [19], a consulting business, has produced a study on the use of AI algorithms and methodologies in the intelligent automation of processes utilizing RPA [20]. In contrast, it should only be utilized in well-established and mature processes, such as those that concentrate on client chores, boosting staff efficiency by optimizing regular operations, and enhancing accuracy in categorizing and routing procedures. When it comes to RPA-based automation, AI and algorithms might be used to aid with any issues or opportunities that may arise. In our opinion, the following commercial and open-source RPA technologies demonstrate how RPA may be applied in the here and now (ideally with the application of some AI techniques or algorithms).

21.4.1 UiPath

Programming scripts may be created and executed using RPA functionalities provided within the UiPath [20–26] framework. If you want to make your business operations more specialized, you may use the program's block-based interface and many plugins. The UiPath platform consists of three parts: UiPath Studio, Robot, and Orchestrator [20]. The Orchestrator is used to coordinate robots. Working with workflows is made easier with the UiPath Studio module [21]. Making and maintaining connections with robots, transferring packages, and managing queues is possible using this tool. RPA process analytics may be improved by using log records coupled with Microsoft's Information Services Server and SQL Server,

Elasticsearch, an open-source search engine based on the Apache License, and a Kibana plugin for data visualization. In [22–24], these characteristics are discussed in further depth. The UiPath tool's UIAutomation module [25] and its official website [26] now include certain AI approaches or algorithms. Recognition, optimization, categorization, and information extraction stand out as some of the most important. In the linked information about AI algorithms, picture and character recognition, optimization, and classification are used.

21.4.2 Kofax

Software and solutions for automating business and organization procedures are made by Kofax [27–33]. In this application, you'll find modules for robotic process automation (RPA), business process orchestration (BPO), document recognition (OCR), and sophisticated data analysis. Many information sources [27–34] were contacted since the tool was proprietary and there was no opportunity to get a test version for this investigation. With this tool, data can be extracted from various documents and other sources (web, e-mail, local files) and using RPA automation and procedural flows may be executed across different computer programs to optimize ERP information system duties. It offers modules for employing AI approaches and algorithms, much like other tools. For example, it can recognize content and context in documents [28] or categorize and distinguish between data found in emails, internet portals, and printed documents [34]. A good example of supervised learning is the use of machine learning algorithms to recognize and categorize optical character recognition (OCR) documents and to analyze the content of e-mails or web pages. A variety of approaches and algorithms for evaluating data via information clustering (a.k.a. "clustering") or density extraction may be employed with natural language processing, however. Kofax's Intelligent Automation platform [32] seems to allow for the usage of certain AI techniques or algorithms based on this information.

Another tool for RPA processes, Automation Anywhere [35–41], provides information on how AI approaches and algorithms might be used. In an ERP setting, RPA may be utilized to cover numerous application areas, including human resources, customer relationship management, and supply chain management. ERPs from SAP and Oracle, as well as those from other organizations, are almost certainly going to be connected. "Digital Workers" is the RPA process that is the most automated or intelligent. A module for cognitive automation is included in the RPA tool, as are tools for analyzing data. As a multipurpose app, it provides a set of data that makes it possible to set up, run, and deploy RPA processes [35–41]. Using methods and algorithms from Artificial Intelligence such as fuzzy logic and artificial neural networks, the Automation Anywhere tool extracts information from documents using its Bot tool [40]. This makes document validation more efficient. In this way, it looks like the IQ Bot platform gives the Automation Anywhere smart word processing app access to some AI approaches or algorithms [40].

21.4.3 WinAutomation

There are a number of RPA-integrated automation functions in the WinAutomation tool [42]. Automating emails and files in various formats (such as PDF and Excel), OCR, and other aspects of the post worker's work environment are among these functions (desktop or web). Softomotive, a provider of RPA technologies, came up with WinAutomation. WinAutomation is designed for desktop environments with integrated process design, desktop automation, web automation, macro recording, multitasking, automatic task execution, mouse and keyboard automation, User Interface designer, email automation, Excel automation, file and folder automation, system monitoring and triggering, auto-login, security, FTP automation, exception handling, repository and control images, command line control, and web data exfiltration. The "processrobot" module includes a number of modules with RPA capability. Using CaptureFast's AI-powered information capture engines, as well as data extraction from documents and systems, as well as hybrid document categorization, RPA capability may be increased. The Cognitive module, according to research [42–45], enables Microsoft services to be incorporated with IBM and Google's Cognitive analytical information processing engines. At the level of AI that is available right now, it seems that the tools are not proven.

21.4.4 AssistEdge

EdgeVerve Systems, a part of Infosys, owns the AssistEdge product. The product is also available to the general public as an "open source" version [49]. To process documents, it can read OCR, depending on the document's context, according to institutional information [46–49]. For autonomous data collecting, data analysis via the study of process changes based on individual process monitoring and combining information for recommendation processes makes use of AI algorithms (such Artificial Neural Networks) [49].

21.4.5 Automagica

There is an open-source version of the closed-source Automagica program [50] available for non-commercial usage on GitHub [51]. It was mostly made with the programming language Python and other community implementations may be able to use it (e.g., AI techniques or algorithms). Basic RPA capabilities include reading OCR, extracting text from PDF files, and automating information in word files and Excel spreadsheets, as well as browser-collected information. It may also be used in conjunction with Google TensorFlow to recognize images and text.

21.5 RPA Tools with IA Support

As a discussion based on a review of the most popular RPA tools, the following Table 21.1 compares the AI aims and algorithms of each technology.

Think about how AI tools like Artificial Neural Networks could be used in computer vision. NA means that there is no information.

A selection of RPA-related task-automation tools was compiled by searching the Internet and digital libraries. These tools do "smart" tasks that are linked to automation, but their main job is to use AI-based approaches and algorithms. It looks like the proprietary solutions have a bigger pool of data and more RPA and AI capabilities. TagUI [52, 53], TaskT [54], and the Robot Framework Foundation [55] are examples of open-source tools that are being made better or getting more features right now.

As a tool, the UiPath has many capabilities and extensive documentation on how to utilize them. Plug-ins may be written to integrate with a variety of different platforms, including PowerShell, the SAP ERP system, Oracle database, and Microsoft Dynamics GP. Several RPA processes are run by Kofax and Automation Anywhere technologies that connect to enterprise resource planning systems, mostly SAP ERP. AssistEdge proved that it could work with Microsoft's cognitive systems (Azure Machine Learning) and Google's cognitive systems (cognitive Services), making the software from these two major tech companies easier to use.

There is always a licensing cost when private enterprises use RPA with AI or combine it with other ERP systems. This is because private companies have access to the most RPA features and data. When it comes to open-source RPA technologies, the number of activities and implementations keeps growing. In the past few years, though, Academic algorithms have been implemented in open-source programming languages like R and Python, and certain robotic process automation (RPA) initiatives might

Table 21.1 Comparison of goals associated to AI and various technologies.

	Goal associated to IA				Artificial intelligence techniques or algorithms used								
Tool	**Recognition**	**Optimization**	**Classification**	**Information extract**	**Computer vision (*)**	**Fuzzy matching**	**Statistic methods**	**Artificial neural networks**	**Decision trees**	**Fuzzy logic**	**Natural language processing**	**Text mining**	**Recomendation system**
UiPath	X	X	X	X	X	X	X	X					
Kofax	X	X	X	X				X				X	
Automation anywhere	X	X	X	X				X			X	X	
WinAutomation	NA	NA	NA	NA	NA	NA	NA	NA	NA	NA	NA	NA	
AssistEdge	X	X	X	X			X	X					X
Automagica	X	X	X	X								X	X

make use of these academic implementations as well. With Microsoft's AI research and the Azure Machine Learning platform, RPA solutions utilizing the .NET framework may be directly explored using the .NET framework as a result of Microsoft's AI research and development efforts.

21.6 Conclusions

Using RPA and AI for ERP-related activities is examined in this article. Using information from business websites and tools, blogs, other web-based digital libraries and scientific digital libraries, it was developed. It created a list of commercial and open-source RPA solutions and tools with information about their RPA capabilities and ERP integration, as well as their ERP support. Our research shows that the majority of proprietary solutions are successful in achieving AI-related objectives including recognizing, optimizing, categorizing, and extracting information from RPA documents or processes. Users of these applications may make better use of and learn more about their data thanks to this feature. Computer vision (such as image recognition using Artificial Neural Networks), statistical methods, decision trees, neural networks for classification and prediction, and fuzzy logic, as well as text mining, natural language processing, and recommendation systems, are all incorporated into these tools' algorithms and AI techniques.

An alternative view is based on how Internet of Things (IoT), intelligent automation (IA), intelligent devices and processes (IDPs), and cyber-physical systems all come together in the present industrial transformation, known as Industry 4.0. When these ideas and technologies come together, they have a big impact on how digital operations work in an organization. To improve these processes, they are now using robots to do some of the work (RPA). Also, RPA now combines intelligent methods and algorithms (AI), as shown in this article, with many of its technologies. This makes it possible for business processes to be automated to intelligent levels.

References

1. Infopédia (2020). Dicionário Infopédia da Língua Portuguesa, 2020. [Online]. Available from : https://www.infopedia.pt.
2. Aguirre, Santiago & Rodriguez, Alejandro. (2017). Automation of a Business Process Using Robotic Process Automation (RPA): A Case Study. 65-71. DOI: 10.1007/978-3-319-66963-2_7. Available from: https://www.researchgate.

net/publication/319343356_Automation_of_a_Business_Process_Using_Robotic_Process_Automation_RPA_A _Case_Study
3. van der Aalst, W. M., Bichler, M., & Heinzl, A. (2018). Robotic Process Automation. Bus Inf Syst Eng 60, pp. 269–272. https://doi.org/10.1007/s12599-018-0542-4
4. Asquith, A., & Horsman, G. (2019). Let the robots do it!–Taking a look at Robotic Process Automation and its potential application in digital forensics. Forensic Science International: Reports, 1, 100007.
5. Moffitt, K. C., Rozario, A. M., & Vasarhelyi, M. A. (2018). Robotic process automation for auditing. Journal of Emerging Technologies in Accounting, 15(1), 1-10.
6. Madakam, S., Holmukhe, R. M., & Jaiswal, D. K. (2019). The future digital work force: robotic process automation (RPA). JISTEM-Journal of Information Systems and Technology Management, 16.
7. Enríquez, J. G., Jiménez-Ramírez, A., Domínguez-Mayo, F. J., & García-García, J. A. (2020). Robotic Process Automation: A Scientific and Industrial Systematic Mapping Study. IEEE Access, 8, 39113-39129.
8. Williams, D., & Allen, I. (2017). Using artificial intelligence to optimize the value of robotic process automation. Available from: https://www.ibm.com/downloads/cas/KDKAAK29
9. Nilsson, N. J. (2014). Principles of artificial intelligence. Morgan Kaufmann Editors.
10. Bahrin, M. A. K., Othman, M. F., Azli, N. N., & Talib, M. F. (2016). Industry 4.0: A review on industrial automation and robotic. Jurnal Teknologi, 78(6-13), pp:137-143.
11. Zheng, P., Sang, Z., Zhong, R. Y., Liu, Y., Liu, C., Mubarok, K., ... & Xu, X. (2018). Smart manufacturing systems for Industry 4.0: Conceptual framework, scenarios, and future perspectives. Frontiers of Mechanical Engineering, 13(2), pp:137-150.
12. Ustundag, A., & Cevikcan, E. (2017). Industry 4.0: managing the digital transformation. Springer Editors. Available from: https://www.springer.com/gp/book/9783319578699
13. Haenlein, Michael & Kaplan, Andreas. (2019). A Brief History of Artificial Intelligence: On the Past, Present, and Future of Artificial Intelligence. California Management Review.
14. Mitchell, T. M. (1997). Machine Learning. New York: McGraw-Hill. ISBN: 978-0-07-042807-2.
15. Leno, V., Dumas, M., La Rosa, M., Maggi, F. M., & Polyvyanyy, A. (2020). *Automated Discovery of Data Transformations for Robotic Process Automation*. https://arxiv.org/abs/2001.01007
16. Huang, F., & Vasarhelyi, M. A. (2019). Applying robotic process automation (RPA) in auditing: A framework. *INTERNATIONAL JOURNAL OF ACCOUNTING INFORMATION SYSTEMS, 35*. https://doi.org/10.1016/j.accinf.2019.100433

17. Agostinelli, S., Marrella, A., & Mecella, M. (2020). *Towards Intelligent Robotic Process Automation for BPMers*. Available from: https://www.researchgate.net/publication/338401505_Towards_Intelligent_Robotic_Process_Automation_for_BPMers
18. FLUSS, D. (2018). Smarter Bots Mean Greater Innovation, Productivity, and Value: Robotic process automation is allowing companies to re-imagine and re-invest in all aspects of their businesses. *CRM Magazine*, *22*(10), 38–39.
19. Delloite (2019). Automation with intelligence Reimagining the organisation in the 'Age of With'. Available from: https://www2.deloitte.com/content/dam/Deloitte/tw/Documents/strategy/tw-Automation-with-intelligence.pdf
20. Tripathi, A. (2018). Learning robotic process automation: Create software robots and automate business processes with the leading RPA tool, UiPath. Packt Publishing Book Series.
21. UiPath (2020a). UiPath Studio: introduction. [Online]. Available from: https://docs.uipath.com/studio/docs/introduction.
22. GitHub (2020a). Open Source, Distributed, RESTful Search Engine. [Online]. Available from: https://github.com/elastic/elasticsearch.
23. GitHub (2020b). Your window into the Elastic Stack. [Online]. Available from: https://github.com/elastic/kibana
24. UiPath (2020b). Prerequisites for Installation. [Online]. Available from: https://docs.uipath.com/orchestrator/docs/prerequisites-for-installation.
25. UiPath (2020c). About the UI automation activities pack. [Online]. Available from: https://docs.uipath.com/activities/docs/about-the-ui-automation-activities-pack
26. UiPath (2020d). Artificial Intelligence RPA Capabilities. [Online]. Available from: https://www.uipath.com/product/ai-rpa-capabilities
27. Kofax (2020a). Developer's Guide Version: 11.0.0 [Online]. Available from: https://docshield.kofax.com/RPA/en_US/11.0.0_qrvv5i5e1a/print/KofaxRPADevelopersGuide_EN.pdf
28. Kofax (2019). Product summary Kofax RPA. [Online]. Available from: https://www.kofax.com/-/media/Files/Datasheets/EN/ps_kofax-rpa_en.pdf
29. Kofax (2020b). Maximize Your ERP with Integrated Accounts Payable Automation. [Online]. Available from: https://www.kofax.com/Solutions/Cross-Industry/Financial-Process-Automation/AP-and-Invoice-Automation/ERP-Integration
30. Kofax (2020c). Power your process. [Online]. Available from: https://www.kofax.com/-/media/Files/E-books/EN/eb_how-rpa-capture-empowers-customer-journey_en.pdf
31. Kofax (2011). Kofax Capture (versão 10.0). [Online]. Available from: https://issues.alfresco.com/jira/secure/attachment/56073/KofaxCaptureDevelopersGuide_10.pdf
32. Kofax (2020e). Kofax intelligent automation platform. [Online]. Available from: https://www.kofax.com/Products/intelligent-automation-platform

33. Kofax (2020d). Cognitive Document Automation. [Online]. Available from: https://www.kofax.com/Blog/Categories/Cognitive-Document- Automation
34. Schmidt, D. (2018). RPA and AI. [Online]. Available from: https://www.kofax.com/Blog/2018/september/rpa-and-ai-the-new-intelligent-digital-workforce
35. Automation Anywhere (2020a). Robotic process automation to ERP. [Online]. Available from: https://www.automationanywhere.com/solutions/robotic-process-automation-to-erp
36. Automation Anywhere (2020b). Automate any ERP process with RPA. [Online]. Available from: https://www.automationanywhere.com/lp/automate-any-erp-process-with-rpa
37. Automation Anywhere (2020c). Actions in the Workbench. [Online]. Available from: https://docs.automationanywhere.com/bundle/enterprise-v11.3/page/enterprise/topics/aae-client/metabots/getting-started/selecting-actions-in-the-logic-editor.html
38. Automation Anywhere (2020d). Bot Execution Orchestrator API. [Online]. Available from: https://docs.automationanywhere.com/bundle/enterprise-v11.3/page/enterprise/topics/control-room/control-room-api/api-deploy-and-monitor-bot-progress.html
39. Automation Anywhere (2020e). Automation Management API. [Online]. Available from: https://docs.automationanywhere.com/bundle/enterprise-v11.3/page/enterprise/topics/control-room/control-room-api/api-bot-deployment.html
40. Automation Anywhere (2020f). IQBot – Intelligent Document Processing. [Online]. Available from: https://www.automationanywhere.com/products/iq-bot
41. E. Global (2017). Automating Content-Centric Processes with Artificial Intelligence. [Online]. Available from: https://www.automationanywhere.com/images/lp/pdf/everest-group-automating-content-centric-processes-with-ai.pdf
42. WinAutomation (2020a) Desktop automation https://www.winautomation.com/product/all-features/desktop-automation
43. WinAutomation (2020b) About Softomotive. Available on: https://www.winautomation.com/about-softomotive/
44. WinAutomation (2020c) Installation Requirements. Available from: https://support.softomotive.com/support/solutions/articles/35000081666-winautomation-installation-requirements
45. WinAutomation (2020d) Softomotive RPA Review. Available from: https://www.rpa-star.com/softomotive-vs-winautomation-rpa-review/
46. AssistEdge (2020a). RPA. Available from: https://www.edgeverve.com/assistedge/robotic-process-automation/
47. AssistEdge (2020b). AssistEdge RPA Brochure. Available from: https://query.prod.cms.rt.microsoft.com/cms/api/am/binary/RE42s9D
48. AssistEdge (2020b). Uso das Redes Neuronais Artificiais para análise de variações dos processos. Available from: https://www.edgeverve.com/assistedge/

assistedge-discover/ and from: https://www.infosys.com/newsroom/press-releases/2019/launches-assistedge-discover-true-value-automation.html
49. AssistEdge (2020c). AssistEdge RPA OpenSource Community. Available from: https://www.edgeverve.com/assistedge/community/
50. Automagica (2020a). Automagica GitHub Repository. Available from: https://github.com/automagica/automagica
51. Automagica (2020b). Automagica Documentação. Available from: https://automagica.readthedocs.io/index.html and https://github.com/automagica/automagica/wiki/Documentation
52. TagUI (2020a). TagUI – AI Singapure Platforma – National institute. Available from: https://makerspace.aisingapore.org/do-ai/tagui/
53. TagUI (2020b). TagUI – GitHub Repository Available from: https://github.com/kelaberetiv/TagUI/tree/pre_v6
54. TaskT (2020a). TaskT RPA .NET Platform. Available from: https://github.com/saucepleez/taskt/wiki/Automation-Commands
55. Robocorp (2020). Robocorp hub. Available from: https://hub.robocorp.com/new-to-robocorp-suite/get-started/quickstart-guide/

22

Advanced Method of Polygraphic Substitution Cipher Using an Automation System for Non-Invertible Matrices Key

Devendra Kuril[1]*, Manoj Dhawan[2] and Gourav Shrivastava[1]

[1]*Department of Information Technology, Shri Vaishnav Vidyapeeth Vishwavidyalaya, Indore, India*
[2]*Department of CSE, Avantika University, Ujjain, India*

Abstract

The polygraphic substitution cipher is the primary cryptosystem (cryptsys) technique that has various advantages in sheltered key encryption (encryp). A substitute enhancement is an inescapable key that is essential for converting cipher text to plain text and it is not appropriate for cryptsys, a readable text comprising of nix. The unmasking of these works is to improve the assessment of polygraphic substitution crypsys to get the better of these complications. Examination of the preceding conclusions presented specifically usual polygraphic substitution cipher techniques that are not yet suitable without automation because there may be many chances of vulnerability. These techniques are especially vulnerable when identifying plaintext. These algorithms also have improved modification attributes and as an outcome, they are advance resistant likened to recognized plaintext attacks using automation systems.

On the other hand, these enhanced polygraphic substitution cipher algorithms have a modulate matrices key problem which is rapidly solved by the system. Moreover, these algorithms are not appropriate for entirely null values for respective blocks of crypsys. In the present report, the evaluation of polygraphic substitution enhanced the advanced technique which applies to the singular key with an automation system. Expanding of the recommended system is substantiated via generated key matrix approaches that are relatively complex to recognize. The preceding exploration concentrated on a polygraphic substitution using a sharable or single key algorithm without an automation system.

**Corresponding author*: dev.kuril19@gmail.com

Romil Rawat, Rajesh Kumar Chakrawarti, Sanjaya Kumar Sarangi, Rahul Choudhary, Anand Singh Gadwal and Vivek Bhardwaj (eds.) Robotic Process Automation, (329–338) © 2023 Scrivener Publishing LLC

During the decoding of the encrypted text, there are raised vulnerabilities. This survey keeps accelerating the usage of the evaluated polygraphic substitution cipher to act in a comparative approach in evaluating the raised vulnerabilities conversely to the key matrix. In this research, an automation system has no issues to identifying the vulnerabilities.

Keywords: Automation system, polygraphic substitution cipher, vulnerability, singular matrix

22.1 Introduction

In an effort to clarify the structure use, chromatic scholars mostly industrialized products for analyzing and determining computer end-user satisfaction. Currently, technology recognition is defined as a sustainable amenability surrounded by a technology set to enlist information technology intended for when the performance is true, therefore the perception isn't utilized for circumstances in which manipulators have the privilege to enlist it apart from furnishing substantiation of usage. Clearly, there was some ambiguity at first as there is always the possibility of a slight deviation from the ideal, planned use for effective operation. The concept of the automation system is that similar deviations aren't valuable and it means the mechanism of the automation system of several data and facts for planned objectives can be augured and modeled. Certainly, on many prototypes, users are unable to access information systems that, if enabled and used, would yield significant performance gains. Therefore, automation systems have been considered the essential aspect in actuating the achievement or downfall of several data and fact security system designs. Figure 22.1 below illustrates the introductory conception underpinning the technology and propositions. Each of them has its own features that will be reviewed and tasks performed in a time scale.

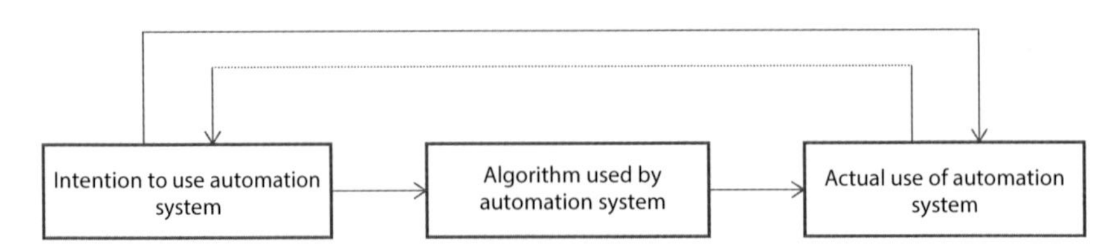

Figure 22.1 Automation system perception.

Crypsys is the method and wisdom of creating a block of readable text that has been encrypted. Caesar's cipher is one of the crypsys which was used by its first owner. Entirely, such encryptions belong to the underpinning for the latest crypsys.

There are two sorts of crypsys systems: secret-key crypsys and public-key crypsys [7, 8]. While in secret-key encryption, the initiator and target serve an identical or single key. In reference with these, a similar key is utilized for cipher and en-cipher, in both techniques. In public-key encryption [11], dissimilar keys are used. In the modern era of cryptographic implementation, both unbalanced and balanced cryptosystems function together with an automation system. In this analysis, we are attentive to the evaluation research on the polygraphic substitution method which has non-regular vulnerabilities [12] during encryption and decryption (decryp). The attack surface of crypsys has made it unfeasible in practice, but it does play a vitally informative part in crypsys and linear algebra [13].

ORGANIZATION OF CHAPTER

The rest of the chapter is outlined as follows. Section 22.2 shows the significance of the advanced polygraphic substitution technique which improves in the Hill Cipher Algorithm, Section 22.3 shows related work, Section 22.4 outlines the proposed methodology, Section 22.5 shows the result analysis, and finally, Section 22.6 concludes this chapter.

22.2 Significance of Advanced Methods of a Polygraphic Substitution Cipher

The Hill Cipher in cryptography existed amongst the basic polygraphic cipher structures that were erected on the applied structure by an additional 3 signs or letters in one. Cryptography shows a pivotal part in considering the practice of the Hill Cipher commonly. Then, the performer has to know that several conceivable matrices in the system don't signify a crucial matrix. While decryption of unreadable text requires an inverse key matrix, after the matrix, the inversion knows whether the inverse occurs or not. So, if the determinant's value is zero or an aspect apart from 1, it denotes that the matrix is not an inverse. In such a case, we discover a random key matrix for converting the readable text into unreadable text. Although, in this exploration we eliminated such types of discrepancy by using an automation system because if the input is given by the user for encryption the user may also be able to produce a mistake and this will reflect the affair.

There's a limit to giving input as a crucial matrix and if it isn't followed by the user, then the output may not give the result as plain text. So, this automation system generates a limit to the given input which generates output as we want. One further issue to overcome with this automation system is that at every three blocks of variable, the stoner may give the input whether it is needed or not, while the automation system of this fashion performs tasks whenever it's needed. So, this is very salutary for performing and adding the efficiency of encryption and decryption of a large amount of information.

22.3 Related Work

Since the polygraphic technique of cryptology functions as an important task in information security and matrix computation algebra, numerous types of exploration were completed to ameliorate this tool. Toorani, and Falahati explored a robust crypsys algorithm for non-invertible matrices performed on the Hill Cipher [9]. The non-invertible crucial matrix discrepancy is answered by converting each readable text character into a couple of ungraspable shared characters. The procedure comprises of the conversion of a couple of ungraspable communication characters into a single readable text and symbols. The Table 22.1 shows about the different studies of encryption/decryption techniques using automation or without an automation system.

Table 22.1 Different studies of encryption/decryption techniques using automation or without an automation system.

S. no.	Study	Description
1.	References (2015) [1]	The exposure to the source of a code system with emphasis on classical encryption techniques, generating mechanisms and systematic techniques to support in floating the authentication effectiveness and to provide specific protection for sensitive information by using various techniques with a distinct focus on confidential information or sensitive data.

(*Continued*)

Table 22.1 Different studies of encryption/decryption techniques using automation or without an automation system. (*Continued*)

S. No.	Study	Description
2.	References (2021) [2]	This research implemented (IARC) Improved Automated Random Cryptography for data blocks that are stored in a cloud system. In this approach, researchers have presented a novel encryption policy by converting the static S-box in the AES algorithm to a dynamic S-box. Additionally, the algorithms Twofish and RSA are used to convert the plain text to cipher text generated keys to enhance privacy issues.
3	References (2020) [4]	This research explains how encryption and decryption are possible with the help of a Playfair cipher and huge mathematical computation for encryption/decryption because here, 95 is used as a modulus operation.
4.	References (2019) [3]	This research focuses on the abilities of a (TPM), i.e., trusted platform module. Moreover, the motive is to confirm the legitimacy of the data spread between two entities on similar (horizontal interoperation) or dissimilar (vertical interoperation) classified levels communicating through Modbus and TCP protocol to utilize the functionalities obtained by integrating trusted platform modules. As of the perception of the optional results, the paper's intentions highlighted the benefit of integrating TPM security in automation/SCADA systems.
5.	References (2017) [5]	This exploration implemented a survey on how various methods are used for encryption and decryption in a perfect manner. In a sense, various methods like the RSA public key cryptosystem used for database record encryption, chip-secured data on an untrusted server, a framework for storage security in RDBMS, database security using mixed cryptography, etc.

(*Continued*)

Table 22.1 Different studies of encryption/decryption techniques using automation or without an automation system. (*Continued*)

S. No.	Study	Description
6.	References (2017) [6]	This research focuses on how encryption/decryption is possible if the key matrix is non-invertible. In this study, when the encryption process applies to plain text characters, then each plain text character converts into two characters of cipher text. Similarly, each two cipher text character converts into a single plain text character when the decryption process has been done. So here, chances of complicated mathematical computation and chaos are increased. This is easy and understandable for 2x2 matrices; if characters or plain text matrices are huge, then it is very complicated.

22.4 Proposed Methodology

The planned crypsys contains a calculating concept that's illustrated in the description of encryp and decryp schemes. The encryp process takes the collaborative configuration of the polygraphic substitution cipher further than the arbitrary presented pattern and to support it in contrast to the participated vulnerability and every knob of information is stated using an arbitrary numeral. Designed to neglect several arbitrary numeral creations, a simple single arbitrary is the number produced at the inception of encrypt and the original arbitrary numeral of the corresponding information knob is constantly produced by regressed matrices.

As recently confirmed, decoding or decryp prefers the invertible matrix. However, in most conditions, the invertible of a matrix is not available. It's a documented statistic which is the part of computation that the complete matrices are non-invertible automating this system.

If the determinant of a matrix is zero in a non-invertible case, then in most developed Hill Ciphers it is vulnerable to decode the encrypted text. According to the statement about the raised vulnerability, it is advised to make the setting a flag and generate its own progressive code for encryption and decryption, which are no longer and no shorter.

The advanced method of evaluation of a polygraphic substitution Hill Cipher is a well-organized system for encryp and decryp because it is chaos in scientific calculation. However, its intended authentic user and authenticate user are generated by making the automation system using an algorithm. This is actually intricate for hackers. The advanced method of evaluation of a polygraphic substitution Hill Cipher is a far more invulnerable technique in a standard manner compared to other available complicated evaluation polygraphic substitution cipher algorithms. This procedure discovers the calculation of removed vulnerability with evaluation code. For the reason that mathematical processing is very difficult to comprehend and execute, the overall performance of the technique implemented in a formal manner with legitimate users can show how it provides a solution for vulnerability in this encryption/decryption technique.

22.4.1 Proposed Encryption Technique Using Automation System

Consider PT = plaintext,
MK = Randomly choose a key in matrix form using an automation system
MKm = Modification of random key matrix
CT = ciphertext
Modulus m = mod m = mod 41 (Contains alphabets, numbers, and some special symbols)

(i) Initially it accepts the readable text PT and random matrix key MK.
(ii) Now, to get ahead it calculates key matrix determinant KM, which is |MK|.
(iii) Set flag = 1 if |MK| > = 0 or if |MK| < 0, making flag = -1
(iv) Make the flag value, then modify the random key KM recently.
(v) Replace modified key MKm instead of random key MK after modification in the matrix key.
(vi) Determine ciphertext CT = MKm × PT mod 41.
(vii) Now, go ahead for completion of all steps, then in half process Cipher-text will be generated.
(viii) Apply decryption process on generated cipher text.

22.4.2 Proposed Decryption Technique

PT = plaintext
MKm = Modification of random key matrix using an automation system
MKm-1 determines the inversion modified matrix key.
CT = cipher text

Let's assume P, Q, x, y, i, j is the variable.
Modulus m = mod m = mod 41 (Contains alphabets, numbers, and some special symbols)

(i) Initially it accepts the ciphertext CT and improved matrix key MK like modified key MKm.
(ii) Compute the modified matrix key determinant MKm set P = |MKm| mod 41.
(iii) Set P = P+ 41 if P < = 0 or Set P = P for P > 0.
(iv) In next step, compute the value of x. For this, the x × P mod 41 = 1 function is helpful regarding the calculation of x.
(v) Determine the value of x and assign to i = x. Find the modified matrix key inversion MKm-1, i.e., MKm-1 = adj MKm × i mod 41.
(vi) Determine the transpose of MKm-1 = (MKm-1)T.
(vii) At last, determine and match the plaintext PT = CT × MKm-1 mod 41.
(viii) Set PTxy = PT_{xy} + 41. If PT_{xy} < = 0, i.e.,, x = 0 to 2 and y= 0.

22.4.3 Planned Advance Codes Regarding Numbers and Some Symbols

@ = 27	. = 28	_ = 29	? = 30	: = 31	0 = 32	1 = 33	2 = 34
3 = 35	4 = 36	5 = 37	6 = 38	7 = 39	8 = 40	9 = 41	

22.5 Exploration

22.5.1 Zero Vulnerability Overcome by Automation System

Similarly, encryp and decryp with a noninvertible matrix key of any plaintext PT includes special symbols and text between the limit of 27-31 and the numbers 0 to 9 at a limit of 32 to 41 according to the proposed codes. If a legitimate user arranges all elements of zero of the random key matrix, then this will also overwhelm zero vulnerability because the automation system will not generate all elements of zero. If, unfortunately, the system generated all elements of zero, then according to the advanced algorithm it overwhelms this vulnerability.

22.5.2 While Used in Noninvertible Matrices

Rendering to the evaluation of a polygraphic substitution Hill Cipher, there are different types of vulnerability raised with a solution. Here is the advanced method of Hill Cipher for a polygraphic substitution vulnerability of an invertible key matrix. The finish of the clarification of the execution of the matrix, which is invertible, provides an encryp/decryp procedure in a formal manner. But, using the advanced method of polygraphic substitution process converts is interesting because this strategy uses mod41, which offers a novel style of plaintext and ciphertext.

Substitution cyphers [10] are susceptible to frequency analysis threats [14, 15] and vulnerabilities [16] posed by cyber threat agents [17], where an expert analyses letter frequencies in ciphertext and substitutes characters for those that arise at a comparable frequency in text in natural language.

22.6 Conclusions

The advanced method of Hill Cipher polygraphic substitution presents a randomly chosen matrix key applied by the automation system, which is evaluated based on the computed prior ciphertext chunks and a multiplying factor. This milestone enhanced the vulnerability of the algorithm to the accepted plaintext attack. The Rectangular Matrix Key for polygraphic substitution of a cipher also implements symmetric and asymmetric key generation algorithms. By comparatively uncertain conclusions, it shows that the advanced method of Hill Cipher is the unique algorithm that asserts both assessment factors, needs an invertible matrix key, and resolves vulnerability when encrypting zeroes in plaintext chunks.

References

1. Dr. Mozamel M. Saeed (2015). Gaps of Cryptography and Their Automatic Treatments with Reference to Classical Cryptography Methods, International Journal of Research Studies in Computer Science and Engineering (IJRSCSE), Volume 2, Issue 1, January 2015, PP 22-28 ISSN 2349-4840 (Print) & ISSN 2349-4859.
2. Noha E. El-Attar, Doaa S. El-Morshedy and Wael A. Awad (2021). A New Hybrid Automated Security Framework to Cloud Storage System, Cryptography 2021, 5, 37. https://doi.org/10.3390/ cryptography 5040037.
3. Alexandra Tidrea, Adrian Korodi * and Ioan Silea, (2019). Cryptographic Considerations for Automation and SCADA Systems Using Trusted Platform Modules, Sensors 2019, 19, 4191; doi:10.3390/s19194191.
4. Tuti Alawiyah, Agung Baitul Hikmah:" Generation of Rectangular Matrix Key for Hill Cipher Algorithm Using Playfair Cipher", 2020. *et al.* 2020 J. Phys.: Conf. Ser. 1641 012094.
5. P.R.Hariharan & Dr. K.P. Thooyamani (2017). Various Schemes for Database Encryption - A Survey, International Journal of Applied Engineering

Research ISSN 0973-4562 Volume 12, Number 19 (2017) pp. 8763-8767 © Research India Publications. http://www.ripublication.com.
6. K. Mani, M. Viswambari: "Generation of Key Matrix for Hill Cipher using Magic Rectangle" Advances in Computational Sciences and Technology, 2017. Advances in Computational Sciences and Technology ISSN 0973-6107 Volume 10, Number 5 (2017) pp. 1081-1090 © Research India Publications.
7. Maxrizal, Baiq Desy Aniska Prayanti (2016) "A New Method of Hill Cipher: The Rectangular Matrix As The Private Key" International Conference on Science and Technology for Sustainability, Proceeding, Volume 2, November 2016 ISSN 2356-542X.
8. Dr. V.U.K. Sastry, K. Shirisha (2012) "A Novel Block Cipher Involving a Key Bunch Matrix" International Journal of Computer Applications, (IJACSA), Vol. 3, No. 12, 2012 October pp.116-122.
9. Toorani, M. and A. Falahati (2009): "A secure variant of the hill cipher". Reprinted from the Proceedings of the 14th IEEE Symposium on Computers and Communications (ISCC'09), pp.313-316, DOI 10.1109/ISCC.2009.5202241]. .
10. Rangel-Romero, Y., G. Vega-García, A. Menchaca-Méndez, D. Acoltzi-Cervantes and L. Martínez-Ramos *et al.* (2006): "Comments on How to repair the Hill cipher". J. Zhejiang Univ. Sci. A., 2006. DOI: 10.1631/jzus.A072143
11. Bibhudendra, A. (2009): Novel methods of generating self-invertible matrix for hill cipher algorithm. International Journal of Security, Volume1: Issue (1).
12. Pour, D.R., M.R.M. Said, K.A.M. Atan and M. Othman: "The new variable-length key Symmetric cryptosystem". Journal of Mathematics and Statistics 5 (1): 24-31, 2009 ISSN 1549-3644 © 2009 Science Publication.
13. Eisenberg, M.(1998): Hill ciphers and modular linear algebra. Mimeographed notes. University of Massachusetts. ©Copyright 1998 by Murray Eisenberg.
14. Mahor, V., Garg, B., Telang, S., Pachlasiya, K., Chouhan, M., & Rawat, R. (2022, June). Cyber Threat Phylogeny Assessment and Vulnerabilities Representation at Thermal Power Station. In *Proceedings of International Conference on Network Security and Blockchain Technology: ICNSBT 2021* (pp. 28-39). Singapore: Springer Nature Singapore.
15. Rawat, R., Gupta, S., Sivaranjani, S., CU, O. K., Kuliha, M., & Sankaran, K. S. (2022). Malevolent Information Crawling Mechanism for Forming Structured Illegal Organisations in Hidden Networks. *International Journal of Cyber Warfare and Terrorism (IJCWT)*, 12(1), 1-14.
16. Rawat, R., Rimal, Y. N., William, P., Dahima, S., Gupta, S., & Sankaran, K. S. (2022). Malware Threat Affecting Financial Organization Analysis Using Machine Learning Approach. *International Journal of Information Technology and Web Engineering (IJITWE)*, 17(1), 1-20.
17. Rawat, R., Mahor, V., Chouhan, M., Pachlasiya, K., Telang, S., & Garg, B. (2022). Systematic literature Review (SLR) on social media and the Digital Transformation of Drug Trafficking on Darkweb. In *International Conference on Network Security and Blockchain Technology* (pp. 181-205). Springer, Singapore.

23

Intelligence System and Internet of Things (IoT) Based Smart Manufacturing Industries

Vinod Mahor[1]*, Sadhna Bijrothiya[2], Ankita Singh[3], Mandakini Ingle[4] and Divyani Joshi[5]

[1]*Department of Computer Science and Engineering, Millenium Institute of Technology and Science, Bhopal, MP, India*
[2]*Department of Computer Science and Engineering, Maulana Azad National Institute of Technology, Bhopal, MP, India*
[3]*Department of Information Technology, Shri Vaishnav Vidyapeeth Vishwavidyalaya, Indore, India*
[4]*Department of Computer Science and Engineering, Medicaps University, Indore, India*
[5]*Department of Computer Science and Engineering, IPS Academy, Indore, India*

Abstract

The Internet of Things (IoT) is often referred to as the internet of everything and the Industrial Internet is a new technological paradigm envisioned as a worldwide network of interconnected equipment and objects. The Internet of Things (IoT) is widely acknowledged as one of the most crucial areas for future technologies and also is attracting significant interest from a variety of businesses. The term "artificial intelligence" has developed. The extensive use of CPS in production contexts makes design and manufacturing smarter. These advanced technologies are penetrating the manufacturing sector and enabling the integration of physical and virtual worlds via cyber-physical systems (CPS), ushering in the fourth stage of industrial production. The first step is to provide a conceptual framework for smart production systems for Industry 4.0. Second, smart design, smart machining, smart control, smart monitoring, and smart scheduling demonstration scenarios are offered. In this chapter, based on these illustrative situations with intelligence systems and IoT, key technologies and their potential applicability to Industry 4.0 smart manufacturing intelligence systems are discussed. Finally, difficulties and potential prospects are highlighted and analyzed.

***Keywords*:** IoT, AI, Industry 4.0, Smart Manufacturing System

**Corresponding author*: vinodengg.mt@gmail.com

Romil Rawat, Rajesh Kumar Chakrawarti, Sanjaya Kumar Sarangi, Rahul Choudhary, Anand Singh Gadwal and Vivek Bhardwaj (eds.) Robotic Process Automation, (339–354) © 2023 Scrivener Publishing LLC

23.1 Introduction

Over the last few years, Big Data has emerged as an emerging topic in AI (artificial intelligence) for both academic and industry populations. Chen *et al.* classified business intelligence and analytics (BI & A) development into three stages: the first is focused on DBMS (database management systems)-based structured content, known as BI & A 1.0; the second is on text and web analytics for unstructured web content, known as BI & A 2.0; and the third is on mobile and sensor-based content, known as BI & A 3.0. The amount of data collected continues to expand dramatically as companies digitalize and use the Internet of Things (IoT). Manufacturing has been designated as one of five industries where Big Data has disruptive potential [1]. Meanwhile, a new production paradigm known as "smart manufacturing" is gaining popularity in academic and industrial areas. It was defined by the Smart Manufacture Leadership Coalition (SMLC) as "the increased use of sophisticated intelligence technologies to enable quick manufacturing of new goods, dynamic reaction to product demand, and real-time decision-making in manufacturing production and supply-chain network optimization." [2].

As we all know, intelligent manufacturing (IM) is the application of artificial intelligence (AI) in manufacturing. So, what is the connection between SM and IM, as well as Big Data and AI? How do they change? This article will attempt to answer these issues by examining the current status of manufacturing intelligence relevant to next-generation AI, as well as identifying domains where planned efforts should be made on next-generation intelligent manufacturing.

23.2 Development of Artificial Intelligence

Since the term "artificial intelligence" was created in 1956, the field has had ups and downs, with two big winters in 1974-80 and 1987-93 [3]. AI may be divided into symbolic and sub-symbolic techniques (e.g., neural networks, fuzzy sets, and evolutionary algorithms). Symbolic intelligence has achieved significant success with small demo programmers for tackling "toy issues" during the first wave in the 1960s. Meanwhile, techniques based on neural networks or cybernetics have been abandoned or pushed to the sidelines [4].

Expert systems (ESs) or knowledge-based systems (KBSs) achieved significant success in the 1980s, with knowledge bases containing high-level subject information gathered from experts and articulated in specialized structured forms. Manufacturing applications have also been explored to tackle complicated machine programming challenges with several tools working on the same workpiece at the same time. Such a necessity prompted the development of knowledge engineering/knowledge representation, which is important to traditional AI research. Meanwhile, the "winter" of connectionists continues. When the work of Hopfield, Rumelhart, and others reignited large-scale interest in neural networks in the mid-1980s, research came to a stop. As a result, academics began to pay attention to sub-symbolic techniques that lack explicit representations of knowledge, such as neural networks (NNs), fuzzy sets, statistics, and genetic algorithms (GA). As a result, sub-symbolic approaches began to appear in AI systems. The loss of the Lisp machine market, the cancellation of new AI investments by the Strategic Computing Initiative, the failure of fifth-generation computers, and the decline of expert systems [5] all contributed to the second AI winter.

In distributed AI (DAI) became more focused on the emergence of multi-agent systems (MAs), which caused previous AI systems in centralized, hierarchical control structures to flatten and be replaced by a set of loosely coupled agents collaborating with each other, interoperating with messages, and mutually learning from experience. Online 2.0, web services, and web intelligence arose in AI systems. Because of the Internet's popularity, enormous amounts of data are now available online. Furthermore, individuals and businesses generate a large amount of structured and unstructured data that must be processed. Having to store and process so much data is both an urgent need and a big problem when it comes to mining and turning this data into knowledge [6].

The ongoing wave of interest in AI began around 2010, fueled by three interconnected factors [7]: the sources of Big Data, which include e-commerce, social media, the research community, organizations, and government; machine learning approaches and algorithms that have been dramatically improved based on raw material provided by Big Data; and powerful computers that support Big Data computing. Traditional AI, or AI 1.0, is being replaced by a new version called Artificial Intelligence 2.0 (AI 2.0), which focuses on machine learning (especially deep learning) and has unstructured content and decentralized (distributed) control structures. Figure 23.1 shows how AI has changed in terms of what it can do and how it is controlled, from AI 1.0 (Symbolic AI) to AI 2.0 via DAI (Distributed AI), which is called AI 1.5D, or Web AI, which is called AI 1.5W.

In the course of AI's progress, more and more sub-symbolic techniques were introduced. Although neural networks were reintroduced in the 1980s, symbolic techniques dominated AI at the time. Later, hybrid expert systems were created by combining traditional expert knowledge with artificial neural networks, evolutionary algorithms, and fuzzy sets in various configurations. By training patterns rather than loading rules, neural networks aid in the process of knowledge acquisition to some extent. Evolutionary algorithms were employed as effective methods for solving complex practical engineering optimization challenges like production scheduling [8]. Deep learning (DL) arose from neural networks, which approaches intelligence from enormous volumes of unstructured data in such a way that it overcomes, to some extent, the symbolic AI bottleneck problem that relies on knowledge extraction, which may be the most difficult component of developing an expert system. Due to the multiplicity of AI approaches and solutions, there is no obvious differentiation between which AI techniques are used. For example, in fuzzy systems, information is retained in symbolic feature styles, but in neural networks, knowledge is implemented as a neuron-like numerical algorithm. Furthermore, intelligent agents can employ both symbolic and sub-symbolic ways. As a result, there is a transition termed 1.5X between AI 1.0 and AI 2.0, which comprises of 1.5D and 1.5W, as shown in Figure 23.1.

Because AI is defined as a highly computerized system whose behaviors need intelligence [8, 9], computing development took over jobs as expert systems that had previously been deployed on expensive special-purpose machines, thereby creating a misleading impression of the so-called "collapse

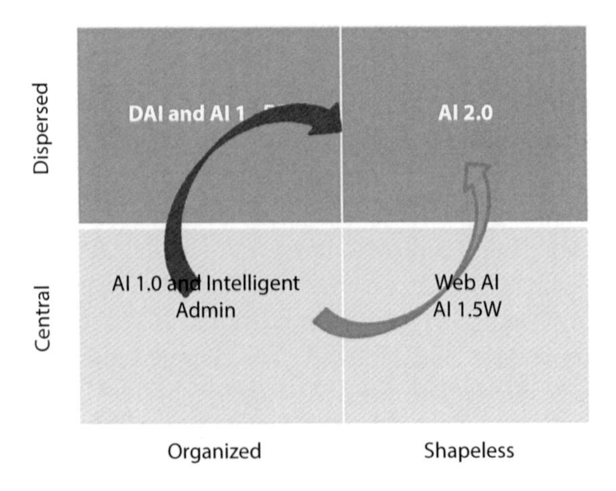

Figure 23.1 AI evolution from perspectives of content and control.

of the Lisp machine market". We are now approaching a new era of IoT, cloud computing, pervasive computing, or ubiquitous computing, which will culminate in the rise of Big Data. New AI technologies are required to collect and evaluate such Big Data. The Table 23.1 shows about the AI development.

Table 23.1 AI development.

AI focus	Computation	Processing content	Focus	Control structure
Symbolic	Workstations	DBMS-based structured content/ Knowledge	Representation	Central
Expert System & Sub-symbolic	PCs	Computational intelligence/ soft computing/ Data analytic	Statistical Methods	Central
Agent	PCs	Distributed computing	Intelligence	Disseminated
Web	Nets	Unstructured user created content/Web analytics	Web Intelligence	Web-service based
Clever	Things + clouds	IoT- based big data/ Context-aware analysis/	Deep Learning	CPS-based distributed

23.3 AI Evolution from Intelligent Manufacturing to Smart Manufacturing

Manufacturing Intelligence: Intelligent manufacturing (IM) may be defined as the junction of artificial intelligence (AI) and manufacturing. As seen in Figure 23.2, IM advances in tandem with the advancement of AI. The first book in the IM field, Manufacturing Intelligence, was published in 1988 [10] and since then, we have seen applications of AI methods, techniques, and paradigms in industrial production, resulting in the emergence of many specific IM systems such as those in design, work schedules, production, inspection, diagnosis, modelling, and control [11–13].

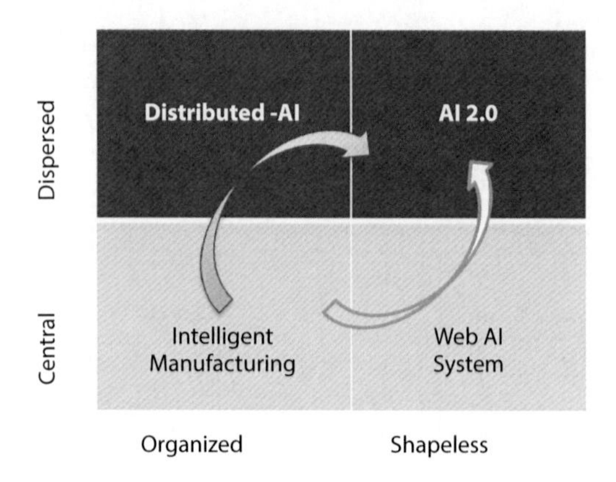

Figure 23.2 Intelligent manufacturing evolution along with AI.

There are research articles on the application of AI in the industrial industry. Teti and Kumara [11], for example, reviewed the important AI approaches established in manufacturing before 1997 and classified them as follows:

Knowledge-based/Expert systems (KBSs/ESs), Neural Networks (NNs), Fuzzy Logic (FL), Multi Agents (MAs), and other techniques such as Evolutionary Algorithms and Simulated Annealing (SA). AI applications resulted in intelligent CIM components such as intelligent CAX (CAD, CAP, CAM, and CAQ) and intelligent robots. KBSs/ESs received significantly greater attention in the early years of intelligent manufacturing system (IMS) development and, subsequently, NNs, case-based reasoning, GA, and FL received attention as well [12]. KBSs/ESs were efficiently introduced in CIM components, whereas IMS was somewhat deployed in the industry, although mostly for big enterprises. The most notable IMS study was the worldwide cooperative research scheme called the Intelligent Manufacturing System, which was founded and initially dated back to Japan and whose participants were from Japan, the United States, the European Union, and other industrial nations [13].

Agent-based systems for intelligent manufacturing [14] appeared, followed by web-services-based systems for manufacturing, Enterprise 2.0, and crowdsourcing. The agent-based strategy appeared to be a viable option since it provided a good paradigm for the intelligent CIM components and IMS, as detailed in [15–17]. Intelligent agents are utilized in distributed AI (DAI) and an agent-based DAI strategy has the potential to address the challenges of current software applications, particularly those with extremely dynamic and unpredictable working conditions [18]. However, most agent-based systems are still in the laboratories for research and prototyping and have not yet been extensively deployed in production.

Intelligent Manufacturing: Instead of intelligent technologies (Symbolic AI) in manufacturing, we are seeing a similar convergence of "smart" technologies (called "smart AI" in contrast to Symbolic AI) in manufacturing in the 2010s, with the potential to radically improve the management of manufacturing enterprises in the product life cycle in order to provide more options for customers [18, 19], as shown in Figure 23.3.

The technologies utilized for smart manufacturing implementation encompass a wide range of disciplines, which are originally referred to as IoT technologies [20], and then many other related methods such as Internet of Services (IoS), Cyber-Physical Systems (CPS), Big Data, and sophisticated robots are added in. These smart technologies are at center stage in intelligent manufacturing 2.0, often known as smart manufacturing. With the growth of IoT/CPS and smart objects (phones), items have become more networked and accessible and the quantity of data collected enables for accurate targeting and proactive management of organizations through informed, timely, and in-depth decision execution. Furthermore, the combination of human, data, and smart/intelligent algorithms has far-reaching implications for production efficiency. Big Data emphasizes data analysis, but CPS offers a broader scope than IoT or IoS and is becoming increasingly essential in the manufacturing setting [21].

Figure 23.3 depicts smart manufacturing as a cyber-physical production system, which is seen as a hybrid of IoT and IoS. Manufacturing-related resources are virtualized and packaged as cloud services in the cyber space, such as the cloud, and may be shared and used on demand via IoS. Because there are many manufacturing services and a single service cannot usually meet complex task requirements, it is necessary to solve a so-called service composition and optimal selection problem in order to form a manufacturing business process using intelligent optimization algorithms such as particle swarm optimization, differential evolution, and bee colony algorithm [22]. Then, by connecting each cloud manufacturing service (virtual machine) to its corresponding physical machine, the streamlined business process produced in the cyber system is transmitted to the actual shop floor for execution. Meanwhile, the state of the physical workshop is sent to the cyber system through IoT to determine whether or not the business process has been completed.

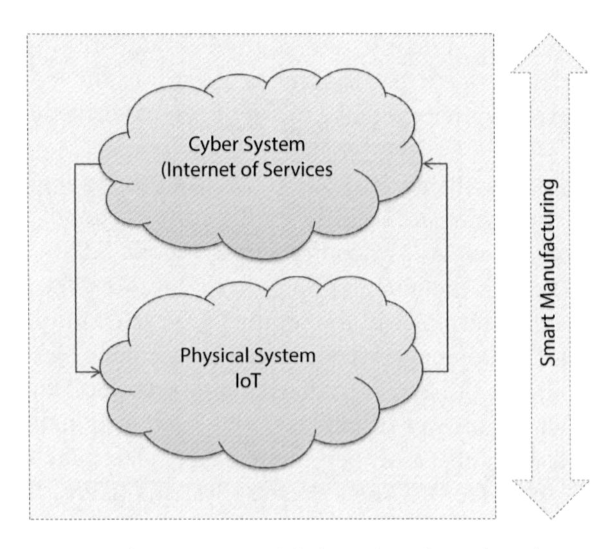

Figure 23.3 Smart manufacturing exemplified as cyber-physical production system.

23.4 IM and SM Comparison

Traditional IM systems are centralized in setup with organized contents such as databases, knowledge bases, intelligent CAD (ICAD), ICAP, and ICAM, and are often utilized on a limited scale in enterprise departments. Meanwhile, agent technology encourages the use of distributed AI in manufacturing. Manufacturing firms have migrated to the Web to sell and advertise their products as the Internet has evolved, resulting in the growth of unstructured data from social media. Later, improvements in the Internet of Things and Services, as well as smart technologies, are rapidly being used by the industry, resulting in the birth of smart manufacturing and forcing organizations to confront the problem of exponentially rising Big Data. As a result, businesses must employ Big Data-related strategies for forecasting, preventative maintenance, and production [22, 23].

However, because of limitations in data collecting and processing capabilities, such Big Data analytics are not available in traditional, even agent-, or web-based manufacturing systems. As previously said, conventional IM is based on symbolic AI, which tries to integrate human experience and expertise in manufacturing, which is often derived from production specialists. Human experience/knowledge is conveyed in classic IM systems using IF-THEN production systems or other structured methods (e.g., frames, objects, and semantic). Big Data created in SM, on the other hand, is beyond the processing capabilities of typical database systems and software tools. As a result, Big Data analytics become crucial for businesses in order to transform raw data into useful information and/or knowledge that aids decision making [23].

IM, as seen in Figure 23.4, is knowledge-based, whereas SM is data-driven and knowledge-enabled. Data and knowledge are linked in the Data, Information, Knowledge, and Wisdom hierarchy (DIKW hierarchy). The introduction of "Big Data (analytics)" causes organizational decision makers to change their attention from knowledge to "data". Because of the massive amounts of data being processed, prediction models and decisions that emerge are based on machine learning, particularly deep learning, which can be used to abstract high-level representations from massive amounts of data [24].

As a result, in the Big Data age, decision making is driven by forecasts, learning from data (experience) to predict, and actions are made in reaction to predictions. Machine learning, which learns from data and uses statistical approaches to assist decision making and works well in practice, contrasts with the older "expert system" approach, which aims to mimic the rules of human experts with the help of programmers who translate the explicit rules into software code.

There is now a growing body of literature on Big Data in manufacturing, as evidenced by surveys conducted in the previous few years. In DBMS, there is a significant difference between Big Data and structured data. Big Data is defined as unstructured large-scale data sets that are difficult to analyze by existing software tools in a reasonable amount of time and is characterized by a high Volume, Velocity, and Variety, necessitating developing new ways for its conversion into Value. Although BI&A

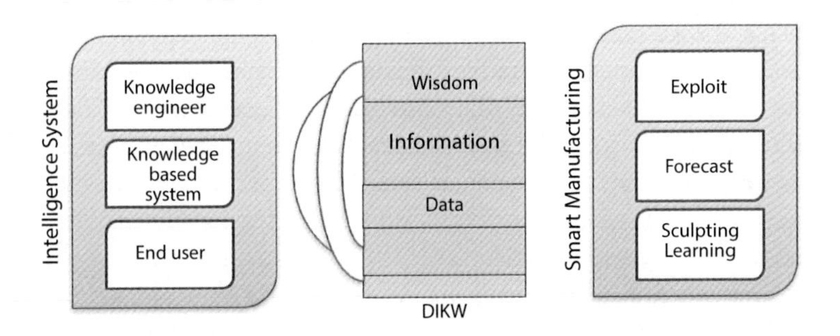

Figure 23.4 Intelligent manufacturing versus smart manufacturing.

is separated into three stages based on data structures/sources, both BI & A 2.0 and 3.0 are connected to Big Data and differ only in the data sources [25].

As previously indicated, we treat AI or intelligent manufacturing in two stages. Prior to 2000, we referred to the application of AI in manufacturing as intelligent manufacturing (IM), but smart manufacturing is now increasingly utilized as a next-generation production model with smart sensing and control techniques. Initially, the smart factory was researched with the use of IoT in production and later became a crucial component of Industry 4.0 (Industrie 4.0). Then, more and more "smart" technology, such as IoS, cloud computing, and CPS, were brought to the smart factory or smart manufacturing [26].

The word "smart" now refers to the generation and use of data across the whole product cycle for the goal of more flexible manufacturing processes that adapt promptly to on-demand modifications at cheap cost without affecting the environment. The advent of these technologies is what distinguishes manufacturing as "smart" and distinguishes it from "older" intelligent manufacturing. IoT and CPS are two acronyms that attempt to connect the cyber and physical worlds, resulting in huge scale data sets, also known as Big Data. SM's strength is in applying manufacturing intelligence (MI) from a complete global perspective with the use of previously unavailable ICT technologies like IoT, IoS, CPS, and Big Data.

As such, Davis *et al.* defined SM as "the use of data-driven MI in multiple real-time applications deployed throughout all operating layers across the factory and supply chain" [26] and the future enterprise as "data driven, knowledge enabled, and model rich with visibility across the enterprise such that all operating actions are executed proactively by applying the best information and performance metrics" [24].

Such widespread use or access to mined information/knowledge from Big Data across the entire product value chain, from product lines to demand-supply networks, enables new services and business models such as "Everything-as-a-Service" and "Pay-per-use," as in cloud-based design and manufacturing [25]. As a result, "Design-as-a-Service" and "Product-as-a-Service" may be accessed on demand. As a result, we are approaching the next stage of intelligent production- smart manufacturing-in which factories can detect, comprehend, think, and respond to our demands. Table 23.2 compares IM and SM and we can see that SM delivers many more benefits than IM.

Table 23.2 Comparison of IM and SM.

Characteristics	Intelligence manufacturing system	Smart manufacturing system
Adaptability		Y
Big Data		Y
Context-Awareness		Y
Deep Learning		Y
Entire Value Chain Support		Y
Everything-as-a-Service		Y
IoS/Cloud Computing		Y
IoT/CPS		Y
Optimal Scale	Usually Local	Global
Proactivity		Y
Self-Organization		Y
Self-Predictiveness		Y
Configuration	Centralized	Distributed
Structured Content (data)	Y	Y
SoS		Y
Ubiquitous Access		Y
Virtualization		Y
Visibility		Y

23.5 Further Smart Manufacturing Development for Industry 4.0

The phrase "Industry 4.0" comes from the German government's high-tech initiative, which is derived from "smart factories". Following the first Industrial Revolution of "Mechanization," the second of "Mass Production," and the third of "Automation," Industry 4.0 arises through the use of CPS,

IoT, and IoS [27]. Smart factories (manufacturing) and Industry 4.0, both of which are frequently depicted in CPS designs [25–27], are enabling one another. However, the CPS design is insufficient for Industry 4.0 or a manufacturing system, which is socio-technical by definition. The "Made in China 2025 Strategy," like Germany's Industry 4.0, focuses on intelligent (smart) manufacturing [28]. Furthermore, with the growth of Enterprise 2.0, socialized firms, crowdsourcing, social manufacturing, and open innovation, as indicated in Figure 23.5, the social dimension should be incorporated in smart manufacturing/smart factories/Industry 4.0. Wisdom manufacturing (or wise manufacturing) in a form of social CPS (SCPS) that was proposed to satisfy such demands [29].

The last three industrial revolutions attempted or centered on mass manufacturing, but Industry 4.0 emphasizes mass customization/personalization. As a result, Industry 4.0 represents a socio-technical revolution, transforming the previous technological revolution into a socio-technical revolution through the introduction of smart manufacturing technologies and convergence with social intelligence and human wisdom, such as social computing, collective intelligence, crowdsourcing, and innovation. And, as seen in Figure 23.6, SCPS-based manufacturing may be considered an extension of CPS-based manufacturing (including smart manufacturing and 3D printing, which naturally revives craft production in the CPS form) by adding the social component [30].

In the manufacturing environment, Wisdom Manufacturing merged IoT, IoS, the Internet of Contents and Knowledge (IoCK), and the Internet of People (IoP). DIKW, Internet of Knowledge, or Internet of Data can all be used to describe IoCK (Big Data). As a result, IoT, IoS, IoCK, and IoP may be combined to form IoTSKP (Internet of Things, Services, Knowledge,

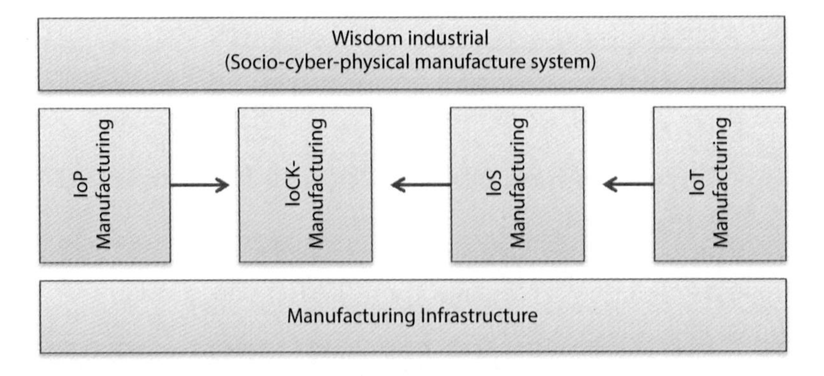

Figure 23.5 Wisdom manufacturing vs. other emerging manufacturing models with big data in common.

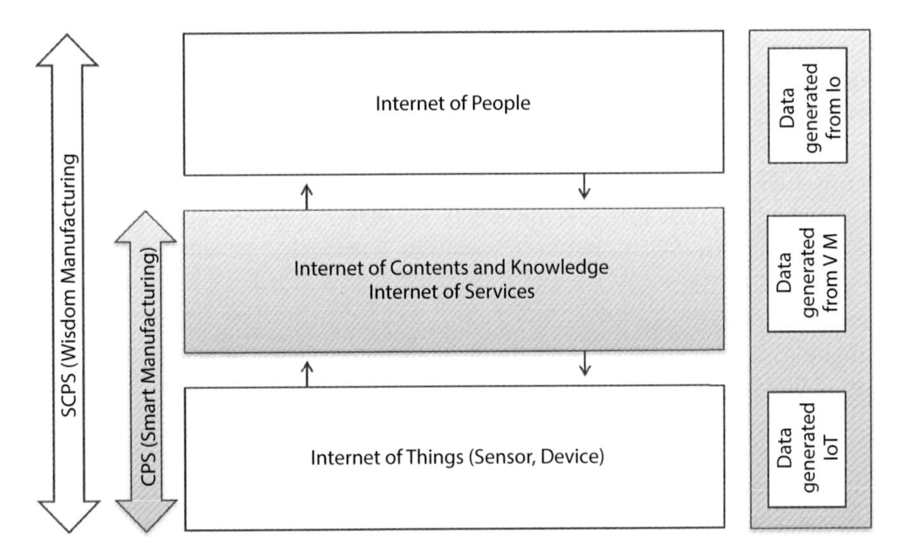

Figure 23.6 Framework for SCPS-based manufacturing.

and People) [30, 31]. Most new emerging manufacturing models concentrate on one or two aspects of IoTSKP, such as smart factories based on IoT, cloud manufacturing based on IoS (cloud computing), Enterprise 2.0/crowdsourcing based on IoP, predictive manufacturing based on Big Data, and CPS-based smart manufacturing. As seen in Figure 23.5, wisdom manufacturing is a model that blends the ideals of these developing models [28].

The integration of IoT, IoS, and IoP in manufacturing results in a rapid increase in available data sets, which overwhelms organizations. These new manufacturing methods all have one thing in common: Big Data. On the one hand, Big Data as data-intensive computing offers us a new paradigm that goes beyond experimental and theoretical study and computer simulations of natural processes [28–31] to reconsider what AI or intelligent (smart) manufacturing entails. On the other hand, Big Data is regarded as one of the biggest difficulties that 21st-century businesses must face. As a result, turning Big Data into usable information/knowledge for these developing production models is an urgent requirement and predictive problem [32].

23.6 Conclusion

SCPS-based manufacturing (wisdom manufacturing) combines the physical, cyber, and social systems as a whole, covering 6 semiotic levels, from physical to social, and generates data from the social system, which includes social media networks, Web 2.0, crowdsourcing communities, and mobiles; the cyber system, which includes the IoCK and IoS, as well as digitalization of manufacturing such as NC/CAD/CAM/CAE/CAPP/PDM/ERP, simulation, and virtual manufacturing; and the physical system, which includes sensors and smart objects.

It is becoming fashionable to combine "symbolic" AI with "clever" AI, resulting in a hybrid AI known as "wise" AI or Artificial Wisdom. As a result, AI progresses from "symbolic" (AI 1.0) to "clever" (AI 2.0), and finally to "wise" (AI 3.0). AI 3.0 can be thought of as a hybrid of symbolic AI, clever AI, and other technologies. We have "sage AI (AI 3.0) = clever AI (AI 2.0) + symbolic AI (AI 1.0) + others."

This chapter to manufacturing is progressing from clever to smart and will eventually become wise (wisdom). Wise (wisdom) manufacturing integrates not only symbolic AI and smart technologies (smart AI), but also human intelligence and wisdom in manufacturing, more specifically integrating humans, computers, and machines, things, ubiquitous, artificial/collective intelligence, and human knowledge and experience as a whole.

There are also recommendations for further study in the areas of data collection, virtualization, and decision-making. This effort is anticipated to give the industrial sector ideas for quickly deploying Industry 4.0.

References

1. Yao, X., Zhou, J., Zhang, J., & Boër, C. R. (2017, September). From intelligent manufacturing to smart manufacturing for industry 4.0 driven by next generation artificial intelligence and further on. In 2017 5th international conference on enterprise systems (ES) (pp. 311-318). IEEE.
2. Rawat, R., Mahor, V., Chirgaiya, S., Shaw, R. N., & Ghosh, A. (2021). Sentiment analysis at online social network for cyber-malicious post reviews using machine learning techniques. In Computationally intelligent systems and their applications (pp. 113-130). Springer, Singapore.
3. Wikipedia. (2017, Mar. 22). AI winter. Available: https://en.wikipedia.org/wiki/AI_winter
4. Wikipedia. (2017, Mar. 20). Artificial intelligence. Available: https://en.wikipedia.org/wiki/Artificial_intelligence
5. National Science and Technology Council. (2016). Preparing for the future of artificial intelligence. Available: https://www.whitehouse.gov/sites/default/files/whitehouse_files/microsites/ostp/NSTC/preparing_for_the_future_of_ai.pdf

6. A. Mostafaeipour and N. Roy, "Implementation of Web based Technique into the Intelligent Manufacturing System," Int. J. Comput. Appl., vol. 17, pp. 38-43, 2011.
7. E. Estellés-Arolas, "Towards an integrated crowdsourcing definition," J. Inf. Sci., vol. 38, pp. 189-200, Apr 2012.
8. H. A. Abbas, S. I. Shaheen, and M. H. Amin, "Simple, Flexible, and Interoperable SCADA System Based on Agent Technology," Intelligent Control and Automation, vol. 06, pp. 184-199, 2015.
9. C. Leiva. (2015). On the Journey to a Smart Manufacturing Revolution. http://www.industryweek.com/systems-integration/journey-smartmanufacturing-revolution.
10. L. H. Wang, M. Torngren, and M. Onori, "Current status and advancement of cyber-physical systems in manufacturing," J. Manuf. Syst., vol. 37, pp. 517-527, 2015.
11. D. Z. Wu, D. W. Rosen, L. H. Wang, and D. Schaefer, "Cloud-based design and manufacturing: A new paradigm in digital manufacturing and design innovation," Comput. Aided. Design., vol. 59, pp. 1-14, Feb 2015.
12. M. M. Najafabadi, F. Villanustre, T. M. Khoshgoftaar, N. Seliya, R. Wald, and E. Muharemagic, "Deep learning applications and challenges in big data analytics," J. Big Data, vol. 2, p. 1, 2015.
13. P. O'Donovan, K. Leahy, K. Bruton, and D. T. J. O'Sullivan, "Big data in manufacturing: a systematic mapping study," J. Big Data, vol. 2, pp. 1-22, 2015.
14. A. D. Mauro, M. Greco, and M. Grimaldi, "A formal definition of Big Data based on its essential features," Libr. Rev., vol. 65, pp. 122-135, 2016.
15. J. Davis, T. Edgar, R. Graybill, P. Korambath, B. Schott, D. Swink, *et al.*, "Smart manufacturing," Annu. Rev. Chem. Biochem. Eng., vol. 6, pp. 141-160, 2015.
16. M. Hermann, T. Pentek, and B. Otto. (2015). Design principles for Industrie 4.0 scenarios: a literature review. Available: http://www.snom.mb.tu-dortmund.de/cms/de/forschung/Arbeitsberichte/Design-Principles-for-Industrie-4_0-Scenarios.pdf
17. H. S. Kang, J. Y. Lee, S. Choi, H. Kim, J. H. Park, J. Y. Son, *et al.*, "Smart Manufacturing: Past Research, Present Findings, and Future Directions," Int. J. Pr. Eng. Man-Gt., vol. 3, pp. 111-128, Jan 2016.
18. Wikipedia. (2017, Mar. 20). Industry 4.0. Available: https://en.wikipedia.org/wiki/Industry_4.0
19. N. Jazdi, "Cyber physical systems in the context of Industry 4.0," in 2014 IEEE International Conference on Automation, Quality and Testing, Robotics, 2014, pp. 1-4.
20. R. F. Babiceanu and R. Seker, "Big Data and virtualization for manufacturing cyber-physical systems: A survey of the current status and future outlook," Comput. Ind., vol. 81, pp. 128-137, Sep 2016.
21. Rawat, R., Rajawat, A. S., Mahor, V., Shaw, R. N., & Ghosh, A. (2021). Surveillance robot in cyber intelligence for vulnerability detection. In Machine Learning for Robotics Applications (pp. 107-123). Springer, Singapore.

22. Rawat, R., Mahor, V., Chirgaiya, S., & Garg, B. (2021). Artificial cyber espionage based protection of technological enabled automated cities infrastructure by dark web cyber offender. In Intelligence of Things: AI-IoT Based Critical-Applications and Innovations (pp. 167-188). Springer, Cham.
23. Mahor, V., Pachlasiya, K., Garg, B., Chouhan, M., Telang, S., & Rawat, R. (2022). Mobile Operating System (Android) Vulnerability Analysis Using Machine Learning. In International Conference on Network Security and Blockchain Technology (pp. 159-169). Springer, Singapore.
24. Rawat, R., Chouhan, M., Garg, B., TELANG, S., Mahor, V., & Pachlasiya, K. (2021). Malware Inputs Detection Approach (Tool) based on Machine Learning [MIDT-SVM]. Available at SSRN 3915404.
25. Mahor, V., Badodia, S. K., Kumar, A., Bijrothiya, S., & Temurnikar, A. (2022). Cyber Security for Secured Smart Home Applications Using Internet of Things, Dark Web, and Blockchain Technology in the Future. In Dark Web Pattern Recognition and Crime Analysis Using Machine Intelligence (pp. 208-219). IGI Global.
26. J. Lee, B. Bagheri, and K. Hung-An, "A cyber-physical systems architecture for Industry 4.0-based manufacturing systems," Manuf. Lett., vol. 3, pp. 18-23, 2015.
27. P. J. Mosterman and J. Zander, "Industry 4.0 as a Cyber-Physical System study," Software and Systems Modeling, vol. 15, pp. 17-29, Feb 2016.
28. J. Zhou, "Intelligent manufacturing - Main Direction of "Made in China 2025""," China Mechanical Engineering, vol. 26, pp. 2273-2284, 2015.
29. X. F. Yao and Y. Z. Lin, "Emerging manufacturing paradigm shifts for the incoming industrial revolution," Int. J. Adv. Manuf. Tech., vol. 85, pp. 1665-1676, Jul 2016.
30. X. F. Yao, H. Jin, and J. Zhang, "Towards a wisdom manufacturing vision," Int. J. Comput. Integ. M., vol. 28, pp. 1291-1312, Dec 2 2015.
31. J. Zhou, X. Yao, M. Liu, J. Zhang, and T. Tao, "A state-of-the-art review on new emerging intelligent manufacturing paradigms," Integr. Manuf. Syst, vol. 23, pp. 624-639, 2017.
32. X. Yao, J. Zhou, C. Zhang, and M. Liu, "Proactive manufacturing - a big-data driven emerging manufacturing paradigm," Integr. Manuf. Syst., vol. 23, pp. 172-185, 2017.

24

E-Healthcare Systems Based on Blockchain Technology with Privacy

Vinod Mahor[1]* and Sadhna Bijrothiya[2]

[1]*Department of Computer Science and Engineering, Millenium Institute of Technology and Science, Bhopal, MP, India*
[2]*Department of Computer Science and Engineering, Maulana Azad National Institute of Technology, Bhopal, MP, India*

Abstract
Healthcare challenges include sluggish access to medical data, poor system interoperability, lack of patient agency, and data quality and quantity for medical research that can be mitigated by blockchain. Doctors can access a patient's whole medical record thanks to the security and convenience provided by blockchain technology, while researchers can only access statistical data and not any personally identifiable information. Distributed Ledger Technology (DLT) and Blockchain (BC) can be extremely beneficial in healthcare systems for access control and massive data management. Putting in place a pure blockchain system, or migrating to one, is a complex task. Certain design and implementation dynamics must be addressed before an efficient solution can be built.

***Keywords*:** Blockchain, healthcare, Electronic Medical Records (EMR), security, privacy

24.1 Introduction

Blockchain has been a popular buzzword in recent years, with the notion that it is a panacea for many (if not all) security issues. There is no arguing that it improves system transparency, traceability, and security, yet it is not a one-size-fits-all technology. Several research papers on the use of

**Corresponding author*: vinodengg.mt@gmail.com

Romil Rawat, Rajesh Kumar Chakrawarti, Sanjaya Kumar Sarangi, Rahul Choudhary, Anand Singh Gadwal and Vivek Bhardwaj (eds.) Robotic Process Automation, (355–370) © 2023 Scrivener Publishing LLC

blockchain in various applications, spanning from industrial automation to automotive networks, and from the Internet of Things to financial markets, are freely available. Practically speaking, they are easier said than done. We examine how the healthcare business might benefit from DLT in general and BC in particular for process automation, digital/electronic medical record management (including big data), access control, and smart contracts in this paper. Several works of literature have addressed these issues (Azaria, A, 2016), however, the majority of them address extremely particular concerns while disregarding the broader picture. In this post, we will first examine and describe how business blockchain may be utilized effectively in healthcare, followed by the specific needs of a healthcare system. In the next sections, we will cover migration issues and possible solutions, the trade-off between unified and multi-chain environments, consensus algorithms for healthcare, users and access rights, smart contracts, and e-healthcare business rules (D. Dwivedi, 2019).

Do not consider crypto currency: Blockchain has attracted substantial interest in recent years, owing mostly to the skyrocketing values of Bitcoin. Since then, hundreds more crypto-currencies have sprouted up all over the world. The most common misperception regarding blockchain is that it is just for crypto-currencies. Blockchain is a Distributed Ledger Technology that has been largely specialized for financial transactions. The generic DLT, on the other hand, is primarily concerned with providing a set of protocols and processes for the dissemination of records among numerous nodes in a collaborative system (D. Dwivedi, 2019, D. C. Nguyen, 2019). The system may be owned by a single company or it may link to a single but shared and distributed Ledger. As a result, blockchain inherits the benefits of DLTs while adding a few more.

Security via smart contracts, which are established agreements between parties to do business, transparency and accountability via immutable records kept at scattered places, and efficiency and cost reduction via process automation are some of the applications of blockchain technology. Blockchains for crypto currencies are based on the notion of tokens, which are transferred between participating users. However, the advantages provided are not restricted to tokens. If the token is viewed as a data element that is produced and exchanged while leaving an audit trail, then any digital asset (or piece of information) that is moved among participants while needing an audit trail might potentially benefit from blockchain. Furthermore, smart contracts may be used to quickly establish access control for such digital assets, while the data itself can be kept in the distributed ledger system, boosting its dependability and validity. Based on these considerations, using blockchain for purposes other than crypto-currency is not only viable, but also extremely practical. Business Blockchain (BBC) is a version of classic blockchain that intends to use BC protocols inside a business process, such as the gathering of authenticated and validated data from asset line sensors, voting and auditing in an e-government system, or asset tracking (S. Biswas, 2019). Another technique for categorizing blockchains is based on the system's openness, such as public, consortium/ federated, or private blockchains. Public blockchains are available to everybody, whilst consortium blockchains are confined to a group of organizations and private blockchains are limited to a single organization. Anyone may become a miner, peer, or trader on public blockchain since they are available to the public. In contrast, consortium/private access is often permissioned, with users first registering and authenticating. Business BC is often consortium/private with restricted access, with a peer in charge of verification consensus creation (https://gdpr-info.eu).

Because Hyper ledger is a Linux foundation solution that may be used as a basis platform for constructing commercial blockchain, most of the discussion in this article revolves around its use and the flexibility it provides. It employs five frameworks and consensus processes designed for various sorts of contexts. Hyper ledger fabric is a significant implementation that provides customizable consensus algorithm implementation, smart contract integration, and Internet of Things (IoT) integration assistance. It is vital to remember that it is merely a platform and does not give a comprehensive blockchain business solution in any specific case.

24.2 Blockchains in Healthcare

Digitization and integration of the Internet of Things (IoT) with E-Healthcare Systems (eHS) have elevated it to one of the fastest-growing fields, paving the way for smart healthcare. (H. Jin, 2019). According to statistics, worldwide healthcare spending will continue to rise in 2020 and beyond, with a strong emphasis on digital transformation (https://hhs.gov/hipaa/). To cut costs and enhance treatment quality, medical service providers will boost their usage of innovative solutions such as cloud computing, 5G, big data analytics, blockchain, artificial intelligence, and so on.

Integrating blockchain with eHS can provide several benefits, including but not limited to the security of electronic medical records (EMR), access control for different types of users, automated execution of services, remote data collection and logging, information unification or standardization, redundancy and fault tolerance, healthcare regulatory enforcement, logistics, and so on (N. Kshetri, 2018). However, implementing such a blockchain is incredibly difficult. To begin, a contemporary eHS is a conglomeration of several distinct technologies at the device, operational, and management system levels. As a result, the blockchain solution should not only address the demands of small-scale sensor devices, but also devices that generate large amounts of data (CT scans). Simultaneously, this data must be exchanged among departments as well as with third-party service providers such as insurance companies. To make matters even more complicated, interoperability across multiple service providers may be impossible owing to very varied automation technologies (P. Pace, 2019).

To be more precise, some of the significant problems are: i) Existing centralized eHS employ relational databases to store data, whereas blockchain uses a file database and DB. There may not be a one-to-one mapping in the schema; ii) Due to transaction size limitations in a block, it is hard to keep entire medical images as part of the chain; iii) Due to real-time transactions on a large scale, it is difficult to transfer all medical history of patients to the blockchain ledger; iv) It is probable that certain medical papers in an eHS are paper-based and as a result, the only method to digitize them is to save them as photographs, which is a slow process; v) Access to patient data must be strictly regulated for various types of internal and external users (P. Sundaravadivel, 2018); vi) An eHS may allow integration of third-party IoT devices (smartwatches, health sensors) to be part of the system, making verification and validation difficult; and vii) Interaction with other non-BC sub-systems of the e-healthcare ecosystem, including regulatory bodies (P. Tasca and C. J. Tessone, 2019).

This is not an entire list of the primary problems that must be overcome when developing a complicated blockchain solution for e-healthcare systems. In the parts that follow, we will go through each and every element of constructing such a system, as well as discuss the technical aspects of various alternatives (R. Jayaraman, 2019). It is crucial to emphasize that the goal of this work is not to give a full solution, but to help the audience understand what the issues are and what the benefits of many viable solutions may be, notwithstanding the authors' preference for certain design choices.

Figure 24.1 depicts a basic company blockchain procedure before advancing. Users create trades (transactions) that contain digital assets that must be shared with other users or devices. A Membership Service Provider (MSP) is made up of an administrator and a Certificate Authority (CA) that is in charge of delivering keys, signatures, certificates, and configuration information. Peers are specialized nodes with the resources to run consensus methods and keep the distributed ledger up to date. Ordering Service is in charge of consolidating all endorsed/approved deals

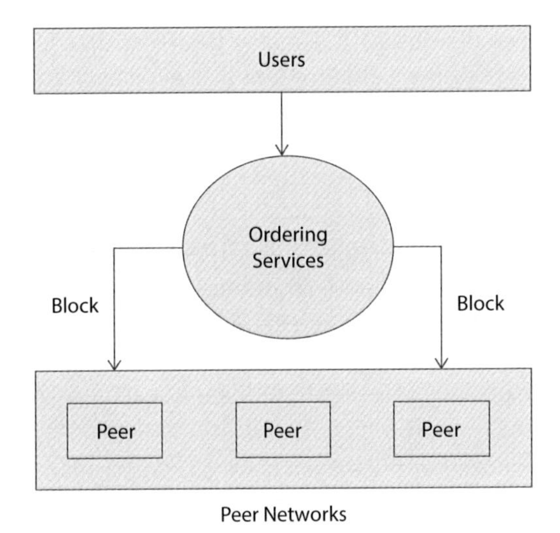

Figure 24.1 Blockchain framework for business.

into a freshly generated block. On peer nodes, smart contracts or chain code are implemented to verify transaction agreements between multiple users (S. Biswas, K. Sharif *et al.*, 2020).

24.3 Regulations and EHR Privacy

The major rationale for incorporating blockchain into any system is to improve security. It is critical to note that BC only adds validation and immutability to the asset exchange process and stored data. These additions, however, have a large and profound effect on the entire security architecture (T. McGhin, K.-K. R. 2019). Validation is accomplished by smart contracts and consensus protocols, which guarantee that no unlawful trade occurs and immutability is ensured through hash connection in the chain, which assures that nothing is altered later. Because BC does not provide any new encryption techniques, signature mechanisms, hash functions, etc., efficient use of current algorithms or creation of new algorithms in this area is critical (V. P. Yanambaka, 2019). Several security and privacy primitives must be evaluated in an e-healthcare use case. Some BC systems (with public miners, for example) allow miners to view transaction payloads for validation and smart contract execution. This payload in e-healthcare might be an EHR that must not be disclosed (even in encrypted format). The compromised (or cooperating) miner/peer cannot be ruled out on a private blockchain.

As a result, the privacy of EHR may be jeopardized. Similarly, if the digital signatures used for validation can be traced back to patients or their physicians, this may lead to a violation of privacy and healthcare standards.

Based on this, two considerations must be made before creating a BC-based e-healthcare system:

(i) Regulation Understanding: HIPAA and GDPR must be obeyed and misunderstandings regarding both must be dispelled. Many researchers, for example, associate GDPR with the right to forget. Nevertheless, the legislation expressly specifies that this is not an absolute right for medical practitioners. As a result, data privacy as defined by HIPAA/GDPR must be incorporated into the system for both internal and external parts (W. Mougayar, 2016).

(ii) Identification of Blockchain Usage Case: Due to privacy and regulatory concerns, it is critical that the use of blockchain within the healthcare system be recognized. Consent management, for example, is a cornerstone of healthcare standards. BC can be utilized well for this purpose. Similarly, access control to EHR, medication control, prescription administration, patient monitoring, insurance, and accounting, among other applications where immutability and accountability are required, might benefit considerably from blockchain (W. Zhang, 2019).

Blockchain cannot be viewed as a one-size-fits-all solution for all healthcare subsystems. Smart contracts will be the most effective way to implement rules such as GDPR and HIPAA. As a result, methods are required to ensure that smart contracts are constructed in such a way that healthcare privacy standards are satisfied.

24.4 Issues with Migration

Crossover time with the old system is always required when designing and deploying any new system for a large or medium-sized firm. The old system is gradually phased out as data and processes are moved to the new. The majority of blockchain research focuses on algorithmic technicalities, ignoring the reality that the startup time for a blockchain system, specifically one designed for healthcare institutions, may render the current option impractical. We tackle this problem from two perspectives (W. Zhang, 2019).

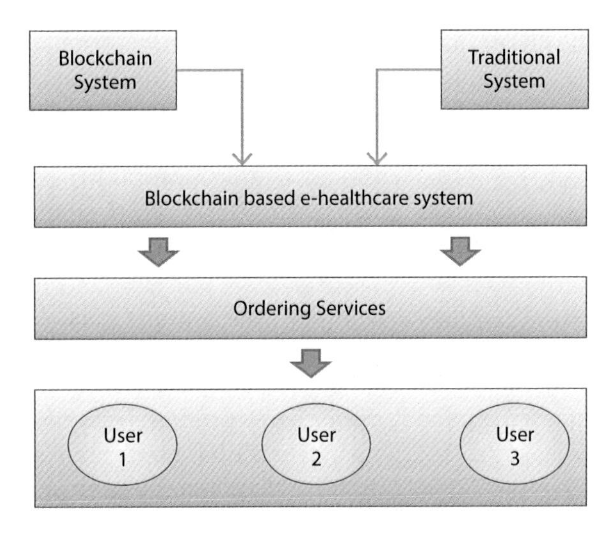

Figure 24.2 Blockchain based framework of e-healthcare model.

Changes in Infrastructure and Architecture: As seen in Figure 24.2, traditional eHS systems are often centralized. A single server hosts the core application, its related database, and maybe the certificate authority. Although the server is on the cloud, it is still a centralized system in terms of implementation. It is also feasible that a major eHS provider has many locations and consequently many centralized systems that communicate at various levels. As a result, the databases may be distributed while the web-based application may be centralized, resulting in a completely new architecture.

In comparison, the blockchain system is completely decentralized. Furthermore, this decentralization is not the same as a decentralized database or distributed systems. As seen in Figure 24.2, a group of peer nodes creates a particular peer network that accomplishes consensus formation, while a specialized ordering service (a group of ordered nodes) is in charge of block generation and distribution back to the peer network. The transition from centralized to distributed blockchain necessitates considerable modifications in infrastructure. This difficulty must be considered when developing solutions (Mahor, V., 2022). It is vital to remember that without an application interface, users or devices cannot begin transactions. Many systems use thin-clients on the user side, which means that the blockchain network must include an application server. The bulk of research articles downplay this feature, assuming that users are transmitting transactions straight to the peer. However, it is possible that the application server is to blame. This assumption may be totally secure for thick-clients, but user devices would still communicate with a system entity that maintains access control. Figure 24.2 depicts how a merged system would look. It is possible.

It has been intuitively noticed that the application server might generate a single point of failure, thus it is vital to understand that simply using blockchain does not make a system temper proof. This creates various additional issues for blockchain security and interoperability with other systems, necessitating the development of safe and standardized APIs for system interaction.

Data Migration and Synchronization: The conversion of old records and databases to the new system is one of the least investigated aspects of blockchain implementation. The most basic reason is that records cannot migrate in their existing state. First, the prior entry with outdated timestamps cannot be accepted by the ledger. A current timestamp is required for each new transaction. Second, because the blockchain record is immutable, any timestamp alteration after the creation of a block is impossible. As a result, any adjustments or updates must be completed prior to the transfer. This is a difficult undertaking that may differ from one eHS to the next (Rawat, R, 2022). Third, the traditional centralized system may have thousands of patient records. At startup time, bootstrapping the blockchain system with all of that data might be a lengthy operation. The same system may be in operation at the same time, producing (or maybe modifying) existing data. This causes a cyclical migration problem that must be addressed throughout the design process. Furthermore, effective migration methods and synchronization strategies are required (Rawat, R, 2022, Mahor, V., 2022).

One alternative option is to move the data only as necessary, rather than during the startup step. As demonstrated in Figure 24.2, a Data Migrator module can be used only when necessary to structure relational database entries into Ledger approved trades. A patient who has visited the eHS, for example, has several records in the old system. Only when the same patient visits the institution after the blockchain transfer are the relevant records synced. The blockchain technology is used to create all new records, whereas the relational database is solely used to store existing ones. This ensures that cyclic record updating is avoided and that the initial bootstrap time is negligible. Efficient designs for such data transfer interfaces and algorithms will be critical to migration success.

24.5 Blockchains: Unified or Multiple

Blockchain solutions must be tailored to individual applications. Multiple service providers with distinct systems may exist in an e-health situation.

Cooperation among these systems is possible, provided an operational level agreement is reached. However, moving EMR from one patient to another or unifying them in a single database is frequently difficult. This problem is multiplied in a blockchain-based solution.

To begin with, if one service provider migrates to a blockchain solution, its operational collaboration with a conventional centralized service provider will be terminated immediately, as there is no default interface between blockchain and non-blockchain systems. The severity of this difficulty is illustrated by the fact that a service provider must simultaneously migrate all of its hospitals to the blockchain system or risk non-cooperation among its own service points (Mahor, V, 2022).

Second, even if all collaborating service providers switch to blockchain solutions, unification concerns may still arise. Figure 24.3 depicts three different sorts of solutions in this regard. All e-healthcare service providers link to a single blockchain in the first Unified blockchain solution, which is maintained by either a consortium or the government. Furthermore, all eHS keep their own local servers and only send exchanges involving multiple eHS. This is a hybrid method that may allow certain eHS to operate a traditional system with a blockchain backbone interface. The other systems depicted in Figure 24.3 comprise a multi-blockchain solution in which each service provider has its own blockchain which is subsequently linked to other blockchains for interoperability. In this case, the eHS can either make their whole peer network part of the unified chain or restrict certain peers from being part of the global chain while the others stay local to the chain. This is a more complicated system, but it does allow individual eHS to have their own blockchain. The solutions for a traditional system to blockchain interface are a major design

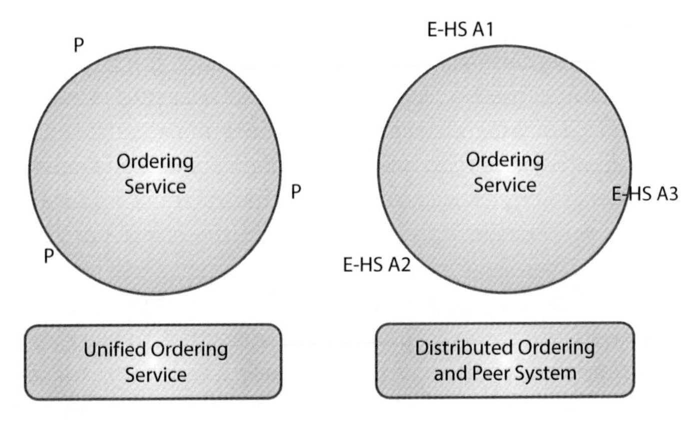

Figure 24.3 Solution for cooperative e-healthcare interoperability.

challenge in this case. The following issues must be addressed in any of the solutions stated above.

Interoperability: This enables one eHS to communicate data with another eHS without the data being interpreted. It increases patient participation, facilitates access, improves efficiency, and, to some extent, permits regulatory compliance. Interoperability may be characterized as follows from an engineering standpoint:

Structural interoperability: Allows data sharing without requiring either system to modify the data format. It is saved and utilized without being interpreted.

Semantic interoperability: Allows systems to understand data without modifying it. This indicates that not only is the data structure the same, but so is its meaning. Temperature, for example, is recorded as an integer but interpreted as Celsius or Fahrenheit.

It's also worth noting that EHR interoperability standards, like Fast Healthcare Interoperability Resources (FHIR) (Mahor, V.,2022, Rawat, R, 2022), are mostly implemented at the application level. Due to storage and query efficiency issues, EHR storage is typically different. However, attempts may be made as part of the trades to store EHRs in native FHIR format, which may lead to enhanced interoperability.

Structure of Trade: Despite the fact that corporate blockchains support unstructured data, the block and trade structure remain set. For example, a block header, transaction payload, and metadata are all included in a blockchain-based Hyperledger. Each component contains multiple parameters that reflect trade-specific information. Other protocols may not be compatible with this format (e.g., Ethereum). To ensure collaboration, all eHS must be able to adhere to the same trade format, which is challenging. A solution to this might be Type-Length-Value (TLV) fields, which represent each portion of a block with a TLV. Furthermore, if all participating systems reach an agreement on

If a block has just the minimum needed TLVs, the order or excess TLVs are irrelevant. TLV usage can also enable FHIR native format for EHR sharing across diverse parties, which could be an interesting study path.

Ledger Storage: After reaching a consensus, the ordering service distributes the block to all peers, who add it to their ledgers and update the global state. This is BBC's final commit procedure (Mahor. V, 2022).

Sharing a block with all peers is an intriguing topic, especially when using a multiple-blockchain approach. Assume that two eHS are participating in a deal that will be included in block B1. In a unified blockchain, B1 must be delivered to all peers following consensus formation. The amount of trades generated by each eHS can be quite vast, resulting in

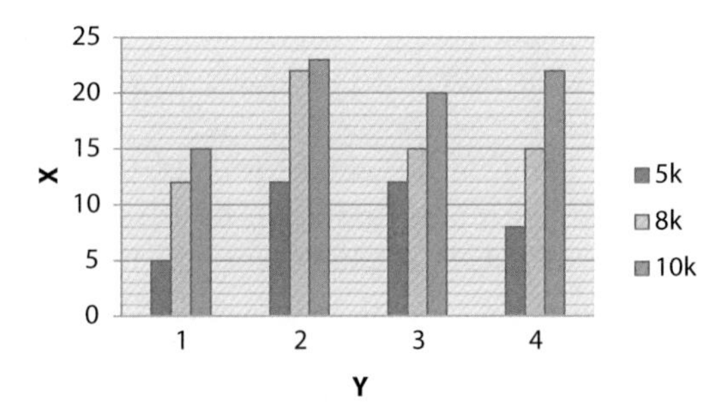

Figure 24.4 Increase in Ledger's RAM use.

astronomically high memory needs for the ledger. Efficient storage solutions have emerged as an intriguing research topic for this challenge. If, on the other hand, a multiple-blockchain solution is used, the participating eHS may choose to keep B1 just in their own peers. This reduces block duplication but may cause access problems if the same patient visits a third eHS that requires the information contained in B1. Figure 24.4 depicts the amount of memory used by Hyperledger Fabric at each peer. In addition to the rigorous access control system, this necessitates an efficient trade/block discovery method across several collaborating blockchains.

24.6 Formation of a Unanimity

The core issues in blockchain for non-cryptocurrency applications are connected to trade verification and consensus creation and this is maybe the most mistaken study topic. The exchange of digital assets is the transaction in a non-crypto system such as healthcare. As a result, any agreement must be reached in order for data pieces to be exchanged validly.

What Should be Verified?: Several IoT devices create data on patients in an e-healthcare system, which must be stored and accessible by various service providers (such as physicians and nurses) as well as third-party services (insurance agencies). A medical test result or a prescription written by a physician, for example, is also considered a digital asset. As soon as a digital asset is generated, its validity, authenticity, and access level must be verified. Here, trade structure becomes an open research challenge once more (Mahor, V, 2022).

Similarly, there may be many more occurrences that must be tracked, such as the administration of a medicine, which must be logged as a transaction.

A trackable event is requesting a patient's old EMR. Although this does not produce digital assets, it is reliant on adequate access control, therefore a query trade is required.

The difficulty is to identify the many forms of trades that might take place in a blockchain for current automated e-health systems. Furthermore, the trading structure must be adaptable enough to accommodate such dynamic information.

Any modification must provide Byzantine Fault Tolerance (BFT), which means that it must be agreed upon by 51% of peers.

Scalability & User Variety

E-healthcare blockchain is a particular situation with a wide range of user types. This type variance is related to their access privileges (Mahor, V., 2022). A patient has full access to all of its trades (for example, medical history), but a minor's guardian may have restricted access. This may vary over time, thus the system must adjust. Similarly, one physician may have total update access, while a consultant physician may just have read access, and the pharmacy may only be able to observe prescription swaps for a patient.

A blockchain solution does not manage user diversity by default, therefore it must be tightly integrated with the entire system's ACL. This tight connection is still a work in progress. Furthermore, in a unified blockchain ecosystem, this relationship should be very scalable.

The important aspects of this research challenge are simplicity and ease of creation.

The next phase is to reach an agreement, which leads to the establishment of a block. It should ideally have two sections; however, some systems simply do one to enhance the transaction rate (TPS). The first is to validate each transaction, which implies that the deal must fulfill the related SC, have valid signatures from all sides, and be supported by peers. The candidate block must be checked for authentic signatures and endorsed by peers in the second step. Because conducting both portions takes time, the problem is to develop consensus-forming techniques that are very efficient without compromising the two-part process (Rawat, R., Garg, B., Mahor, V, 2022).

It is also worth noting that Hyperledger Fabric merely does transaction verification and does not require a majority vote of 51 percent of peers. Peers who participate might be as few as two. This boosts TPS greatly but affects BFT. However, Hyperledger fabric also supports pluggable consensus mechanisms, making it a versatile platform. This opens the door to study towards replaceable or dynamically changeable algorithms in which peers can choose which algorithms are best suited for consensus for certain sorts of trades or blocks. As a result, in multiple-blockchain systems, different blockchains involved in a trade must either follow the same consensus building procedure (which is excessively restrictive) or dynamically pick one (which should be interoperable). To work on several chains, the consensus method itself should be very scalable. Similarly, smart contracts should be compatible with several blockchain systems, which is a difficult research task.

24.7 Access Control & Users

All users are made equal in a public blockchain, particularly for cryptocurrencies. They can originate transactions or serve as peers to build consensus. This is not the case with healthcare systems, which require stringent access control. It is critical to distinguish between access to the whole system and access to the blockchain. The former may only allow the user to be locally logged in by verifying a local certificate authority (CA) or Access Control List (ACL), whilst the latter allows the user to generate transactions or query the ledger. The specific issues in this area are discussed further below.

Channel Management and Access Control: The previously outlined connection is a complex approach since it needs every deal to be cross-checked by the ACL, which negates the point of utilizing blockchain. Smart contracts and channels are a better answer. The notion of channels was inspired by Hyperledger Fabric, which assigns each user/device a logical way for connecting to a peer. These pathways may then be used to construct the access control solution (channel). The channel should be associated with the patient and may have many versions. A new version is produced whenever the patient wishes to update the access privileges. Because the channel information is kept as part of the blockchain network, it is not subject to the ACL. Furthermore, the smart contract and channels have distinct roles and should be used wisely.

Although this is still a research topic, the answer must be controlled within the blockchain network.

Large Data: In a blockchain, information is stored or digital assets are exchanged through transactions, which include all pertinent data (pictures, etc.). The transactions (called blocks) are recorded in a file-based ledger, which cannot contain big pictures. The normal size of a single block in any blockchain system is a few gigabytes, as it has a direct influence on the system's speed.

As previously stated, e-healthcare systems rely largely on medical images (x-rays, CT scans, etc.). This factor alone may render blockchain deployment in e-healthcare systems impractical. Off-chain storage provides a solution, but it necessitates many changes in the way deals are executed and data is kept. For starters, off-chain storage should only enable access permitted by the peer or the ordering service. Off-chain storage must provide immutability, distributed nature, and access by verified users only, as the goal of a distributed ledger was to have duplicated immutable copies of data (Rawat, R., Mahor, V., & Telang, S. 2021.). As a result, protocol adjustments for query trades are required. Second, in order to add data to off-chain storage and link it to a specific transaction, the trade must include a pointer to the data's storage location. This pointer can lead to a hash value or a string. Another effective method for storage is third-party, off-chain storage, but security must be ensured. Because a CA is believed to be safe and trustworthy, practical storage assurances should be provided.

All of these criteria translate into design issues for blockchain implementations in e-healthcare (and large data) applications. Consider Blockchain as a shell that surrounds current and traditional database systems. This would essentially enable the usage of database storage and query efficiency while safeguarding them within the working concept of a blockchain system.

Finishing the Ecosystem: Finally, EHR management is not the primary process in a hospital. Accounting, human resources, pharmaceutical logistics, emergency services, and other departments are all connected into the ecosystem (https://www2.deloitte.com/global/en/insights.html). As previously said, switching to the blockchain will have a significant influence on inter-departmental communication. The majority of healthcare research in BC is focused on EHRs, but the elements and their interaction, as indicated in Figure 24.2, are vitally significant. A viable and deployable BC solution can only function if all entities in the ecosystem are on the same page. As a result, the research community must focus on BC and non-BC system interfaces while ensuring that none compromises the other.

24.8 Conclusion

The purpose of this paper was to help the reader understand the complexities of establishing a blockchain solution for e-healthcare systems and to seek potential solutions. Because healthcare is not an isolated network, blockchain solutions applied by separate health service providers must be interoperable, necessitating the development of new protocols for trade and consensus management. Off-chain data management and security must be included into the ecosystem. Finally, blockchain is an exciting and efficient solution for many security and accountability challenges; nevertheless, transfer of old systems is a lengthy process and the first step is to understand the application domain's demands. Many of the design issues discussed in this article may also apply to other domains.

References

Azaria, A. Ekblaw *et al.*, "Medrec: Using blockchain for medical data access and permission management," in International Conference on Open and Big Data, Aug 2016, pp. 25–30.

D. Dwivedi, G. Srivastava *et al.*, "A decentralized privacypreserving healthcare blockchain for iot," Sensors, vol. 19, no. 2, 2019.

D. C. Nguyen, P. N. Pathirana *et al.*, "Blockchain for secure ehrs sharing of mobile cloud-based e-health systems," IEEE Access, vol. 7, pp. 66 792–66 806, 2019.

Fast Healthcare Interoperability Resources (FHIR) R4. (Online). Available: https://www.hl7.org/fhir/

S. Biswas, K. Sharif *et al.*, "A Scalable Blockchain Framework for Secure Transactions in IoT," IEEE Internet of Things Journal, vol. 6, no. 3, pp. 4650–4659, June 2019.

General Data Protection Regulation. (Online). Available: https://gdpr-info.eu

H. Jin, Y. Luo *et al.*, "A review of secure and privacy-preserving medical data sharing," IEEE Access, vol. 7, pp. 61 656–61 669, 2019.

Health Insurance Portability and Accountability. (Online). Available: https://hhs.gov/hipaa/

Hyperledger: Hyperledger business blockchain technology. (Online). Available: https://www.hyperledger.org/projects

N. Kshetri, "Blockchain and electronic healthcare records (cybertrust)," Computer, vol. 51, no. 12, pp. 59–63, Dec 2018.

P. Pace, G. Aloi *et al.*, "An edge-based architecture to support efficient applications for healthcare industry 4.0," IEEE Transactions on Industrial Informatics, vol. 15, no. 1, pp. 481–489, Jan 2019.

P. Sundaravadivel, E. Kougianos *et al.*, "Everything you wanted to know about smart health care," IEEE Consumer Electronics Magazine, vol. 7, no. 1, pp. 18–28, Jan 2018.

P. Tasca and C. J. Tessone, "A taxonomy of blockchain technologies: Principles of identification and classification," Ledger, vol. 4, Feb 2019.

R. Jayaraman, K. Salah *et al.*, "Improving opportunities in healthcare supply chain processes via the internet of things and blockchain technology," International Journal of Healthcare Information Systems and Informatics, vol. 14, pp. 49–65, Feb 2019.

S. Biswas, K. Sharif *et al.*, "PoBT: A lightweight consensus algorithm for scalable IoT business blockchain," IEEE Internet of Things Journal, vol. 7, no. 3, pp. 2343–2355, Mar 2020.

T. McGhin, K.-K. R. Choo *et al.*, "Blockchain in healthcare applications: Research challenges and opportunities," Journal of Network and Computer Applications, vol. 135, pp. 62–75, 2019.

V. P. Yanambaka, S. P. Mohanty *et al.*, "PMsec: Physical unclonable function-based robust and lightweight authentication in the internet of medical things," IEEE Transactions on Consumer Electronics, vol. 65, no. 3, pp. 388–397, Aug 2019.

W. Mougayar, The Business Blockchain: Promise, Practice, and Application of the Next Internet Technology. Wiley, 2016.

W. Zhang, Y. Lin *et al.*, "Inference attack-resistant e-healthcare cloud system with fine-grained access control," IEEE Transactions on Services Computing, pp. 1–1, 2019.

Mahor, V., Badodia, S. K., Kumar, A., Bijrothiya, S., & Temurnikar, A. (2022). Cyber Security for Secured Smart Home Applications Using Internet of Things, Dark Web, and Blockchain Technology in the Future. In *Dark Web Pattern Recognition and Crime Analysis Using Machine Intelligence* (pp. 208-219). IGI

Rawat, R., Mahor, V., Chouhan, M., Pachlasiya, K., Telang, S., & Garg, B. (2022). Systematic literature Review (SLR) on social media and the Digital Transformation of Drug Trafficking on Darkweb. In *International Conference on Network Security and Blockchain Technology* (pp. 181-205). Springer,

Mahor, V., Bijrothiya, S., Bhujade, R. K., Mandloi, J., Mandloi, H., & Asthana, S. (2022). Enthusiastic Cyber Surveillance for Intimidation Comprehension on the Dark Web and the Deep Web. In *Using Computational Intelligence for the Dark Web and Illicit Behavior Detection* (pp. 257-271). IGI GlobalSingapore.

Rawat, R., Garg, B., Mahor, V., Chouhan, M., Pachlasiya, K., & Telang, S. (2021). Cyber Threat Exploitation and Growth during COVID-19 Times. In *Advanced Smart Computing Technologies in Cybersecurity and Forensics* (pp. 85-101). CRC Press.

Rawat, R., Mahor, V., & Telang, S. (2021). Study of Phylogenetic for Computational Analysis of Sleep Apnea Syndrome for Patient (Healthcare & Treatment) Using Machine Learning (Robot Vision) (No. 6202). EasyChair.Global.

S. Allen, "2020 global health care outlook," 2019. (Online). Available: https://www2.deloitte.com/global/en/insights.html

25

An Intelligent Machine Learning System Based on Blockchain for Smart Health Care

Vinod Mahor[1]*, Rahul Choudhary[2], Sadhna Bijrothiya[3], Jitendra Jatav[1] and Harsh Dubey[1]

[1]*Department of Computer Science and Engineering, Millenium Institute of Technology and Science, Bhopal, MP, India*
[2]*Department of Computer Science and Engineering, Shri Vaishnav vidhyapeeth vishwavidhyalaya, Indore, MP, India*
[3]*Department of Computer Science and Engineering, Maulana Azad National Institute of Technology, Bhopal, MP, India*

Abstract
Blockchain and Machine Learning work together to provide the best solutions for various jobs in the Smart Healthcare system. The emergence of these two new technologies has risen in the recent decade. We suggested secure, transparent, and intelligent approaches in this research. Machine learning models and blockchain technologies are being used in the internet of medical things market to increase security and train our models in order to better diagnose, prevent, and treat disease in the healthcare system, patients, patient rights, patient autonomy, and patient equality, which are all important.

***Keywords*:** Machine learning, blockchain technology, security, smart structure

25.1 Introduction

According to Frost & Sullivan, healthcare firms will have access to integrated health IT systems based on a growing collection of information management technologies such as blockchain [1] and the Internet of

**Corresponding author*: vinodengg.mt@gmail.com

Romil Rawat, Rajesh Kumar Chakrawarti, Sanjaya Kumar Sarangi, Rahul Choudhary, Anand Singh Gadwal and Vivek Bhardwaj (eds.) Robotic Process Automation, (371–382) © 2023 Scrivener Publishing LLC

Things (IoT) [2] over the next five to ten years. Privacy is a major issue when it comes to storing and transferring health data and with existing healthcare data storage systems lacking top-tier security, blockchain might provide a solution to weaknesses such as hacking and data theft. Interoperability is a characteristic of blockchain technology in healthcare that enables the secure exchange of medical data among the many systems and staff involved, resulting in a variety of advantages such as enhanced communication, time savings, and operational efficiency. According to the survey, the usage of blockchain technology for claims adjudication and billing management applications is predicted to grow by 66.5 percent by 2025 due to challenges such as mistakes, duplications, and improper billing. All of these concerns may be resolved with blockchain [3].

Machine learning is a computer-based approach for analyzing free-form text or speech input using a predefined set of ideas and technology, such as linguistic and statistical methods. In terms of patient care, knowledge and experience are two highly important aspects for physicians. Nevertheless, humans are restricted in terms of obtaining knowledge through data accumulation, whereas machine learning shines in this area. Machine learning is classified into two types: supervised learning and unsupervised learning [4].

25.2 Review of Literature

The World Health Organization (WHO) has proposed that governments prepare a "Pandemic Plan" in light of the increased risk of pandemic. A Pandemic Plan is frequently constructed in line with the World Health Organization's pandemic stages, with the purpose of attaining unambiguous outcomes in pandemic management from the outset [5].

In healthcare, many approaches for preparing for an emergency may be recognized. In fact, each crisis is separated into four phases: mitigation, preparation, reaction, and recovery [6]. The "tabletop exercise" is a useful strategy for modeling the emergence of a crisis situation; it provides a scenario that benefits from communication and collaboration across numerous sectors and areas, such as management, employees, logistics, communication, and finance. A suitable strategy can provide a wide framework as well as a mental picture of the ideal environment for future decision-making.

Each transaction in a network block is validated via a procedure based on consensus dispersed over all nodes (that is, the devices/users connected to the network). Blockchain technology is a subset of Distributed Ledger

Technology, whose operation is primarily based on a register structured in blocks linked in a network and each transaction performed in a network block is validated through a process based on consensus distributed across all nodes (that is, the devices/users). Transactions are the results of operations that occur among the network's subjects. The notion of blockchain stems from the fact that each block uses a cryptographic mechanism to preserve a reference to the preceding one. Blockchain, unlike traditional internet services, is kept on network devices (computers) called nodes, each of which has a copy of the complete blockchain [7].

Furthermore, two critical aspects of this form of technology should be underlined for our consideration: i) consensus decentralization and ii) ledger decentralization. Because of the decentralization of consensus, the presence of trustworthiness among persons participating in any type of transaction and the presence of a central authority may no longer be required. Similarly to the second point, replicating and storing different copies of different blockchains across network nodes ensures greater system security and equity among users who can access the same information at the same time, as well as the traceability and immutability of the validated transactions contained in the blocks. As a result, blockchain is a peer-to-peer network in which all network members may trust the system but not each other [8, 9].

25.3 Use of Blockchain in the Healthcare System

The blockchain may offer a single transaction layer where organizations may submit and exchange data through one safe channel by retaining a specific set of standardized data on the chain, along with private encrypted connections to separately stored information such as radiographic or other pictures. Smart contracts and uniform authorization standards can substantially assist in the provision of seamless connection [10]. There were roughly 176 million data breaches involving healthcare records between 2009 and 2017 [10, 11]. The secure qualities of blockchain can considerably improve the security of health data. Each individual has a public identifier or key, as well as a private key that can only be opened when and for how long is necessary. Furthermore, the obligation to personally target each user in order to gain sensitive information would limit hacking. As a result, blockchain can provide a medical data audit trail that is immutable [12]. Figure 25.1 depicts a schematic. We proposed a blockchain management solution for the healthcare business.

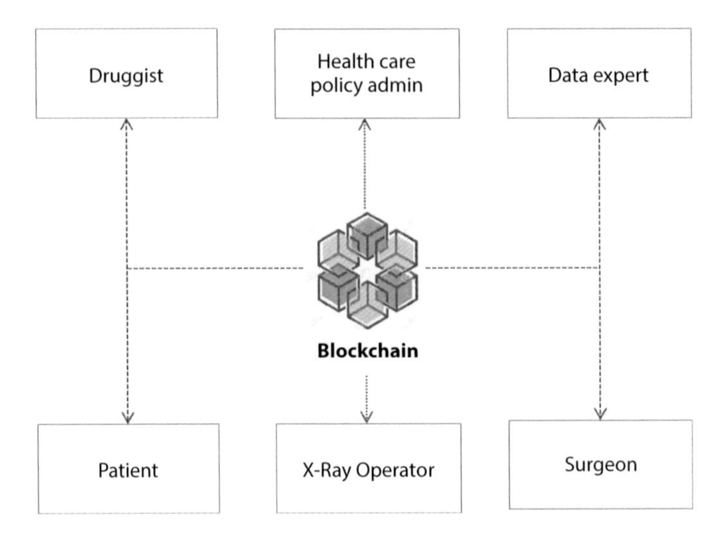

Figure 25.1 Blockchain healthcare management system overview.

25.3.1 Locating and Obtaining Medical Supplies

With total transparency, blockchain can help secure and identify the pharmaceutical supply chain. It can even keep track of the labor costs and carbon emissions related with the manufacture of these things.

25.3.2 Storing Medical Information

Despite significant improvements in the preservation of medical records, its transactions and security remain poor. Health data breaches are prevalent and with the increase of hackers all over the world, hacking of this data has surpassed earlier ways as the most common means of breaching. Meditab [11] bills is one of the major electronic medical records software developers for hospitals. Electronic faxes are part of the company and this method is still commonly used to share patient data with other data seekers. However, this kind of data exchange has shown to be hazardous and untrustworthy, making it an unwise decision. A fax server at Spider Silk [4], a cybersecurity firm located in Dubai, was hosting an Elastic search database containing at least 6 million healthcare records. The server lacked password security, allowing anybody to view the transmitted faxes in real time. The faxes sent contained personally identifiable information about the patients, such as their medical history, past treatment, Social Security numbers, and other details.

25.3.3 Disease and Outbreak Tracking

Blockchain's unique properties can help in real-time disease reporting and disease pattern study, which can aid in understanding the disease's genesis and transmission aspects.

25.3.4 Machine Learning and Artificial Intelligence in the Healthcare System

The growth of modern computing technology is intrinsically linked to Artificial Intelligence (AI). System learning has a long history, with Alan Turing's work cracking the German Enigma machine during WWII, laying the groundwork for most of current computer science. He is also the topic of the Turing Test, which attempts to evaluate if artificial intelligence has become indistinguishable from human intellect [5].

25.3.5 Machine Learning in the Healthcare System

Machine Learning, a subset of artificial intelligence, has seen widespread use in the medical field. Patient information, medical treatment records, and medication status have all been digitized, and a massive amount of data has been created in the area of medicine and healthcare as a result of the growth of ICT technology and the advent of the big data age [12]. Machine learning is utilized in the medical industry to interpret complex medical data, which has become the primary focus of machine learning research.

The paper investigates how the medical industry uses machine learning technologies such as deep learning, neural network learning, and feature fusion to perform data analysis and mining. The objective is to enhance the area of smart medicine by realizing human activity detection, health monitoring, illness prediction, and diagnosis. Gumaei *et al.* [13] recognized human behavior using machine learning. They proposed a multisensor hybrid deep learning model for recognizing human activities. This method can handle multisensor data more intelligently, allowing medical institutions to better care for the old and sick. Souri *et al.* [14] integrated machine learning to the student health monitoring system, gathered data via the Internet of Things, and assessed data using machine learning to continuously monitor the physical state of students. Ali *et al.* [15] created a disease prediction system based on deep learning feature fusion and the information gain approach to forecast the incidence of heart illness and to provide conditions for successful heart disease therapy in patients. Chui *et al.* [16] reviewed recent research in the field of intelligent medicine on sickness diagnosis and proposed novel machine learning techniques. Learning algorithms were explored, as well as the challenges of implementing illness detection in the future.

25.3.6 Oversee Learning

Training data is often used as labelled data in supervised learning. Training data consists of one or more inputs "labelled" output. Models employ these labelled discoveries to evaluate themselves during training, with the objective of improving future data prediction (i.e., a set of test data). Typically, supervised learning models focus on classification and regression approaches. Classification problems are quite prevalent in medicine. A doctor diagnoses a patient in most clinical settings by identifying the ailment based on a collection of symptoms. Regression tasks often focus on projecting numerical outcomes based on a collection of data, such as vital signs and expected duration of stay in the hospital.

Supervisory learning approaches include random forests (RF), decision trees (DT), Nave Bayes models, linear and logistic regression, and support vector machines (SVM), among others [17]. Random forests are a form of decision tree composed of a set of decision trees that have been separately trained. The projections of the trees are frequently averaged to give a better end result and forecast [18]. Each tree is built using a random sample of the data with replacement and a random subset of features is selected at each candidate split. This prevents each learner or tree from focusing too much on the purportedly predictive qualities of the training set, which may or may not be predictive on incoming data. In other words, it enhances the generalization of the model. With noisy data, random forests, which might include hundreds or even thousands of trees, perform well [19]. The model created by merging the results of numerous trees trained on data will yield a prediction that can be evaluated using test data [Figure 25.1]).

25.3.7 Unsupervised Learning

Unsupervised machine learning employs unlabeled data to discover patterns within the data [19]. These algorithms often excel in grouping data into relevant categories, allowing for the discovery of hidden traits that are not immediately apparent. They are, however, more computationally expensive

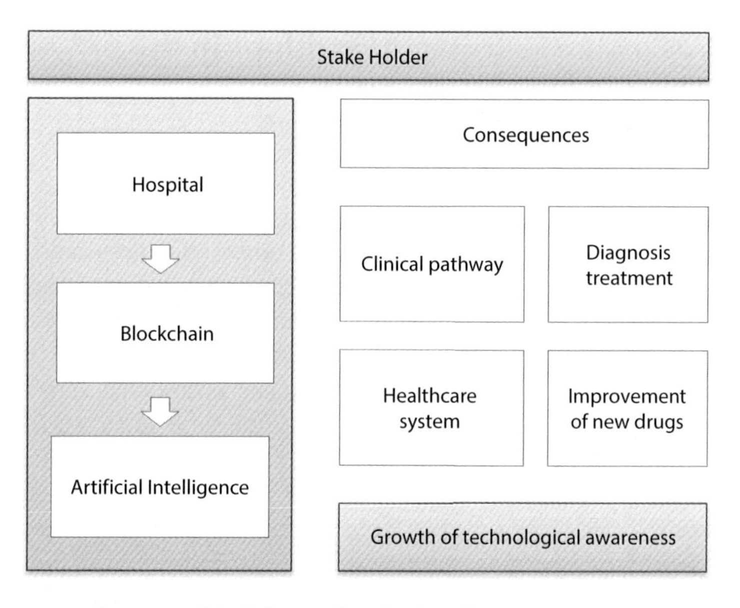

Figure 25.2 Architecture of blockchain and artificial intelligence in smart healthcare system.

and require a greater quantity of data to perform. The most well-known and extensively used approaches are K-means clustering and deep learning, however deep learning may also be used unsupervised [12, 13]. Such algorithms also do association tasks, which are similar to clustering. These algorithms are referred to as unsupervised since no human input is provided as to which set of attributes the clusters will be focused on. Although there are other variations on the traditional k-means method, such as k-medians and k-medoids, the basic concept stays the same. In the case of kclusters, Euclidean distance is used to find the "nearest" center or mean. The center of the current cluster is computed and updated for the following data point [14]. The most significant drawback of this strategy is that it needs the initialization of an expected number of "means" or "centers." Incorrect kvalue selection can lead to poor clustering [15]. The Figure 25.2 shows about the architecture of blockchain and artificial intelligence in smart healthcare system.

25.3.8 The Effort Movement

The proposed design is intended to provide for the safe management of activities in an intelligent technology environment. These tasks include system setup, data gathering from sensors and actuators, and dependable client control over the device to accomplish required operations. All of these techniques are detailed briefly in the sections that follow.

25.4 Machine Learning Algorithms in the Medical Industry

25.4.1 Support Vector Machine

The Support Vector Machine is the most often used machine learning approach in the healthcare business. It utilizes a supervised learning model for classification, regression, and detection outliners. Algorithms have been employed in recent years to estimate cardiac patient medication adherence, which has helped millions of patients escape catastrophic consequences such as hospital re-admission and even death. Applications include protein classification, image segregation, and text categorization [15, 16].

25.4.2 Logistic Regression

This machine learning approach predicts the current scenario of the category dependent variable using predictor variables. It is widely used to categorize and forecast the possibility of an occurrence, such as sickness risk management, which assists physicians in making vital decisions. It also aids medical institutions in identifying high-risk individuals and implementing behavioral health initiatives to help them improve their daily health behaviors [17]. It may be described mathematically, as a data point is then given to that cluster.

$$p = \frac{exp(a0 + a1x1 + a2x2 + a3x3 + \ldots + auxu)}{1 + exp(a0 + a1x1 + a2x2 + a3x3 + \ldots + auxu)}$$

The elements in the equation are identified as expected probability p, independent variables from, and regression coefficients from $a0 + a1$. It may also be compressed as:

$$In\left(\frac{p}{1-p}\right) = a0 + a1x1 + a2x2 + a3x3 + \ldots\ldots + auxu$$

The defined equation is referred to as a basic logistic regression model.

Random Forest: This method is used to generate numerous training trees for classification and regression during training, as well as to solve the problem of over-fitting decision trees. Based on a patient's medical history,

Random Forest is used to forecast sickness risk and analyze ECG and MRI data [17, 18].

Discriminant Analysis: Discriminant Analysis is a machine learning technique for assessing the accuracy of object categorization and categorizing an item into one or more categories. In the healthcare industry, discriminant analysis is utilized for anything from early identification of diabetic peripheral neuropathy to improving the diagnostic qualities of blood vessel imaging. It's also used to detect signs of mental illness and in electronic health record management systems.

25.5 Blockchain and Artificial Intelligence Solutions

In industrialized nations, healthcare accounts for a large amount of the gross domestic product (GDP). Hospital costs, on the other hand, continue to climb, as do unnecessary operations and health-data breaches [19–21]. Blockchain technology has the ability to improve things in this sector. It is capable of a wide range of functions, such as safe encryption of patient data and epidemic control. Estonia was a forerunner in this field, implementing blockchain technology in healthcare in 2012. Blockchain is being utilized to manage its whole healthcare billing system, 95% of health data, and 99% of prescription information [20–22]. Figure 25.1 depicts the Blockchain and Artificial Intelligence application paradigm in the healthcare system. Each user in the Blockchain network will be issued an authorised certificate by the Certificate Authority. It will provide an identity to everybody who transacts on the network. The identification will be provided through a digital certificate. Using the digital certificate, the user will sign the transaction and submit it to the blockchain [23, 24]. The following are the advantages of signing:

- Verify the identity of the individual doing or requesting a transaction on the blockchain to ensure that they are a real user
- Ensure that the user has authorization to view the ledger for the transaction in question

How can blockchain be used to address problems?

- Each authenticated user will have a copy of the shared ledger. This will solve the data collecting problem. The machine learning models may be given incredibly trustworthy data directly, and the results can be derived.

- The model may be trained using real-world data. As a consequence, model efficiency and accuracy will increase, cutting the central authority's additional costs.
- The patient can receive guidance on how to live a healthy lifestyle. The model may be taught by observing the counsel given to other patients experiencing similar issues or symptoms (by clinicians).
- When a patient asks a basic health query, a trained model with a test is uploaded to the blockchain network; the model can forecast epidemics and give medical suggestions. Natural Language Processing can be used to detect illnesses and provide treatments.
- Based on the patient's symptoms, the trained model can give clinical suggestions to professionals.
- The model has been trained to predict outbreaks. For instance, if a patient undergoes a test and the findings are uploaded to the blockchain network, the model can forecast an outbreak and give advice to the doctor.
- In the healthcare profession, we employ a range of devices and equipment to execute any medical test. Every machine or machine component has a set lifetime. It may also predict when the machine or a component of the machine will need to be replaced or removed.

25.6 Conclusions

Researchers and developers have been lured to recent advancements in the use of blockchain and artificial intelligence in smart healthcare systems. Internet of Medical Things researchers and developers are collaborating to blend various technologies on a wide scale in order to serve society to the greatest extent feasible. However, advancements are only attainable if we analyze the numerous challenges and weaknesses in today's clever technical techniques. In this research study, we create a system that uses two separate technologies to improve efficiency and results in the healthcare system. Our solution is capable of providing transparency, security, and intelligence in the Internet of Medical Things (IoMT).

References

1. Ahmad, M.A.; Teredesai, A.; Eckert, C. Interpretable Machine Learning in Healthcare. *2018 IEEE International Conference on Healthcare Informatics (ICHI)* 2018, pp. 447–447.
2. Ali, F.; El-Sappagh, S.; Islam, S.M.R.; Kwak, D.; Ali, A.; Imran, M.; Kwak, K. A Smart Healthcare Monitoring System for Heart Disease Prediction Based On Ensemble Deep Learning and Feature Fusion. *Information Fusion* 2020, *63*. doi:10.1016/j.inffus.2020.06.008.
3. Best EMR Software Company | Medical Software | EHR | IMS | Meditab. https://www.meditab.com/. (Accessed on 10/28/2021).
4. Blockchain Applications in Healthcare. https://www.news-medical.net/health/Blockchain-Applications-in-Healthcare.aspx. (Accessed on 10/28/2021).
5. Chui, K.; Alhalabi, W.; Pang, S.; Pablos, P.; Liu, W.; Zhao, M.; Moraunti, L. Disease Diagnosis in Smart Healthcare: Innovation, Technologies and Applications 2017.
6. Fekih, R.B.; Lahami, M. Application of Blockchain Technology in Healthcare: A Comprehensive Study. *The Impact of Digital Technologies on Public Health in Developed and Developing Countries* 2020, *12157*, 268 – 276.
7. Gagan Kumar, R.K. A survey on machine learning techniques in health care industry. *International Journal of Recent Research Aspects* 2016, *3*, 128–132.
8. Gumaei, A.; Hassan, M.; Alelaiwi, A. A Hybrid Deep Learning Model for Human Activity Recognition Using Multimodal Body Sensing Data. *IEEE Access* 2019, *7*, 99152–99160. doi:10.1109/ACCESS.2019.2927134.
9. Hassan, M.M.; Jincai, C.; Iftekhar, A.; Cui, X. Future of the Internet of Things Emerging with Blockchain and Smart Contracts. *International Journal of Advanced Computer Science and Applications* 2020, *11*.
10. He Warned of Coronavirus. Here's What He Told Us Before He Died. - The New York Times. https://www.nytimes.com/2020/0 2/07/world/asia/Li-Wenliang-china-coronavirus.html. (Accessed on 10/28/2021).
11. Jamison, D.T. Disease Control Priorities, 3rd edition: improving health and reducing poverty. *The Lancet* 2018, *391*, e11–e14.
12. Kumar, S.; Tiwari, P.; Zymbler, M.L. Internet of Things is a revolutionary approach for future technology enhancement: a review. *Journal of Big Data* 2019, *6*, 1–21.
13. Rawat, R., Mahor, V., Chirgaiya, S., & Garg, B. (2021). Artificial Cyber Espionage Based Protection of Technological Enabled Automated Cities Infrastructure by Dark Web Cyber Offender. In Intelligence of Things: AI-IoT Based Critical-Applications and Innovations (pp. 167-188). Springer, Cham.

14. Rawat, R., Mahor, V., Chirgaiya, S., & Rathore, A. S. (2021). Applications of Social Network Analysis to Managing the Investigation of Suspicious Activities in Social Media Platforms. In Advances in Cybersecurity Management (pp. 315-335). Springer, Cham.
15. Rawat, R., Mahor, V., Rawat, A., Garg, B., &Telang, S. (2021). Digital Transformation of Cyber Crime for Chip-Enabled Hacking. In Handbook of Research on Advancing Cybersecurity for Digital Transformation (pp. 227-243). IGI Global.
16. Rawat, R., Rajawat, A. S., Mahor, V., Shaw, R. N., & Ghosh, A. (2021). Surveillance Robot in Cyber Intelligence for Vulnerability Detection. In Machine Learning for Robotics Applications (pp. 107-123). Springer, Singapore.
17. Rawat, R., Rajawat, A. S., Mahor, V., Shaw, R. N., &Ghosh, A. (2021). Dark Web—Onion Hidden Service Discovery and Crawling for Profiling Morphing, Unstructured Crime and Vulnerabilities Prediction. In Innovations in Electrical and Electronic Engineering (pp. 717-734). Springer, Singapore.
18. scikit-learn: machine learning in Python — scikit-learn 1.0.1 documentation. https://scikit-learn.org/stable/. (Accessed on 10/28/2021).
19. sklearn. ensemble .Random Forest Classifier — scikit-learn 1.0.1 documentation.https://scikit-learn.org/stable/modules/ generated/sklearn.ensemble. RandomForestClassifier.html. (Accessed on 10/28/2021).
20. Souri, A.; Ghafour, M.; Ahmed, A.; Safara, F.; Yamini, A.; Hoseyninezhad, M. A new machine learning-based healthcare monitor- ing model for student's condition diagnosis in Internet of Things environment. *Soft Computing* **2020**, *24*. doi:10.1007/s00500-020- 05003-6.
21. spiderSilk - Home. https://spidersilk.com/. (Accessed on 10/28/2021).
22. Toh, C.; Brody, J.P. Applications of Machine Learning in Healthcare. 2021.
23. Wendelboe, A.M.; Miller, A.; Drevets, D.A.; Salinas, L.; Miller, E.J.; Jackson, D.; Chou, A.F.; Raines, J. Tabletop exercise to prepare institutions of higher education for an outbreak of COVID-19. *Journal of emergency management* **2020**, *18 2*, 183–184.
24. Zarzaur, B.; Stahl, C.C.; Greenberg, J.A.; Savage, S.A.; Minter, R.M. Blueprint for Restructuring a Department of Surgery in Concert With the Health Care System During a Pandemic: The University of Wisconsin Experience. *JAMA surgery* **2020**.

26

Industry 4.0 Uses Robotic Methodology in Mechanization Based on Artificial Intelligence

Vinod Mahor[1]* and Sadhna Bijrothiya[2]

[1]Department of Computer Science and Engineering, Milleniume Institute of Technology and Science, Bhopal, MP, India
[2]Department of Computer Science and Engineering, Maulana Azad National Institute of Technology, Bhopal, MP, India

Abstract

The early promise of machine intelligence's influence did not include the division of the fledgling area of Artificial Intelligence. The concept of embedded intelligence was merged between observation, cognition, and actuation, according to AI's inventors. Given the technological advancements of recent decades and the widespread use of information systems in society, the great majority of services supplied by businesses and organizations are now digital. Industry 4.0 is the fourth industrial revolution, and significant advancements in technology and mechanization characterize it. Robotic process mechanization (RPM) offers several benefits in terms of automating organizational and corporate procedures. In addition to these benefits, applying Artificial Intelligence (AI) algorithms and techniques in conjunction with RPM processes improves the accuracy and execution of RPM processes in data extraction, recognition, classification, forecasting, and process optimization. In this context, this research aims to study the RPM tools connected with AI that can help the development of Industry 4.0-related organizational processes. This chapter discusses Artificial Neural Network algorithms, Text Mining techniques, and Natural Language Processing techniques for the extraction of information and the subsequent process of optimization and forecasting scenarios in improving the operational and business strategies of organizations that appear to be extending the functionality of RPM tools.

**Corresponding author*: vinodengg.mt@gmail.com

Romil Rawat, Rajesh Kumar Chakrawarti, Sanjaya Kumar Sarangi, Rahul Choudhary, Anand Singh Gadwal and Vivek Bhardwaj (eds.) Robotic Process Automation, (383–394) © 2023 Scrivener Publishing LLC

Keywords: Industry 4.0, robotic methodology, artificial intelligence

26.1 Introduction

The primary contribution of this study is an overview of AI and RPM contributions to industry 4.0, as well as the functional examination and comparison of numerous proprietary and open-source tools. The structure of this chapter is as follows. The overall idea of Robotic Process Mechanization is described in Section 26.2 and the general concept of Artificial Intelligence and Industry 4.0 is offered in Section 26.3. In Section 26.4, numerous proprietary and open-source tools are introduced, along with their essential qualities so a discussion of the devices may occur in Section 26.5. The conclusions are offered in Section 26.6, followed by the sources that support this study.

26.2 Mechanization of Robotic Processes

Robotic Process Mechanization (RPM) is the mechanization of service processes that mimic human labor. The mechanization is carried out with the assistance of software robots or AI employees capable of doing repetitive jobs properly. The developer creates the task instructions by screen capturing and setting variables. These duties involve, among other things, entering into programs, copying and pasting data, opening emails, and filling out forms. According to Van der Aalst *et al.* [1], "RPM is an umbrella word encompassing tools that work on the user interfaces of other computer systems." Although conventional kinds of process mechanization (such as screen recording, scraping, and macros) rely on the computer's user interface, the primary purpose of the RPM is element recognition rather than screen coordinates or XPath choices. In most circumstances, this results in a more intelligent interaction with the user interface, commercial providers of RPM technologies reporting an increase in demand [2], and we have seen some studies where these tools are utilized to automate digital forensics, auditing and industrial. The fourth industrial revolution (Industry 4.0) opens the way for new approaches to automate boring rules-based business activities by leveraging RPM technologies and data from intelligent devices [3]. RPM is the extrapolation of a human worker's repetitive chores by a robot in corporate operations (where those tasks are done quickly and profitably). This intends to replace people with mechanization from the outside in. Unlike traditional approaches, RPM does not constitute part of the information infrastructure but instead sits on top of it, meaning a low degree of intrusiveness [4] and the potential for cost savings. According to specific estimates, the usage of RPM technology reduces operating expenses of transactional tasks inside shared services by 35% to 55%.

26.3 Industry 4.0 and Artificial Intelligence

AI was previously a concept that was separated into critical sectors of application. Natural language processing, automatic programming, robotics, computer vision, automatic theorem proving, intelligent data retrieval, and other domains are examples. These application fields are now so diverse that each may be regarded as a separate field in and of itself. AI is presently best defined as a collection of fundamental principles underpinning many of these applications. The key idea of intelligent factories and industry 4.0 is the application of AI by robots to fulfill complicated activities, cut costs, and enhance the quality of goods and services [5]. AI technologies are penetrating the manufacturing industry and, with the support of cyber-physical systems, fusing the physical and virtual worlds. The usage of AI in the industrial business makes it intelligent and capable of tackling current difficulties such as configurable needs, shorter time to market, and a rising number of sensors in equipment [6]. The usage of flexible robots in conjunction with AI facilitates the production of various items. AI approaches (such as data mining) can analyze massive amounts of real-time data collected from numerous sensors, as shown in Figure 26.1.

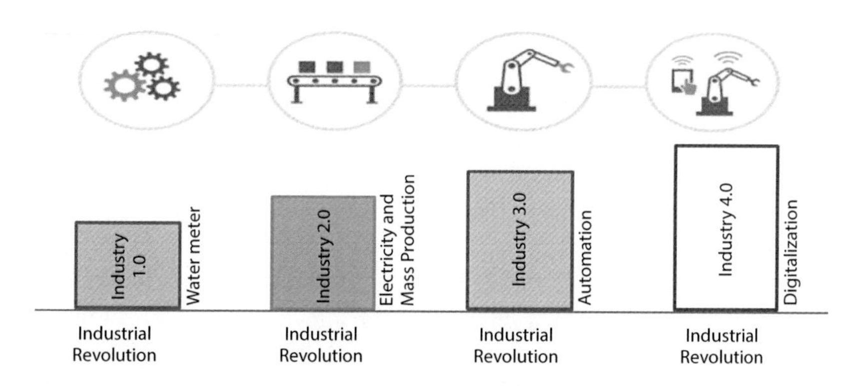

Figure 26.1 Structural design of industry 4.0.

26.4 RPM Outfits that Sustenance Artificial Intelligence

AI algorithms [7] and Machine Learning (ML) techniques have been effectively deployed in real-world settings such as commerce, industry, and digital services. ML [8] is used to "teach" machines how to deal with data more efficiently, simulating the learning concept of rational beings, and can be implemented with AI algorithms (or techniques), reflecting paradigms/approaches of good characteristics such as connectionist, genetics, statistics and probabilities, based on cases, and so on. It is possible to investigate and extract information using AI algorithms and the ML technique to classify, associate, optimize, group, forecast, find trends, and so on. Given the breadth of AI's application, RPM has steadily included implementations of algorithms, or AI approaches used in specific settings (e.g., Enterprise Resource Planning, Accounting, Human Resources) to classify, recognize, categorize, and so on for its mechanization capabilities. Some academic studies on challenges and potential, as well as case studies of the applicability of RPM and AI, have been published in recent years, as are cases of articles [8] in the field of automatic discovery and data transformation in the audit area, in the application of Business Process Management, and productivity optimization processes. Other studies on intelligent mechanization of processes using RPM have been published, for example the consultancy of Delloite [9], which presents the potentialities of the applicability of AI algorithms and techniques, but they should be applied in well-defined, stabilized mature processes, such as in strategic areas focused on customer tasks, increasing employee productivity (optimizing routine tasks), improving accuracy in categorizing and routing approaches, and improving overall process efficiency. In this context, and based on the above, if there are obstacles and potentialities to the RPM idea, they may be strengthened further by applying algorithms and AI approaches. The sections that follow present commercial and free source technologies that we believe are typical of RPM's recent applicability (ideally with the application of some AI techniques or algorithms) [10–12].

26.4.1 UiPath

UiPath [13–16] is a tool for developing RPM features in its framework to generate and run programming scripts, allowing it to be programmed using an interface of blocks and numerous plugins for business process changes. The RPM UiPath platform is now divided into three modules:

UiPath Studio, UiPath Robot, and UiPath Orchestrator, the latter of which allows for robot orchestration [13]. The UiPath Studio module is a tool that allows you to create, model, and execute workflows [13], as well as assist in the establishment and maintenance of robot connections, package transfer, and queue management. In turn, by storing log records and connecting them to Microsoft's Information Services Server and SQL Server, as well as Elasticsearch (an open-source search engine built on the Apache License), a Kibana data visualization plugin can improve the view of analytical data associated with RPM process execution. These characteristics are discussed in further depth in [12–14]. Some Artificial Intelligence algorithms or methodologies are now accessible through the UiPath tool via its UI Mechanization module [15] and revealed on its official page [16], the most notable of which are recognition, optimization, categorization, and information extraction. In terms of AI algorithms, picture and character recognition, optimization, and classification are used for the information examined.

26.4.2 Kofax

Kofax is a firm that creates software for process mechanization in businesses and organizations. The product includes modules for RPM, business process orchestration via software activity flows, document recognition (using Optical Character Recognition - OCR procedures), and sophisticated data analysis. Because this was a proprietary product and there was no way to access a test version for this study, numerous sources of information were examined in order to gather as much information on the tool as possible [17]. This tool, like RPM mechanization processes, allows for the extraction of data from documents and other sources (web, e-mail, local files) in various formats and designs, as well as the execution of procedural flows between computer applications to optimize tasks associated with Enterprise Resource Planning (ERP) information systems. As with other tools, it includes modules for the implementation of AI-related techniques or algorithms. The tool, which can be more or less profound in its application of these approaches, can recognize the content and context of a document [18] or classify and recognize information in emails, online portals, and chapters [17, 18]. Because a set of previous knowledge is necessary to classify and verify the contents, the employment of ML techniques combined with the recognition and classification of OCR documents and the analysis of the contents of e-mails or web pages can be considered types of supervised learning. Natural language processing, on the other hand, may be utilized in supervised learning for classification or unsupervised learning to evaluate material through information clustering ("clustering") or density extraction, depending on the approach or algorithms employed. In this regard, it appears that various AI approaches or algorithms are now available via the Kofax tool via the Intelligent Mechanization platform and its Cognitive Document Mechanization module [18].

26.4.3 Wherever Mechanization

Mechanization wherever is another RPM-focused solution that also provides information on the application of AI approaches and algorithms. As an RPM tool applied to ERP settings, it covers numerous areas of application such as human resources, Customer Relationship Management, and Supply Chain, and it is especially liable to be linked or interconnected with ERPs from SAP and Oracle, as well as other ERPs from other organizations. RPM is allied with the most automated or intelligent procedure known as "Digital Workers." The RPM solution includes a cognitive mechanization module as well as analytical data analysis tools for RPM procedures. As a multifunctional application, it provides a collection of information that enables the configuration, operation, and deployment of RPM processes [19, 20]. Through its Bot tool, the Mechanization Wherever tool internally provides the execution of some Artificial Intelligence techniques and algorithms such as fuzzy logic, Artificial Neural Networks, and natural language processing for the extraction of information from documents and, as a result, improves document validation efficiency. In this regard, it appears that various AI approaches or algorithms are presently available through the IQ Bot platform by the Mechanization Wherever intelligent word processing application [20].

26.4.4 Win Mechanization

The Win Mechanization program provides a collection of features connected with mechanization processes that are included in RPM processes, specifically, mechanization of emails, files in various formats (e.g., PDF and Excel), OCR, and other aspects linked with the work environment of post employees (desktop or web). Soft motive, on the other hand, is an RPM solutions provider that developed Win Mechanization. Win Mechanization is intended for desktop environments with built-in process design, desktop mechanization, web mechanization, macro recording, multitasking, automatic task execution, mouse and keyboard mechanization, a User

Interface designer, email mechanization, excel mechanization, file and folder mechanization, system monitoring and triggering, auto-login, security, File Transfer Protocol mechanization, exception handling, repository, and other features.

Control pictures, command line control, web data extraction, PDF mechanization, scripting, OCR capabilities, computer vision, sophisticated synchronization, auditing and logging, web recorder, inactive and non-interactive execution, database and SQL, and cognitive and terminal emulation [21]. In terms of RPM functionalities, the tool provides a set of modules through the "process robot" module, and a partnership with the company Capture Fast allows it to extend its RPM functionalities with information capture engines that use AI, data extraction in documents, and systems that automatically and hybrid classify documents. According to the reviewed literature], the module enables the integration of capabilities with the analytical information processing engines from Microsoft, IBM, and Google's Cognitive. However, it appears that at the level of availability of AI features, the instruments do not give proof.

26.4.5 Assist Edge

The Assist Edge tool, owned by Edge Verve Systems (an Infosys subsidiary), is a proprietary utility with an "open source" community version. According to institutional data, its functions include OCR reading for document processing based on the context associated with the kind of document. It employs AI algorithms (e.g., Artificial Neural Networks) [49] based on information from automated processes for automatic data capture, data analysis through the study of process changes based on individual process monitoring, and information categorization for recommendation processes [21].

26.4.6 Automatic

The Automatic Programmer is proprietary, although there is an open-source version (for non-commercial use) accessible on GitHub [22]. It was created primarily in the Python programming language, although it may be used by other community implementations (e.g. of AI techniques or algorithms). RPM's basic functions include reading OCR, extracting texts from PDF files, automating information in word files, excel files, information obtained via the web, and developing mechanization processes. It also supports picture and text recognition using Google TensorFlow.

26.5 Discussion

Based on an examination of the most typical RPM tools covered in this article, a table of technologies is shown below, indicating which Artificial Intelligence and AI algorithms they employ:

Analyzing material on the web and in digital libraries, a set of technologies that implement RPM mechanization processes was found. These tools perform "smart" processes connected with mechanization routines, although they are primarily associated with the application of AI approaches and algorithms. The proprietary tools look to have a larger collection of information and RPM features with AI. Open source tools rely on a developer community and are still in development or expanding in terms of functionality, such as the tools TagUI, TaskT, and Robot Framework Foundation [22].

UiPath is a program with a plethora of functionality and documentation. It contains a number of programmable plugins that allow it to be adapted to other systems such as PowerShell, SAP ERP, Oracle, and Microsoft Dynamics. The Kofax and Mechanization technologies execute numerous RPM procedures that are linked to ERPs, mostly SAP. The Assist Edge tool exhibited the capacity to interact with Microsoft (Azure Machine Learning) and Google (cognitive Services) cognitive systems, enhancing the usability of these two significant technical firms' implementations [23] and various goals, compared in Table 26.1.

Adoption of RPM processes with AI or connectivity with other ERPS will always be connected with a license fee in the case of private enterprises (where the greatest number of RPM features and information is more readily available) [24]. Initiatives and implementations for open-source RPM technologies are still expanding. However, various implementations (at the academic level) of algorithms have been utilized in free programming languages (such as R and Python) in recent years and certain RPM projects may benefit from implementations of Artificial Intelligence approaches and algorithms. On the other hand, as a result of Microsoft's research and availability of AI algorithms, the ability to use the .NET framework from Azure's Machine Learning platform provides for a direct route to investigate RPM implementations [25].

Although the adoption of Industry 4.0 [26] seems to address many of the industrial automation difficulties, new cybersecurity [27, 28] risks can be raised. Hackers [29], cyberterrorists [29], and business rivals may be able to access networks through the use of sensors and remote monitoring.

Table 26.1 Comparison of AI Goals.

ID	Tool	UiPath	Kofax	Win mechanization	Assist edge	Auto magica
T1	ANN	Y	Y	NA	Y	Y
T2	Taxonomy	Y	Y	NA	Y	Y
T3	Computer vision	Y		NA		
T4	Decision trees			NA		
T5	Fuzzy logic			NA		
T6	Fuzzy corresponding	Y		NA		
T7	Information extraction	Y	Y	NA	Y	Y
T8	NLP		Y	NA		Y
T9	Optimization	Y	Y	NA	Y	Y
T10	Recognition	Y	Y	NA	Y	Y
T11	Recommendation systems				Y	Y
T12	Statistic methods	Y		NA		
T13	Text mining					

26.6 Conclusions

This chapter covers research on RPM using AI for ERP-related processes. It was based on a study of information gathered from web-based digital libraries (business websites and tools, blogs, etc.) and scientific digital libraries. A collection of commercial products (UiPath, Kofax, Mechanization Wherever, and Win Mechanization) and open-source tools (Assist Edge and Automagica) were discovered and each was characterized in terms of RPM functionality, ERP integration, and ERP support. We find that the majority of proprietary products employ AI-related methods, such as recognition, optimization, classification, and knowledge extraction from RPM documents or processes. It also improves the optimization and exploration of information by these apps' users. These tools' AI techniques and algorithms focus on computer vision (for example, image recognition using Artificial Neural Networks), statistical methods, decision trees, neural networks for classification and prediction, fuzzy logic, and the implementation of techniques related to text mining, natural language processing, and recommendation systems.

Industry 4.0, on the other hand, is based on the convergence of the Internet of Things, intelligent mechanization, intelligent devices and processes, and cyber-physical systems. The confluence of all these concepts and technology causes a dramatic shift in industrial processes, altering the workflow of digital operations throughout the organization. Robots are being used to automate various phases in these processes to enhance those (RPM). Furthermore, as demonstrated in this study, RPM now combines intelligent approaches and algorithms (AI) in many systems, allowing for higher degrees of intelligence in the mechanization of activities inside a corporation.

References

1. Asquith, A., & Horsman, G. (2019). Let the robots do it!–Taking a look at Robotic Process Automation and its potential application in digital forensics. Forensic Science International: Reports, 1, 100007.
2. Didi Faouzi, Nacereddine Bibi Triki and Ali Chermitti, (2016) Optimizing The Greenhouse Micro– Climate Management by The Introduction of Artificial Intelligence Using Fuzzy Logic, International Journal of Computer Engineering and Technology, 7(3), 2016, pp. 78–92.
3. Enríquez, J. G., Jiménez-Ramírez, A., Domínguez-Mayo, F. J., & García-García, J. A. (2020). Robotic Process Automation: A Scientific and Industrial

Systematic Mapping Study. IEEE Access, 8, 39113-39129. Williams, D., & Allen, I. (2017). Using artificial intelligence to optimize the value of robotic process automation. Available from: https://www.ibm.com/downloads/cas/KDKAAK29

4. FLUSS, D. (2018). Smarter Bots Mean Greater Innovation, Productivity, and Value: Robotic process automation is allowing companies to re- imagine and re-invest in all aspects of their businesses. *CRM Magazine*, *22*(10), 38–39.
5. GitHub (2020a). Open Source, Distributed, RESTful Search Engine. [Online]. Available from: https://github.com/elastic/elasticsearch.
6. GitHub (2020b). Your window into the Elastic Stack. [Online]. Available from: https://github.com/elastic/kibana
7. Huang, F., & Vasarhelyi, M. A. (2019). Applying robotic process automation (RPA) in auditing: A framework. INTERNATIONAL JOURNAL OF ACCOUNTING INFORMATION SYSTEMS, 35. https://doi.org/10.1016/j.accinf.2019.100433
8. Madakam, S., Holmukhe, R. M., & Jaiswal, D. K. (2019). The future digital work force: robotic process automation (RPA). JISTEM-Journal of Information Systems and Technology Management, 16.
9. Moffitt, K. C., Rozario, A. M., & Vasarhelyi, M. A. (2018). Robotic process automation for auditing. Journal of Emerging Technologies in Accounting, 15(1), 1-10.
10. Mukul Anand Pathak, Kshitij Kamlakar, Shwetant Mohapatra, Prof. Uma Nagaraj, (2016) Development of Control Software For Stair Detection In A Mobile Robot Using Artificial Intelligence and Image Processing, International Journal of Computer Engineering and Technology, 7(3), pp. 93–98.
11. Rawat, R., Mahor, V., Chirgaiya, S., Garg, B.: Artificial cyber espionage based protection of technological enabled automated cities infrastructure by dark web cyber offender. In: Intelligence of Things: AI-IoT Based Critical-Applications and Innovations, pp. 167–188. Springer, Cham (2021)
12. Rawat, R., Rajawat, A.S., Mahor, V., Shaw, R.N., Ghosh, A. (2021). Surveillance Robot in Cyber Intelligence for Vulnerability Detection. In: Bianchini, M., Simic, M., Ghosh, A., Shaw, R.N. (eds) Machine Learning for Robotics Applications. Studies in Computational Intelligence, vol 960. Springer, Singapore. https://doi.org/10.1007/978-981-16-0598-7_9.
13. UiPath (2020a). UiPath Studio: introduction. [Online]. Available from: https://docs.uipath.com/studio/docs/introduction.
14. UiPath (2020b). Prerequisites for Installation. [Online]. Available from: https://docs.uipath.com/orchestrator/docs/prerequisites-for-
15. UiPath (2020c). About the UI automation activities pack. [Online]. Available from: https://docs.uipath.com/activities/docs/about-the-ui-automation-activities-pack
16. UiPath (2020d). Artificial Intelligence RPA Capabilities. [Online]. Available from: https://www.uipath.com/product/ai-rpa-capabilities.

17. Kofax (2020a). Developer's Guide Version: 11.0.0 [Online]. Available from: https://docshield.kofax.com/RPA/en_US/11.0.0_qrvv5i5e1a/print/KofaxRPADevelopersGuide_EN.pdf
18. Kofax (2019). Product summary Kofax RPA. [Online]. Available from: https://www.kofax.com/-/media/Files/Datasheets/EN/ps_kofaxrpa_en.pdf
19. Umachandrani. (2017). How tech is making life easier for differently-abled. The Times of India. Retrieved from https://timesofndia.indiatimes.com/india/how-tech-is-making-lifeeasier-for-differently-abled/articleshow/61538902.cms
20. Vempati, Shashi Shekhar. (2016). India and the Artificial Intelligence Revolution. Carnegie India.
21. WinAutomation (2020a) Desktop automation https://www.winautomation.com/product/all-features/desktop-automation
22. WinAutomation (2020a) Desktop automation https://www.winautomation.com/product/all-features/desktop-automation
23. WinAutomation (2020b) About Softomotive. Available on: https://www.winautomation.com/about-softomotive/
24. WinAutomation (2020b) About Softomotive. Available on: https://www.winautomation.com/about-softomotive/
25. Zheng, P., Sang, Z., Zhong, R. Y., Liu, Y., Liu, C., Mubarok, K., ... & Xu, X. (2018). Smart manufacturing systems for Industry 4.0: Conceptual framework, scenarios, and future perspectives. Frontiers of Mechanical Engineering, 13(2), pp:137-150.
26. Rawat, R., Chakrawarti, R. K., Vyas, P., Gonzáles, J. L. A., Sikarwar, R., & Bhardwaj, R. (2023). Intelligent Fog Computing Surveillance System for Crime and Vulnerability Identification and Tracing. *International Journal of Information Security and Privacy (IJISP)*, 17(1), 1-25.
27. Rawat, R., Sowjanya, A. M., Patel, S. I., Jaiswal, V., Khan, I., & Balaram, A. (Eds.). (2022). *Using Machine Intelligence: Autonomous Vehicles Volume 1.* John Wiley & Sons.
28. Rawat, R., Bhardwaj, P., Kaur, U., Telang, S., Chouhan, M., & Sankaran, K. S. (2023). *Smart Vehicles for Communication, Volume 2.* John Wiley & Sons.
29. Mahor, V., Bijrothiya, S., Rawat, R., Kumar, A., Garg, B., & Pachlasiya, K. (2023). IoT and Artificial Intelligence Techniques for Public Safety and Security. *Smart Urban Computing Applications*, 111.

27

RPA Using UiPATH in the Context of Next Generation Automation

Shadab Pasha Khan[1*] **and Rehan Khan**[2†]

[1]*Department of Information Technology, Oriental Institute of Science & Technology, Bhopal, India*
[2]*Department of Computer Science Engineering - Data Science, Oriental Institute of Science & Technology, Bhopal, India*

Abstract
In today's world of technology digitization, automation is the need of the hour. Next Generation applications which do not depend upon human intervention are all about automation. Automation plays a crucial role in all areas where clearly defined or well-defined, mundane tasks are executed to achieve any meaningful objective. RPA establishes a set of processes for all mundane tasks depending upon the domain. It enables the user to configure or customize processes according to need. Automation will become the world's most essential technology for all industries where the execution of repetitive tasks is needed. The revolution has already begun. RPA reduces the time, effort, and cost of doing any well-defined job. RPA offers multiple benefits over traditional methods used in businesses. In this paper, we explore various applications of RPA along with the issues and challenges involved. Email automation, web scraping, and weather forecasting are illustrated using UiPATH. IPA, RPA, RPA Security, and how RPA works in Blockchain are some upcoming trends.

***Keywords*:** IPA, RPA, Automation, Blockchain, UiPATH

27.1 Introduction

The IEEE (Institute of Electrical and Electronics Engineers) Standards Association defines Robotic Process Automation (RPA) as "A preconfigured

**Corresponding author*: shadabpasha@gmail.com
†*Corresponding author*: dayel.rehan@gmail.com

Romil Rawat, Rajesh Kumar Chakrawarti, Sanjaya Kumar Sarangi, Rahul Choudhary, Anand Singh Gadwal and Vivek Bhardwaj (eds.) Robotic Process Automation, (395–422) © 2023 Scrivener Publishing LLC

software instance that uses business rules and predefined activity choreography to complete the autonomous execution of a combination of processes, activities, transactions, and tasks in one or more unrelated software systems to deliver a result or service with human exception management" (IEEE Corporate Advisory Group 2017) [1].

Robotic Process Automation, abbreviated as RPA, is used to increase speed and efficiency. It is achieved by automating basic tasks through hardware or software systems in a way similar to what humans do [2].

The software or robot can be shown a work process with different advances and applications, for example, taking received forms, sending a receipt message, checking the form for completeness, filling the form in a file, and refreshing a calculation sheet with the name of the structure, the date documented, etc. RPA programming is intended to lessen the weight for workers by finishing redundant, essential undertakings [3]. Mechanical cycle mechanization devices are not substitutes for hidden business applications, instead they robotize the all-around manual assignments of human laborers [4]. They take a gander at the screens that workers today check out and fill in and update similar boxes and fields inside the UI by pulling the vital information from the significant area. It joins APIs and UI (UI) cooperation to incorporate and perform redundant undertakings among big business and efficiency applications. By sending scripts that imitate human cycles, RPA apparatuses complete independent execution of different exercises and exchanges across irrelevant programming frameworks. This can extraordinarily diminish work expenses and increment proficiency by speeding things up and limiting human mistakes. The Table 27.1 shows about the comparison of traditional approach vs RPA approach.

27.2 Traditional Approach vs RPA Approach

Table 27.1 Comparison of traditional approach vs RPA approach.

S. no.	Parameters	Traditional methods	RPA methods
1	Processing Speed	Low	Fast
2	Time Taken	More	Less
3	Efficiency	Low	Improved
4	Error Probability	More	Less
5	Accuracy	Less	More
6	Cost Reduction	Less	More

(Continued)

Table 27.1 Comparison of traditional approach vs RPA approach. (*Continued*)

S. no.	Parameters	Traditional methods	RPA methods
7	Error Reduction	Less	More
8	Productivity	Less	Improved
9	Tasks	Better for constantly changing, non-standardized tasks	Better of well-defined rule-based tasks
10	Tools Available	No tools	Many tools such as UiPATH, Blueprism, etc.

27.3 Related Work

RPA technology enables all future generation users of the network to automate the repetitive tasks as per the requirement of the industry. Usage of RPA is almost in all domains, including production, IT, Healthcare, Banking Sector, Education, and many more. I have reviewed many papers for this paper that led in multiple directions. Some of them are listed below.

Authors [1] concluded that RPA is much more used in the corporate world than in academics. They also focused on the tool analysis of RPA. They analyzed the impact performed by RPA in various domains. Authors [2] emphasized RPA and its usage in multiple fields. Authors [3] analyzed the working principle RPA and explored the architecture of RPA and [4] also studied the widely used RPA technology in the context of Next Generation Applications and Authors [5] provided "an overview of various RPA challenges and issues of privacy in RPA. They also explored different threats and technologies to secure it. Authors [6] emphasized software robots along with their usage. Authors [7] explained the use of RPA technology to bring changes to the Education System and make it a Smart Education System. It also explored how the RPA model helps teachers, educators, and parents in mundane tasks. Authors [8] examined the RPA in the education sector to do regular repetitive tasks more efficiently. Authors [9] studied the implementation of RPA in the banking sector and how it helps reduce operational and labor costs and improves efficiency. They also discussed the application market of RPA in the banking sector. Authors [11] explored the applications of RPA in the Banking and Finance Sector, especially in the case of loan processing and fraud detection. Also,

RPA can be used to mitigate fraud risks through various methods and eliminate human errors. Authors [12] mentioned the capabilities of RPA to reduce the burden of Business Process Management (BPM) in the context of the banking and finance sector. With the help of AI and machine learning, how do we reduce the burden of business processes and improve efficiency? Authors [13] emphasized the application of RPA in various domains of the Healthcare Industry. Automating workflow processes in the healthcare industry is necessary to improve efficiency and reduce the work pressure on the healthcare workforce [14].

Authors [16, 17] emphasized RPA usage in Logistics and its various subdomains. Authors [18] studied RPA and its use cases in the Business Sector by BPM. Authors [20] mentioned the new dimension of Intelligent Process Automation, AI, and ML in RPA to improve business process outcomes. Authors [21] explained the combination of AI and RPA in Industry 4.0. There are various challenges and issues when anyone deploys RPA in an industry and the authors [23] mentioned all of them. Blockchain is the upcoming technology that would play a key role in introducing transparency in every contract. The concept of distributed ledger is introduced and using RPA, mundane tasks should be replaced. Blockchain challenges and various opportunities in this sector are also explored by authors [24, 25]. Authors [29, 30] explained web scrapping as an implementation of RPA.

27.4 Applications of RPA

27.4.1 Role of RPA in Education

In this day and age, for the cutting-edge school system innovation is significant in robotizing repetitive, rules-based, and irritating assignments. Robotized schooling systems dispose of desk work and manual cycles with programming for affirmations strategies, planning gatherings, checking participation, surveying tasks and grades, and overseeing funds. Achieving these undertakings while giving students the proper teaching can be difficult for educators and heads. Furthermore, administrators have numerous obligations including the smooth and practical activity of instructive exercises and offices, which are monotonous [7].

RPA or Robotic Process Automation is a product of advanced mechanics innovation modified to assume control over a portion of our everyday undertakings. They ordinarily perform tedious and trifling cycles. This lessens representative responsibility and amplifies task precision.

In education, where the educator understudy proportion is frequently lopsided, RPA can decrease the responsibility of school staff, permitting them to give reasonable and widespread admittance to understudy guiding, mentoring, and course enlistment, and that is only the tip of the iceberg [5].

- Enrollment Process and Course Registration
  Perhaps the most tedious assignment that instructive organizations need to manage is the enlistment of students. Enlistment includes a ton of desk work, which is troublesome and blunder-inclined. It may be tiring to survey, concentrate on reports, and straightforwardly confirm the appropriateness of the course applied for in the wake of auditing the application structure.

  Admission to a school or college includes a ton of desk work and agendas, for example, sending letters of suggestion, old records, financial reports, letters of the plan, migration records (for worldwide understudies), and state-sanctioned test scores. From there, the sky is the limit.

  To have the option to achieve this multitude of undertakings productively and carefully with a brought together manual methodology, instructive foundations need to present RPA. RPA bots are prepared to decipher circumstances and settle on choices autonomously, diminishing disturbance to manual cycles. The instrument can examine structures, break down structures, audit reports, evaluate qualification rules, and eventually choose whether to concede understudies.

  RPA bots can be booked to watch out for every understudy's record and send them periodic updates when the containers on the agenda are not ticked. Alternatively,, they can be sorted out to give students cutoff times and plans for the confirmation cycle. This limits the risk of students passing up enlistment by failing to remember straightforward arrangements or cutoff times.
- Attendance Management
  It is critical to keep exact time audits of the workforce and school organization to give sensible pay. Monitoring the time spent gainfully by instructive accomplices is essential too, which is preposterous through manual time enlistment. Setting up a legitimate participation global positioning

framework assists associations with social occasion data about when they are signed into work.

A few teachers dole out a part of a student's grade because they participated in the course. RPA bots can be modified for online courses to pick the student's name from the participation list at every meeting. They could then put a mark close to their name for class participation that day.

Attendance management is a principal and tedious component of study classroom management. Automation is a straightforward yet precise method for smoothing out processes. Since the information is computerized, instructors can immediately move attendance information into accounting sheets to acquire remarkable experiences, for example, how students are doing in class contrasted with the earlier year. This data can help that person make shifts in the direction to make it seriously captivating. Alternatively, it tends to be utilized for execution assessment by college organizations [6].

- Fee Management
  Managing the charges of countless students utilizing manual power is drawn-out, complex, and challenging. With RPA [6] being carried out, an organization can uninhibitedly make different expense bunches with no rush, make grant applications, and decide any discretionary charges to be gathered (educational expenses, college expenses, and so forth.)
- Grading and Monitoring Students' Performance
  Particularly for huge electives with an enormous number of students, evaluation of tests can take a great deal of significant investment. RPA bots can be customized in a standard-based system for evaluating on the web evaluations.

  The bot can be customized to utilize NLP to scan the response for hints to the response key in the case of open-ended questions. For instance, it tends to be told to stamp a response as "right" on the off chance that it contains a significant catchphrase from the response key or something almost identical. RPA bots can plan student grades and progress, giving an account of the premise of their imprints in assessments and different exercises. This way, it screens and assesses the students' exhibition in scholastics and co-curricular exercises. This empowers students and their folks to perceive weak gatherings and perform better. For the other end of the mentor's responsibility, RPA devices can separate scores and different subtleties saved in the data set, robotizing the whole cycle [8].

27.4.2 RPA in Banking

RPA in the financial business fills in as a valuable device to address the requests of the financial area and assist them with expanding their effectiveness by diminishing administrative expenses through the programming/software model. In an undeniably immersed banking and monetary area, it has become fundamental for banks and other monetary establishments to persistently develop, stay competitive, and give extraordinary client experience.

According to a report by Forrester, it is expected that the industry will be worth $22 billion by 2025 [10]. That is considerable growth and a sharp rise compared to $250 million in 2016. This shows the exponential growth of this industry.

Rising working costs, intensified by administrative fines alongside furious administrative prerequisites, slow cycles down and result in an unfortunate client experience. Tossing more individuals at the issue of tracking down new and better ways of overseeing consistency while chopping down functional costs is certainly not the response. This calls for a better solution so that these problems can be handled. Robotic Process Automation (RPA) can empower banks and money organizations to decrease manual endeavors, offer better consistency, alleviate dangers, and improve the general purchaser experience. Besides, automation is generally appropriate for banks and monetary establishments because there are no extra foundation prerequisites.

There are several benefits of RPA in the banking and financial sector. Robots are exceptionally versatile and allow you to oversee high volumes during top business hours by adding more robots and answering any circumstance in record time. Once accurately set up, banks and monetary establishments can make their cycles much quicker, more valuable, and more proficient. Furthermore, RPA execution permits banks to put more spotlight on innovative systems to develop their business by liberating workers from doing ordinary assignments. Like some other industries, cost-saving is basic to the financial business also. Banks and monetary organizations can take a gander at saving handling time and cost.

Whether you are hoping to diminish manual mistakes or accomplish high exactness for a minimal price, robots work 24/7 to follow through with the jobs allotted to them. Accordingly, repeating consistently present accessibility. With RPA devices giving an intuitive innovation to robotize banking processes, it is extremely simple to carry out and keep up with computerization work processes with next to no (or negligible) coding prerequisites [11, 12].

- Automated Reports Generation
  Undoubtedly, creating consistent reports for fake exchanges as dubious movement reports is a standard prerequisite at banks and monetary establishments. Expectedly, consistence officials should peruse every one of the reports and fill in the fundamental subtleties in the structure of suspicious activity reports. This makes it an incredibly redundant undertaking that requires some investment and exertion.

  RPA innovation, with NLP abilities, can peruse these extensive consistency reports before separating the expected data and documenting the suspicious activity report. For ideal outcomes, the RPA programming can be prepared with inputs from the consistency officials on the pieces of each record which best fit each segment of the report.

  In addition to the fact that this assistance decreases the functional expenses, it saves the time taken to play out the undertaking.
- Incorporating New Customers
  Client onboarding in banks is a tedious, long cycle because a few reports require manual confirmation. RPA can make the interaction much more straightforward by catching the information from the KYC reports utilizing the optical character recognition strategy (OCR). This information can then be matched against the data given by the client in the structure.

  Assuming no inconsistencies after the robotized coordinating, the information is naturally placed into the client at the board entrance. RPA robotization in client onboarding assists in staying away from manual mistakes and recovering a ton of time and exertion by the representatives.
- Know Your Customer (KYC)
  KYC can be easily automated with the help of Robotic Process Automation, which will save financial organizations a lot of time and manual effort. Whether automating the manual cycles or getting dubious financial exchanges, RPA execution is demonstrated to be instrumental in saving

both time and cost when contrasted with typical financial arrangements [9].

- Account Opening
  Robotic Process Automation (RPA) makes the account opening process much easier and quicker when compared to the traditional approach. Automation deliberately disposes of the information record mistakes between the center financial framework and the new record opening solicitations, in this way upgrading the information nature of the general framework.
  An excellent illustration of this is worldwide banks involving robots in their record opening cycle to separate data from input structures, taking care of it in various host applications. The outcome was the disposal of a mistake-inclined, tedious, manual information passage process and a sharp decrease in TAT while simultaneously keeping up with complete functional precision and relieved costs.
- Mortgage Lending and Loan Processing
  Loans are one of the basic assistance regions for any monetary foundation. The course of home-loan loaning is incredibly process-driven and tedious, which makes it very reasonable for RPA automation. RPA innovation can easily take care of the cycle with plainly characterized rules.
  RPA takes into consideration the simple automation of different errands essential to the home loan loaning process, including advance commencement, record handling, monetary examinations, and quality control. Subsequently, the advances can be supported much quicker, prompting upgraded consumer loyalty [9, 11].
- Customer Services
  The volume of ordinary customer questions in banks is tremendous, making it hard for the staff to answer them with a low completion time. RPA devices can permit banks to computerize such every day, rule-based cycles to successfully answer questions continuously, in this manner diminishing the completion time significantly.

27.4.3 Role of RPA in Healthcare

Medical service frameworks contain different difficult undertakings and severe guidelines that require a significant measure of assets like caseloads and patient bookings. This prompts failures, significant expenses of tasks, and slow cycles.

By utilizing the force of robotization and RPA, medical service suppliers can resolve these issues and make medical service frameworks more productive, medical service processes quicker, and work on the general degrees of patient fulfillment easier.

With the contribution of RPA innovation, patients can plan their arrangements without mediation from emergency clinic representatives. Alongside dispensing with the requirement for asset designation for planning, this application can likewise further develop client relations since patients can organize an arrangement quicker. After a medical care administration is given, charging could find an opportunity to be handled due to the manual and dreary undertakings in the case load cycle, for example, contributing, handling, and assessing records and information [13]. Alongside robotizing time-serious errands, an RPA-drive claims the executives' framework can likewise wipe out human mistakes during claims handling.

RPA empowers medical care suppliers to track and report each cycle step in organized logs documents so the organization can consent to outer reviews. Since these cycles will be taken care of by bots, RPA could further develop information privacy too. An effective RPA execution in medical services requires all-encompassing premonition to determine where mechanization can benefit, what assets are accessible to help with the cycle, and whether an organization can help facilitate with ranges of abilities or assets [14].

27.4.4 Role of RPA in Logistics and Transportation

According to the report [15] by the World Economic Forum and Accenture, the computerized change in the field of strategies is supposed to result in $1.5 trillion worth by 2025. Furthermore, this is not easy at all, on the off chance that we consider that the transportation and coordinated operations industry is at present defaced overwhelmingly of ineffectual ways of behaving, for example, a portion of the trucks voyaging void returning in the wake of making a conveyance, which adversely affects modern efficiency. Automatically, the following merchandise is only one of the advantages for organizations involving RPA in transportation.

Clients can access their records to determine where their products are found. They can likewise see when a shipment is supposed to show up and if there are any delivery delays. This similarly applies backward.

This implies organizations could utilize RPA to oversee inbound and take a look at calls. This recovers endless assets and supports proactive preparation. For instance, the client can try to be accessible to get any conveyance requiring a mark or bundles that ought not to be left unattended.

With the utilization of mechanical cycle computerization in transportation and strategies, by putting information at the core of coordinated factors, organizations can smooth out the work process and hence diminish, generally speaking, working expenses and significantly work on functional efficiency. But there is something other than monetary benefit for individual organizations. The above-referred report conceives that the positive repercussions of transportation and coordinated operations going advanced at the society level will reach $2.4 trillion. So, the capability of digitization is the figure to cross the business borders and positively affect society and the environment. RPA in transportation keeps on changing the game continuously. This innovation permits an organization to mechanize the capacity to peruse and answer clients and merchants. Organizations can then refresh framework data in light of IVR or email communications to work with the effective and exact delivery of items [16]. Programming robots are the ideal decision to do this interaction because of their integrative limit, which can smooth the work process. Start-to-finish computerization of request-to-cash processes for some enormous outsider planned operation transporters is attainable since bots can coordinate with the business transportation of merchandise [17, 18].

27.5 Intelligent Process Automation

IPA is an arising set of innovations at its center that consolidates critical process updates with robotic process automation and AI. It is a set-up of business-process enhancements and cutting-edge devices that help the information specialist eliminate dull, replicable, and routine undertakings. Furthermore, it can further develop client ventures by improving associations and accelerating processes.

Intelligent Process Automation (IPA) alludes to using Artificial Intelligence and related new advances, including Computer Vision, Cognitive Automation, and Machine Learning to Robotic Process Automation.

Joining of innovative advancements like Robotic Process Automation (RPA) Solutions, Artificial Intelligence (AI), Intelligent Character Recognition (ICR) Analytics, Optical Character Recognition (OCR), and Process Mining are brought about Intelligence Automation (IA). The fundamental point is to join the most cutting-edge innovations to fortify a quick start-to-finish business process automation. Intelligent automation is called Intelligent Process Automation (IPA) and Hyper Automation.

After some time, IPA emulates exercises completed by people and figures out how to improve. Conventional switches of rule-based automation are expanded with dynamic abilities due to advances in profound learning and mental innovation. The commitment of IPA is drastically upgraded effectiveness, expanded laborer execution, decreased functional dangers, and further developed reaction times and client venture encounters.

Generally, there is disarray between advanced IPA and RPA. However, both are not quite the same as one. Robotic Process Automation incorporates the utilization of inventive apparatuses and systems that computerize and complete responsibilities that require some investment, but a lot lesser than human specialists. These activities are many times rule-based, tedious, and direct. In some cases, RPA can be problematic when clients give wrong information. This is where Intelligent Process Automation enters. At the point when RPA is not sufficient to meet specific prerequisites, IPA is utilized to finish complex systems utilizing AI rationale and dynamic methodologies. The Figure 27.1 shows about the process automation structure. The Figure 27.2 shows about the destination folder. The Figure 27.3 shows about the email automation. The Figure 27.4 shows about the mail sequence and the Figure 27.5 displays the results after the bot is executed.

Digital Process Automation (DPA), Robotic Process Automation (RPA), and Artificial Intelligence are the technologies that make up IPA (Intelligent Process Automation). DPA portrays the spry arrangement of canny process

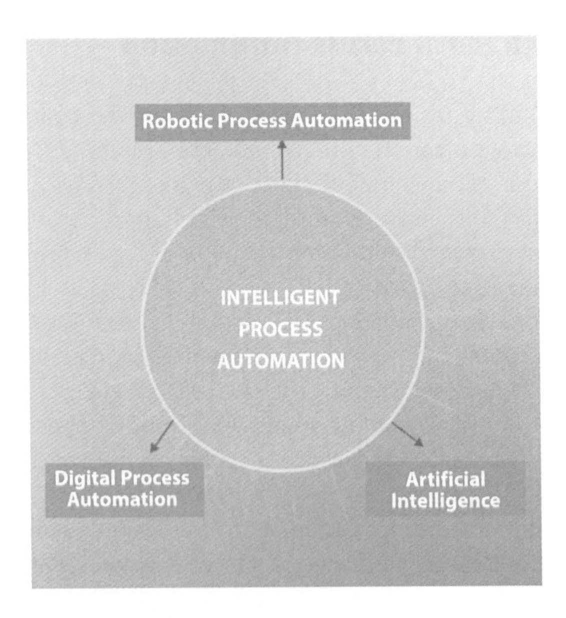

Figure 27.1 Process automation structure.

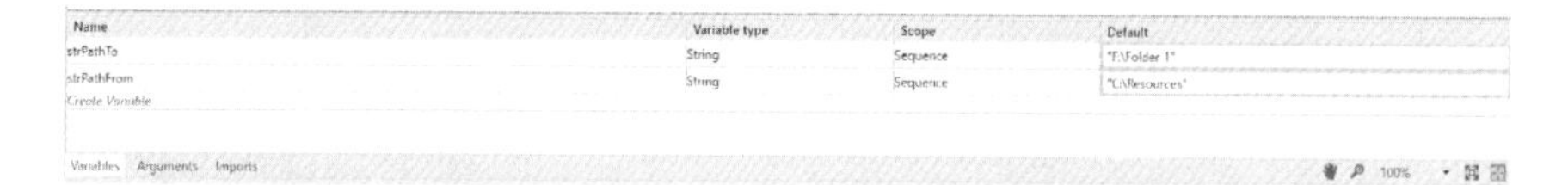

Figure 27.2 Destination folder.

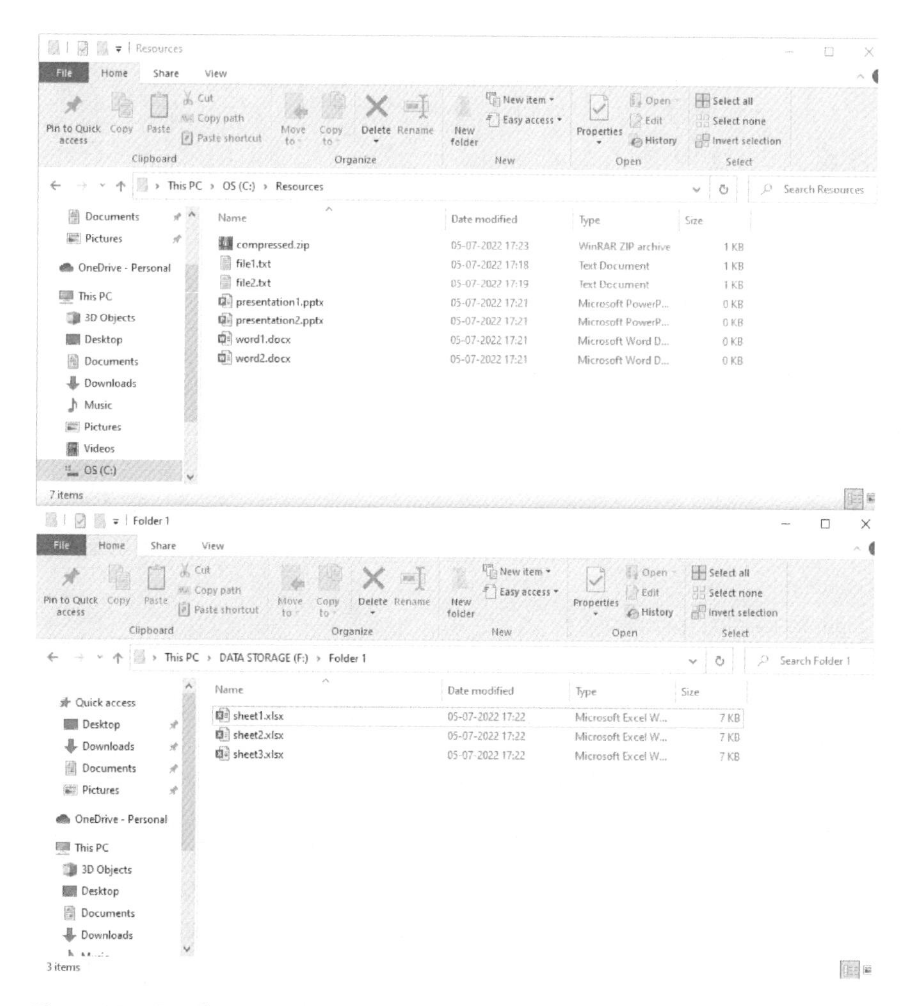

Figure 27.3 Email automation.

automation advances developed from their underlying foundations in BPM innovation. DPA gives the dexterity and understanding expected to empower an all-encompassing way to deal with robotizing business processes. It empowers you to deal with the progression of information across

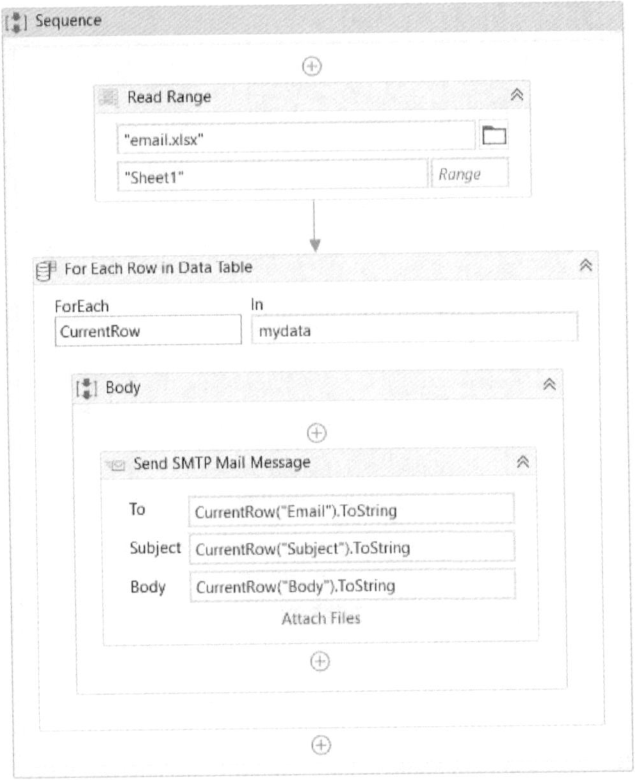

Sequence
Read Range
"email.xlsx"
"Sheet1"
Range
For Each Row in Data Table
ForEach
CurrentRow
In
mydata
Body
Send SMTP Mail Message
To
CurrentRow("Email").ToString
Subject
CurrentRow("Subject").ToString
Body
CurrentRow("Body").ToString
Attach Files

Figure 27.4 Mail sequence.

BEFORE
AFTER

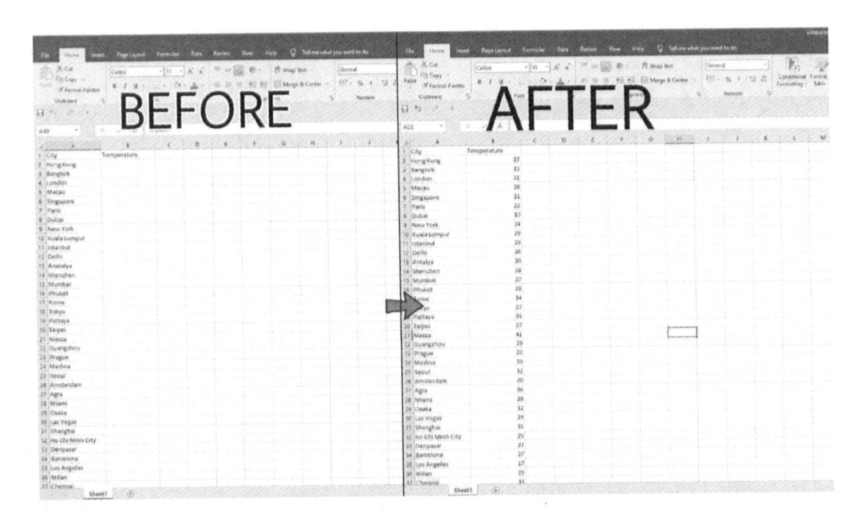

Figure 27.5 Results after the bot is executed.

your venture and makes it more straightforward to distinguish regions for development and roll out deft improvements.

Machine Learning utilizes a calculation that can be advanced by perceiving designs in organized information. Primarily, there are two kinds of learning: supervised and unsupervised. Supervised algorithms rely upon a bunch of organized information data sources and results and afterward gain from them while pushing ahead. Unsupervised algorithms give experiences and distinguish designs by organized information perception.

On the other hand, RPA offers speed and productivity that might be of some value. Sending robots that impersonate human activities assists with decreasing extremely manual, work concentrated undertakings, for example, re-keying information starting with one framework and then onto the next. Simulated intelligence then, at that point, contributes extraordinary knowledge and choices to the blend. This carries one more degree of reasoning to the automation as AI can dissect information so that a human proved unable, perceiving designs in information and gaining from past choices to settle on progressively shrewd decisions [19–21].

27.5.1 RPA Security

Robotic process automation is being utilized in little to medium-sized medical care callings and worldwide monetary administration enterprises. It essentially manages a great deal of private business information. RPA's product robots process data from various organization information bases and sign into various records utilizing provided accreditations to robotize day-to-day business errands like moving documents, processing requests, and leading finance. In this methodology, the automation stage approaches an organization's workers, clients, and merchants' data (stock records, passwords, etc.).

Robotic process automation certifications are generally traded to reuse them. Since these records and certifications are left unmodified and unstable, a digital assailant can snatch them, use them to escalate privileges, and move rapidly to gain admittance to basic frameworks, applications, and information. Administrators, then again, can separate credentials from vulnerabilities because numerous organizations that utilize robotic process automation have numerous bots underway out of nowhere; the risk is exceptionally critical.

The use of Robotic Process Automation (RPA) enables companies to reduce the risk of errors in manual work, resulting in high-quality, high-precision work. It records, analyses, and optimizes all activities of real-time robotic processes and allows you to monitor business processes carefully. Let us see the most significant security risks and how to prevent them [22, 23].

- Abuse of Privileged Access
  The term applies to any organization's internal systems and databases and primarily refers to privileged accounts, such as those with greater access to an organization's sensitive data. This may include the accounts of IT team members or employees who handle sensitive data, such as the accounts of B. system administrators and financial managers. Privileged access to RPA bots can be abused and carry the same risks associated with abusing human access. The best example is:
  Attackers can use special access to RPA bots to infiltrate systems and steal sensitive business data.
  The vulnerability of bots increases with privileged access and attackers can exploit this vulnerability to disrupt critical business activities related to customers, orders, or transactions.
- Framework Weaknesses
  A weakness is breaking a data framework that permits a digital aggressor to acquire unapproved access to the framework and perform malignant activities. Any imprudent way workers behave, including admittance to confined or noxious sites, is one way a weakness can emerge. In this situation, the site is a dangerous source that represents a gamble. Frail passwords, vindictive document transfers, and SQL injections are instances of weaknesses. With regards to weaknesses, there are two primary risk factors:
  Harm to the backend of the RPA programming might permit digital assailants to invade the corporate organization.
  While moving information, most present-day RPA frameworks use encryption; however, there are negligible security-level RPA devices that can cause light information spills while exchanging information that is not encrypted.
- System Outage
  System Outage/Framework Outage (or inactive time) alludes to when a framework/organization can never do its essential capability again. There are many purposes for a framework blackout. Some of them incorporate mistakes from the side of human laborers, outdated equipment, minor or significant bugs in server OS, and joining issues. The risk situations include:
  The working of the RPA bot can be upset by unanticipated organizational disappointments, which prompts less efficiency.
  Rapid sequencing of bot tasks can bring about framework disappointment or interference.

- Divulgence of Confidential Information
  Classified data is any data connected with an organization's business and undertakings that are not accessible to people in general and have business esteem. The unapproved revelation of an organization's financial data, showcasing plans, forthcoming ventures, and some other materials stamped secret might have crushing outcomes.

  Now and then, even a simple call to a colleague during noon or somebody indiscreetly sending an email from a corporate email box to a companion to share some organizational news, might be viewed as a revelation of personal data. This is notwithstanding plenty of situations when such a revelation is finished deliberately with the assistance of additional modern methods.

  In RPA, a risk scenario connected with exposure of classified data might occur when purposeful, careless, or ill-advised preparation of an RPA bot has caused spillage of secret information, for example, credit card data or Visa information to the web.

27.6 RPA and Blockchain

A blockchain is a developing rundown of records, called blocks, that are safely connected together utilizing cryptography. Each block contains a cryptographic hash of the past block, a timestamp, and exchange information. The timestamp demonstrates that the exchange information existed when the block was distributed to get into its hash. As blocks each contain data about the block before it, they structure a chain, with each extra block supporting the ones preceding it [24]. Consequently, blockchains are impervious to adjustment of their information on the grounds that once recorded, the information in some random block can't be changed retroactively without modifying every single ensuing block. In customary unified exchange frameworks, every exchange should be approved through the focal confided in an organization, for instance, the central bank, definitely coming about to the expense and the exhibition bottlenecks at the central servers. In contrast to the unified model, an outsider is not generally required in the blockchain. Agreement calculations in blockchain are utilized to keep up with information consistency in the appropriated network. Each client can collaborate with the blockchain with a generated address, which doesn't uncover the genuine personality of the client. Note that blockchain can't ensure the ideal protection of privacy due to the inborn limitation. Open blockchains are easier to use than some conventional proprietorship records, which, while open to general society, actually require admittance to see. Since all early blockchains were permissionless, contention has emerged over the blockchain definition. An issue in this continuous discussion is whether a confidential framework with verifiers entrusted and approved by a focal authority ought to be considered a blockchain. A benefit to an open, permissionless, or public blockchain network is that making preparations for agitators isn't needed and no entrance control is required. This implies that applications can be added to the organization without the endorsement or trust of others, utilizing the blockchain as a transport layer [25, 26].

Robotic Process Automation (RPA) can assist with working with the reception of blockchain with regard to data trade between various IT frameworks. Blockchain and RPA could work together to follow and perform exchanges across the framework. RPA can work in any environment effortlessly and with practically no worries. The exchanges will be checked and communicated utilizing the blockchain. The data will be put away in a decentralized record, making it difficult to modify or eradicate it.

Blockchain and RPA each enjoy their own benefits, for example, expanded functional viability, the decreased time required to circle back, and upgraded consumer loyalty. A basic activity like finishing up an online form or sending an email has advanced into utilizing AI/ML in mechanizing visual discernment and dynamic cycles. Initially, Blockchain was utilized for Cryptocurrency exchanges. However, it has since developed into an independent framework that permits secure and dependable transactions. Blockchain is viewed as an ideal innovation for resource tracking, records management, and exchanges. Up until this point, the most exceptional use of Blockchain has been for payment handling, claims to handle, and lawful proprietorship management. RPA can empower Blockchain to trade data with existing IT frameworks, as well as assist with laying out checking around exchange handling exercises. Blockchain can guarantee trust to RPA organized process streams by giving a tied down environment to approve all exchanges and keep up with the review trail in a permanent common record. Assuming coordination is planned genially, RPA and Blockchain can completely mechanize start-to-finish automated process streams to accomplish enterprise objectives.

RPA-driven processes communicate with end clients and frameworks to catch information utilizing bots and arrange the work process for handling. Blockchain can give a ledger to record deals and cycles for automated choices like installment handling and the personality of the board.

RPA can automate repetitive consistency undertakings for administrative streams as well as mechanizing process controls. Blockchain can be utilized to make a permanent record of the administrative occasions, which can be gotten to in the event of inner or outer reviews [27, 28].

27.7 Implementation

The following is an implementation of Robotic Process Automation using the tool UiPath. This is a very basic example in which a small bot is created which can move/copy the desired files from one location to the other.

In Figure 27.6, the source folder here is named "Resources" and is located in the C drive, while the destination folder is named "Folder1" and is located in the F drive. As we can see, there are a number of files in our

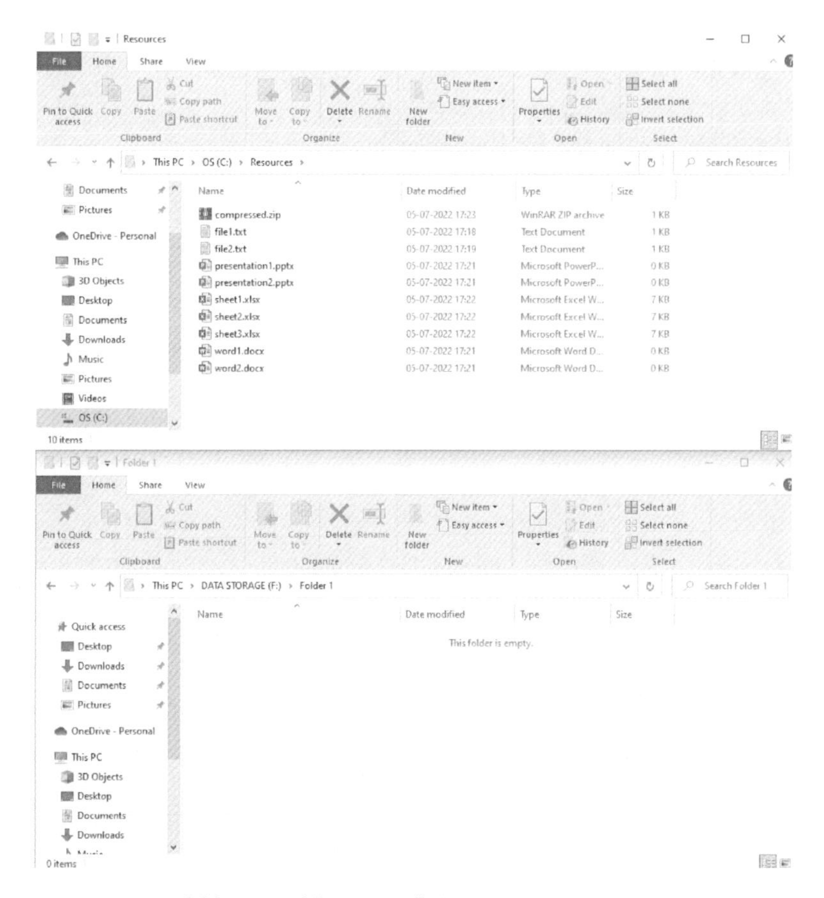

Figure 27.6 Source folder named "Resources".

source folder. Let's create a uibot that will move the excel/spreadsheet files (having the extension as .xlsx) to the destination folder.

Two variables are created named "strPathTo" and "strPathFrom" which store the path of the Destination Folder and Source Folder respectively. Both of these variables store the location/path in the form of a string.

Figure 27.7 shows the Uibot for moving a file.

After running this bot, all of the files having the extension .xlsx are moved to the destination folder from the source folder.

This is a very basic implementation of the Robotic Process Automation (RPA) done using Uipath.

Email Automation

Using Robotic Process Automation, we can automate the sending of emails to multiple users at the same time. We can have an excel file which contains the list of emails (the ones to which some message is to be sent), the subject, and the message body. The bot will read the data from that excel file and deliver the messages when executed.

The following is a bot for the same.

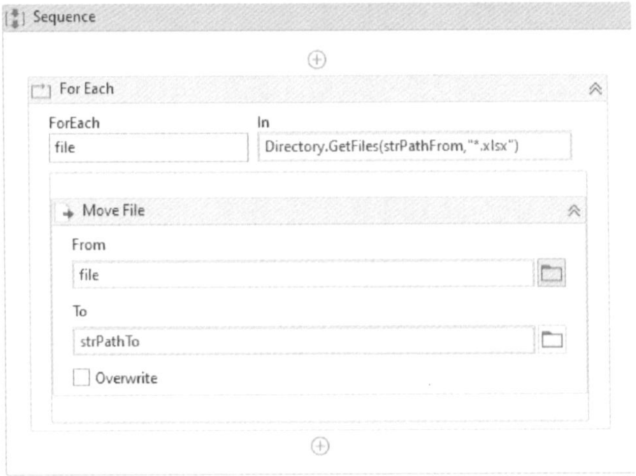

Figure 27.7 Moving file path.

27.7.1 Weather Report of Top 50 Cities Visited by International Visitors

In this example, a weather report is generated using Robotic Process Automation (RPA) with the help of the UiPath Tool. In an excel file, the cities having the most international visitors are added from the list available on Wikipedia [36]. Web scraping is used to collect the cities' data (temperature in this case) from a database of cities from the web. The Uibot scrapes the data from the web of each city in the excel file and updates the temperature of that city right next to it.

Figure 27.8 shows the RPA bot designed in UiPath. After the execution of the bot, it will update the excel file with the city's respective temperature.

The following screenshot shows the results after the above bot is executed.

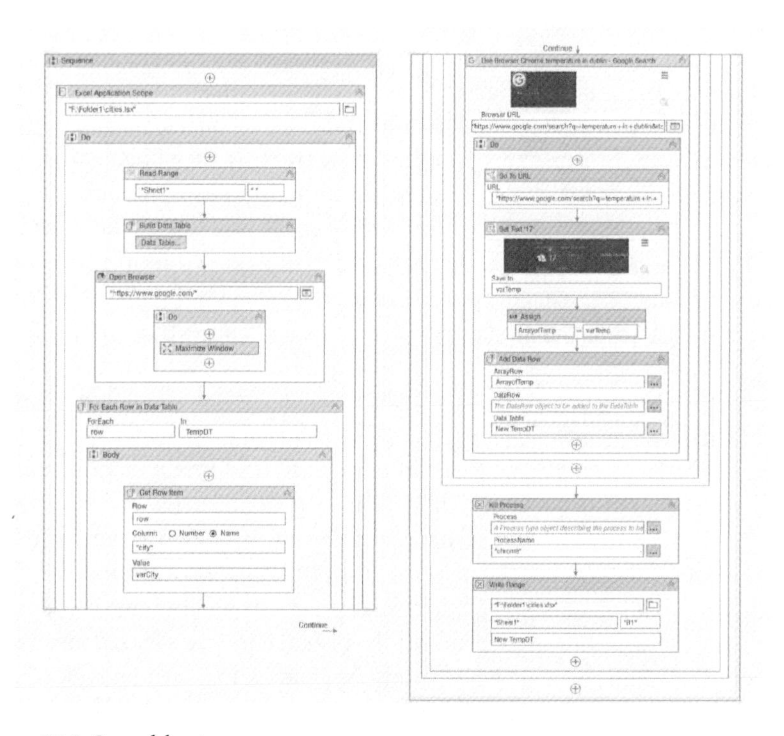

Figure 27.8 Spreadsheet

27.7.2 Use Cases

27.7.2.1 Website Scraping

Web scraping is a programmed technique to get a lot of information from sites. A large portion of this information is unstructured information in an HTML design which is then changed over into organized information in a spreadsheet or a data set with the goal that it very well may be utilized in different applications. There are various ways of performing web scraping to get information from websites [29]. The web scraping programming may straightforwardly get to the World Wide Web utilizing the Hypertext Transfer Protocol or an internet browser. While web scraping should be possible physically by a product client, the term commonly alludes to robotized processes executed utilizing a bot or web crawler. It is a type of replicating in which explicit information is accumulated and duplicated from the web, commonly into a focal neighborhood data set or bookkeeping sheet, for later recovery or analysis. There are techniques that a few sites use to forestall web scraping, like recognizing and refusing bots from crawling their pages. Accordingly, there are web scraping frameworks that depend on involving procedures in DOM parsing, PC vision, and normal language handling to reenact human perusing to empower gathering page content for disconnected parsing [30]. When we do web scraping manually, it can be a very tedious task as it includes many clicks, scrolls, copy and paste/move repetitions of texts or files, or extracting the objected data. That is why switching to Robotic Process Automation (RPA) is done to automate web scraping. Using RPA, errors that occur due to manual entries can be eliminated. Moreover, it reduces the time of data extraction and entries. We can extract text, images, videos, and even objects using RPA. RPA bots do scraping without writing codes or scripts for data extraction [31].

27.7.2.2 Data Migrations

Data migrations are typical work in enormous and medium undertakings that are going through a consolidation, a procurement, an execution, a modernization effort, a retirement of a legacy framework, an ERP redesign, or a transition to the Cloud. Robotic Process Automation (RPA) capacities position this innovation for thought as a data relocation instrument on the grounds that profoundly organized, rules-based movement exercises fit the profile of what robots specialize in. Various robots could be conveyed to relocate and approve consistently, which can surely speed up data movement. RPA can likewise help in diminishing the expenses of movement, as it doesn't need coding and is easy to utilize. RPA for data extraction and movement is an easy-to-understand, basic, and cost-effective answer for handling such difficulties of data tasks. The work approach for data movement won't just take additional time, however, it will likewise build the expense [32].

27.7.2.3 Call Centre Operations

With the assistance of RPA innovation, it is not difficult to help client demands get through call places: normal questions raised by clients and arrangements can be gone to specialists through a dashboard. This reduces human work along with minimization of errors. At the point when an issue is raised to client support specialists (the human workers), RPA assists with consolidating each important data point related to a client on a solitary screen, empowering specialists to have all the data expected from different frameworks for ideal and flawless help, thereby increasing the quality of the provided service [33].

27.7.2.4 Compliance Reporting

With the extension of the business, there will possibly come a time when it will not be easy to monitor the activities of the working staff. For instance, if an employee working in an organization tries to install some unauthorized or restricted application or software, Robotic Process Automation will ensure that all the employees work and do their tasks as per the compliance requirements in case the owner won't be able to keep an eye on them.

27.7.2.5 Customer Order Processing

Whenever a customer places an order, every single time an almost identical cycle is followed with just some minor variations. Dealing with those errands manually, with human power, can be drawn-out and tedious and will be more prone to errors and blunders. RPA arrangements can be customized to oversee client request handling to decrease the blunder from the side of human specialists, further develop Return on Investment, lower costs, and further develop consumer loyalty and a better experience [34].

27.7.2.6 Employee Onboarding

The RPA helps with mechanizing the onboarding system of representatives by enlisting their names, booking gatherings, and acceptance classes, and sending those automated messages and emails in regard to their obligations. RPA gives the best answer for guaranteeing that every representative's onboarding is as per the norm and pre-set interaction and that they get all the important data to work with organization rules.

27.7.2.7 Credit Card Applications

With regards to credit card handling, RPA can help by speeding up the entire cycle and decreasing the long holding-up periods. Significant delay costs the monetary organizations robustly, as well as leaves the clients disappointed. RPA bots are the behind-the-stage entertainers of most MasterCard application processes. RPA empowers the banks to give credit cards to clients in practically no time. They are modified to deal with each interaction stage. An RPA can speak with different frameworks immediately, approve the expected data, lead historical verifications, and settle on a choice in view of the principles to support and oppose the application [9].

27.7.2.8 Scheduling Systems

With RPA technology, online health care appointment scheduling can be improved. Scheduling optimization is a typical battle, especially when it includes thinking about a large number of variables. This can prompt income spillage, unfortunate client support, and employee dissatisfaction, and that's just the beginning. With RPA innovation, planning quality can get to the next level. For instance, RPA bots can gather all data with respect to the patient, for example, insurance details, appointment requests, location inclinations, and so on, to make the arrangement booking process easy and productive. The RPA bot plans patients' arrangements in view of diagnosis, specialist accessibility, area, and different contemplations, including budget reports and insurance data, and gives smoothed-out front-office support and computerized information assortment and handling [35].

The two main risks related to RPA are data theft [37, 38] and theft. If proper security protocols are not in place, sensitive data, such as RPA bot credentials [39] or customer data processed by RPA, may be exposed to attackers [40, 41]. UiPath, an RPA software company, has experienced a cyberattack. The business tracks emails sent to customers informing them that their private details have been disclosed online due to a security breach.

27.8 Conclusion

Automation of mundane and repetitive tasks is of prime importance. RPA would play a key role in all aspects of Next Generation Applications. In this chapter, we have tried to incorporate the role of RPA in various sectors like education, banking, healthcare, logistics, and transportation. RPA Security is also discussed. We have implemented a few use cases using the RPA tool UiPATH and demonstrated how RPA reduces the burden of doing repetitive tasks in every domain. The implementation work motivates other researchers to explore different aspects of automation.

References

1. "IEEE Guide for Terms and Concepts in Intelligent Process Automation," in IEEE Std 2755-2017, vol., no., pp.1-16, 28 Sept. 2017, doi: 10.1109/IEEESTD.2017.8070671.
2. (2018) Robotic process automation - Wikipedia. *Robotic process automation - Wikipedia.*
3. Hofmann, P., Samp, C., and Urbach, N. (2020) Robotic process automation. *Electron Markets*, **30** (1), 99–106.
4. Robotic Process Automation in the Real World: How 3 Companies are Innovating with RPA | IRPAAI. Robotic Process Automation in the Real World: How 3 Companies are Innovating with RPA | IRPAAI.
5. Lasso-Rodríguez, G., and Gil-Herrera, R. (2019) ROBOTIC PROCESS AUTOMATION APPLIED TO EDUCATION: A NEW KIND OF ROBOT TEACHER? 2531–2540.
6. Neethu V Joy, Sreelakshmi P G, 2020, Robotic Process Automation role in Education Field, INTERNATIONAL JOURNAL OF ENGINEERING RESEARCH & TECHNOLOGY (IJERT) NSDARM – 2020 (Volume 8 – Issue 04),
7. Kuppusamy, Palanivel & Joseph K, Suresh. (2020). Robotic Process Automation to Smart Education. 3775.
8. (2020) Robotic Process Automation role in Education Field. *International Journal of Engineering Research*, **8** (04), 2.
9. Vijai, C., Suriyalakshmi, S.M., and Elayaraja, M. (2020) The Future of Robotic Process Automation (RPA) in the banking sector for Better Customer experience. *Shanlax International Journal of Commerce*, **8** (2), 61–65.
10. FED The RPA Market Will Grow To $22 Billion By 2025 | Forrester. The RPA Market Will Grow To $22 Billion By 2025 | Forrester.
11. Thekkethil, M.S., Shukla, V.K., Beena, F., and Chopra, A. (2021) Robotic Process Automation in banking and finance sector for loan processing and

fraud detection. *2021 9th International Conference on Reliability, Infocom Technologies and Optimization (Trends and Future Directions) (ICRITO).*

12. M. Romao, J. Costa and C. J. Costa, "Robotic Process Automation: A Case Study in the Banking Industry," 2019 14th Iberian Conference on Information Systems and Technologies (CISTI), 2019, pp. 1-6, doi: 10.23919/CISTI.2019.8760733.
13. Bhatnagar, R., and Jain, R. (2019) Robotic Process Automation in healthcare-A Review. *International Robotics & Automation Journal*, **5** (1), 12–14.
14. Ratia, M., Myllärniemi, J., and Helander, N. (2018) Robotic process automation - creating value by digitalizing work in the private healthcare? *Proceedings of the 22nd International Academic Mindtrek Conference.*
15. (2022) Creating More Value With Car Circularity | Accenture. *Creating Business Value Through Car Circularity | Accenture.*
16. Gružauskas, V., and Ragavan, D. (2020) Robotic Process Automation for document processing: A case study of a logistics service provider. *Journal of Management*, **Vol. 36** (No. 2), 119–126.
17. Sullivan, M., Simpson, W., and Li, W. (2021) The role of Robotic Process Automation (RPA) in Logistics. *The Digital Transformation of Logistics*, 61–78.
18. Devarajan, Y. (2018) A Study of Robotic Process Automation Use Cases Today for Tomorrow's Business. **5** (6), 7.
19. Kholiya, P.S., Kapoor, A., Rana, M., and Bhushan, M. (2021) Intelligent Process Automation: The future of digital transformation. *2021 10th International Conference on System Modeling & Advancement in Research Trends (SMART).*
20. Chakraborti, T., Isahagian, V., Khalaf, R., Khazaeni, Y., Muthusamy, V., Rizk, Y., and Unuvar, M. (2020) From Robotic Process Automation to Intelligent Process Automation: – Emerging Trends –, in *Business Process Management: Blockchain and Robotic Process Automation Forum*, vol. 393, Springer International Publishing, Cham, pp. 215–228.
21. Ribeiro, J., Lima, R., Eckhardt, T., and Paiva, S. (2021) Robotic Process Automation and Artificial Intelligence in Industry 4.0 – A Literature review. *Procedia Computer Science*, **181**, 51–58.
22. Choi, D., R'bigui, H., and Cho, C. (2021) Robotic Process Automation Implementation Challenges, in *Proceedings of International Conference on Smart Computing and Cyber Security*, vol. 149, Springer Singapore, Singapore, pp. 297–304.
23. Syed, R., Suriadi, S., Adams, M., Bandara, W., Leemans, S.J.J., Ouyang, C., ter Hofstede, A.H.M., van de Weerd, I., Wynn, M.T., and Reijers, H.A. (2020) Robotic Process Automation: Contemporary themes and challenges. *Computers in Industry*, **115**, 103162.
24. Nofer, Michael & Gomber, Peter & Hinz, Oliver & Schiereck, Dirk. (2017). Blockchain. Business & Information Systems Engineering. 59. 10.1007/s12599-017-0467-3.

25. Zheng, Zibin & Xie, Shaoan & Dai, Hong-Ning & Chen, Xiangping & Wang, Huaimin. (2018). Blockchain challenges and opportunities: A survey. International Journal of Web and Grid Services. 14. 352. 10.1504/IJWGS.2018.095647.
26. Yaga, D., Mell, P., Roby, N., and Scarfone, K. (2018) Blockchain technology overview. NIST IR 8202. https://doi.org/10.6028/NIST.IR.8202
27. Sibanyoni, N.A. (2021) A Blockchain-Based Robotic Process Automation Mechanism in Educational Setting:, in *Advances in Data Mining and Database Management* (eds.Mahmood, Z.), IGI Global, pp. 17–41.
28. Mendling, J., Decker, G., Hull, R., Reijers, H.A., and Weber, I. (2018) How do Machine Learning, Robotic Process Automation, and Blockchains Affect the Human Factor in Business Process Management? *CAIS*, 297–320.
29. Zhao, B. (2017) Web Scraping, in *Encyclopedia of Big Data* (eds.Schintler, L.A., and McNeely, C.L.), Springer International Publishing, Cham, pp. 1–3.
30. Singrodia, V., Mitra, A., and Paul, S. (2019) A Review on Web Scrapping and its Applications. *2019 International Conference on Computer Communication and Informatics (ICCCI)*, 1–6.
31. Inc., U. (n.d.). Web scraping software. Retrieved July 18, 2022, from https://www.uipath.com/solutions/technology/web-scraping-software
32. Eddy, D. (2021, September 28). Benefits of RPA for Data Migration: Uipath. Retrieved July 18, 2022, from https://www.uipath.com/blog/rpa/how-rpa-transforms-data-migration
33. Inc., U. (n.d.). Contact Center Automation - improve customer experience. Retrieved July 19, 2022, from https://www.uipath.com/solutions/department/contact-center-automation
34. Kajrolkar, Asmita & Pawar, Shivani & Paralikar, Prasad & Bhagat, Narendra. (2021). Customer Order Processing using Robotic Process Automation. 1-4. 10.1109/ICCICT50803.2021.9510109.
35. Matonya, M., Kocsi, B., Pusztai, L., and Budai, I. Production Planning, Scheduling and Risk Analysis in Manufacturing Operations by Robotic Process Automation. 10.
36. List of cities by international visitors. (2022, July 13). Retrieved July 19, 2022, from https://en.wikipedia.org/wiki/List_of_cities_by_international_visitors
37. Rawat, R. (2023). Logical concept mapping and social media analytics relating to cyber criminal activities for ontology creation. *International Journal of Information Technology*, 15(2), 893-903.
38. Rawat, R., Mahor, V., Álvarez, J. D., & Ch, F. (2023). Cognitive Systems for Dark Web Cyber Delinquent Association Malignant Data Crawling: A Review. *Handbook of Research on War Policies, Strategies, and Cyber Wars*, 45-63.
39. Rawat, R., Chakrawarti, R. K., Vyas, P., Gonzáles, J. L. A., Sikarwar, R., & Bhardwaj, R. (2023). Intelligent Fog Computing Surveillance System for Crime and Vulnerability Identification and Tracing. *International Journal of Information Security and Privacy (IJISP)*, 17(1), 1-25.

40. Rawat, R., Sowjanya, A. M., Patel, S. I., Jaiswal, V., Khan, I., & Balaram, A. (Eds.). (2022). *Using Machine Intelligence: Autonomous Vehicles Volume 1.* John Wiley & Sons.
41. Rawat, R., Bhardwaj, P., Kaur, U., Telang, S., Chouhan, M., & Sankaran, K. S. (2023). *Smart Vehicles for Communication, Volume 2.* John Wiley & Sons.

About the Editors

Romil Rawat, PhD, is an assistant professor at Shri Vaishnav Vidyapeeth Vishwavidyalaya, Indore. With over 12 years of teaching experience, he has published numerous papers in scholarly journals and conferences. He has also published book chapters and is a board member on two scientific journals. He has received several research grants and has hosted research events, workshops, and training programs. He also has several patents to his credit.

Rajesh Kumar Chakrawarti, PhD, is a professor and the Dean of the Department of Computer Science & Engineering, Sushila Devi Bansal College, Bansal Group of Institutions, India. He has over 20 years of industry and academic experience and has published over 100 research papers and chapters in books.

Sanjaya Kumar Sarangi, PhD, is an adjunct professor and coordinator at Utkal University, Coordinator and Adjunct Professor, Utkal University, Bhubaneswar, India. He has over 23 years of academic experience and has authored textbooks, book chapters, and papers for journals and conferences. He has been a visiting doctoral fellow at the University of California, USA, and he has more than 30 patents to his credit.

Rahul Choudhary, PhD, is an assistant professor at the Shri Vaishnav Institute of Information Technology, Indore, India. He has over nine years of academic experience.

Anand Singh Gadwal, is an assistant professor at the Shri Vaishnav Institute of Information Technology, Indore, India, has a masters of engineering degree in computer engineering, and is pursuing a PhD in this area.

Vivek Bhardwaj, PhD, is an assistant professor at Manipal University Jaipur, Jaipur, India. He has over eight years of teaching and research experience, has filed five patents, and has published many articles in scientific journals and conferences.

Index

Also of Interest

From the same editors

AUTONOMOUS VEHICLES VOLUME 1: Using Machine Intelligence, Edited by Romil Rawat, A. Mary Sowjanya, Syed Imran Patel, Varshali Jaiswal, Imran Khan, and Allam Balaram. ISBN: 9781119871958. Addressing the current challenges, approaches and applications relating to autonomous vehicles, this groundbreaking new volume presents the research and techniques in this growing area, using Internet of Things, Machine Learning, Deep Learning, and Artificial Intelligence.

AUTONOMOUS VEHICLES VOLUME 2: Smart Vehicles for Communication, Edited by Romil Rawat, Purvee Bhardwaj, Upinder Kaur, Shrikant Telang, Mukesh Chouhan, and K. Sakthidasan Sankaran, ISBN: 9781394152254. The companion to *Autonomous Vehicles Volume 1: Using Machine Intelligence*, this second volume in the two-volume set covers intelligent techniques utilized for designing, controlling and managing vehicular systems based on advanced algorithms of computing like machine learning, artificial Intelligence, data analytics, and Internet of Things with prediction approaches to avoid accidental damages, security threats, and theft.

Check out these other related titles from Scrivener Publishing

FACTORIES OF THE FUTURE: Technological Advances in the Manufacturing Industry, Edited by Chandan Deep Singh and Harleen Kaur, ISBN: 9781119864943. The book provides insight into various technologies adopted and to be adopted in the future by industries and measures the impact of these technologies on manufacturing performance and their sustainability.

AI AND IOT-BASED INTELLIGENT AUTOMATION IN ROBOTICS, Edited by Ashutosh Kumar Dubey, Abhishek Kumar, S. Rakesh Kumar, N. Gayathri, Prasenjit Das, ISBN: 9781119711209. The 24 chapters in this book provide a deep overview of robotics and the application of AI and IoT in robotics across several industries such as healthcare, defense. education, etc.

SMART GRIDS FOR SMART CITIES VOLUME 1, Edited by O.V. Gnana Swathika, K. Karthikeyan, and Sanjeevikumar Padmanaban, ISBN: 9781119872078. Written and edited by a team of experts in the field, this first volume in a two-volume set focuses on an interdisciplinary perspective on the financial, environmental, and other benefits of smart grid technologies and solutions for smart cities.

SMART GRIDS FOR SMART CITIES VOLUME 2: Real-Time Applications in Smart Cities, Edited by O.V. Gnana Swathika, K. Karthikeyan, and Sanjeevikumar Padmanaban, ISBN: 9781394215874. Written and edited by a team of experts in the field, this second volume in a two-volume set focuses on an interdisciplinary perspective on the financial, environmental, and other benefits of smart grid technologies and solutions for smart cities.

SMART GRIDS AND INTERNET OF THINGS, Edited by Sanjeevikumar Padmanaban, Jens Bo Holm-Nielsen, Rajesh Kumar Dhanaraj, Malathy Sathyamoorthy, and Balamurugan Balusamy, ISBN: 9781119812449. Written and edited by a team of international professionals, this groundbreaking new volume covers the latest technologies in automation, tracking, energy distribution and consumption of Internet of Things (IoT) devices with smart grids.

DESIGN AND DEVELOPMENT OF EFFICIENT ENERGY SYSTEMS, edited by Suman Lata Tripathi, Dushyant Kumar Singh, Sanjeevikumar Padmanaban, and P. Raja, ISBN 9781119761631. Covering the concepts and fundamentals of efficient energy systems, this volume, written and edited by a global team of experts, also goes into the practical applications that can be utilized across multiple industries, for both the engineer and the student.

INTELLIGENT RENEWABLE ENERGY SYSTEMS: Integrating Artificial Intelligence Techniques and Optimization Algorithms, edited by Neeraj Priyadarshi, Akash Kumar Bhoi, Sanjeevikumar Padmanaban, S. Balamurugan, and Jens Bo Holm-Nielsen, ISBN 9781119786276. This collection of papers on artificial intelligence and other methods for improving renewable energy systems, written by industry experts, is a reflection of the state of the art, a must-have for engineers, maintenance personnel, students, and anyone else wanting to stay abreast with current energy systems concepts and technology.

SMART CHARGING SOLUTIONS FOR HYBRID AND ELECTRIC VEHICLES, edited by Sulabh Sachan, Sanjeevikumar Padmanaban, and Sanchari Deb, ISBN 9781119768951. Written and edited by a team of experts in the field, this is the most comprehensive and up to date study of smart charging solutions for hybrid and electric vehicles for engineers, scientists, students, and other professionals.

Printed in the USA/Agawam, MA
August 18, 2023

850030.023